INTRODUCTORY SYSTEM ENGINEERING

McGRAW-HILL SERIES IN ELECTRONIC SYSTEMS

John G. Truxal and Ronald A. Rohrer, · Consulting Editors

CHUA · *Introduction to Nonlinear Network Theory*

DIRECTOR AND ROHRER · *Introduction to System Theory*

HUELSMAN · *Theory and Design of Active RC Circuits*

MEDITCH · *Stochastic Optimal Linear Estimation and Control*

PEATMAN · *The Design of Digital Systems*

RAMEY AND WHITE · *Matrices and Computers in Electronic Circuit Analysis*

ROHRER · *Circuit Theory: An Introduction to the State Variable Approach to Network Theory*

SCHULTZ AND MELSA · *State Functions and Linear Control Systems*

STAGG AND EL-ABIAD · *Computer Methods in Power System Analysis*

TIMOTHY AND BONA · *State Space Analysis: An Introduction*

TRUXAL · *Introductory System Engineering*

INTRODUCTORY SYSTEM ENGINEERING

JOHN G. TRUXAL

Dean of Engineering
State University of New York at Stony Brook

McGRAW-HILL BOOK COMPANY

New York   St. Louis   San Francisco   Düsseldorf   Johannesburg
Kuala Lumpur   London   Mexico   Montreal   New Delhi
Panama   Rio de Janeiro   Singapore   Sydney   Toronto

To
Doris, Carol, and Brian

INTRODUCTORY SYSTEM ENGINEERING

Copyright © 1972 by McGraw-Hill, Inc. All rights reserved.
Printed in the United States of America. No part of this publi-
cation may be reproduced, stored in a retrieval system, or
transmitted, in any form or by any means, electronic, mechanical,
photocopying, recording, or otherwise, without the prior written
permission of the publisher.
Library of Congress Catalog Card Number 76-172265

07-065317-8

1234567890 KPKP 798765432

This book was set in Press Roman by Scripta-Technica, Inc., and
printed and bound by Kingsport Press, Inc. The designer
was Scripta-Technica, Inc. The editor was Charles R. Wade.
Alice Cohen supervised production.

PREFACE

Strong forces encourage change in the undergraduate electrical engineering curriculum. The technology itself is changing rapidly, with the increasingly widespread acceptance of integrated circuits, the advantages of digital circuitry, and the use of computers for straightforward analysis and design. These major trends alone would be sufficient to demand curricular reexamination, particularly with the objective of reducing the total time devoted to the routines of classical network analysis and electromagnetics.

In parallel with this technological change, two other forces have appeared, impacting higher education: pressures to control the cost (primarily by increasing productivity of the individual faculty member), and student pressures for a program with greater relevance and more concern with problems that are clearly significant. These forces are encouraging a general reevaluation of educational processes, with attempts to apply system techniques to a university and to capitalize on our growing understanding of the learning process.

In this environment, the Department of Electrical Engineering of the Polytechnic Institute of Brooklyn (PIB) has been revising its undergraduate curriculum. The introductory sophomore circuits course is followed by junior-level courses in computer engineering, electronics, and electromagnetics. In addition, a one-semester course has the objective of presenting the basic concepts of electrical system engineering. This book has been developed during the past three years when the notes were used as a text for this required, three-credit course.

Thus, the goal of the course is a presentation to all EE undergraduates of the basic concepts in the field of system engineering as it overlaps electrical engineering. For students interested in control and communication, this course is followed with more advanced senior electives; for students with primary interest in computer engineering or electrophysics, it represents a terminal required course.

In the actual course offered at PIB, portions of the book have been covered in self-study or omitted. Chapters 3 and 4 are intended as a review of elements of the sophomore course and are included in an effort to develop reasonable uniformity of student background. These two chapters are covered primarily through self-study activities, including audiocassettes available for assistance in self-evaluation of the student's own problem work. (Cassettes are also available on problems of other chapters and for overall course review.) The discussions of state models (Chap. 7), nonlinear simulation (Chap. 10), and digital simulation (Chap. 14) are covered rather superficially in the course, with more careful consideration reserved for the senior offerings. Associated laboratory work is included within the single, separate laboratory course offered to EE juniors.

During the three years, the author has been particularly fortunate in teaching this material in parallel with Professors Hunt, Drenick, Griesmann, Braun, and D. Miller. These individuals have made major contributions to the content of this book, and have participated enthusiastically in experimentation with flexible scheduling and the use of audiovisual aids, in attempts to develop patterns to control per student costs while improving educational quality.

Primary acknowledgments of the author's indebtedness are to the staff of the Polytechnic Institute of Brooklyn for the exciting environment in which this effort was possible, and to the 500 students over the three years, whose enthusiasm for the excitement and challenges of electrical engineering has never wavered. Above all, the author is indebted to his wife, Doris.

John G. Truxal

CONTENTS

Preface   v

1   SYSTEM ENGINEERING   1

   1.1   System Engineering   2
   1.2   Difficult System Engineering Problems   3
   1.3   Cardiac Pacemaker   7
   1.4   Concepts Underlying System Engineering   11
   1.5   Final Comment   13

2   SYSTEM DIAGRAMS   14

   2.1   The System of Urban Apartments   15
   2.2   The Block Diagram   17
   2.3   Signal Flow Diagrams   25
   2.4   Mason's Theorem for Reduction of Systems   33
   2.5   Complex Systems   40
   2.6   Algebra of Signal Flow Graphs   42
   2.7   Isolation of One Element   44
   2.8   Final Comment   50
          Problems   50

3   TRANSFER FUNCTIONS.   54
    3.1   What is the Transfer Function?   55
    3.2   Critical Frequencies   58
    3.3   Total Response   62
    3.4   Steady-State Response   67
    3.5   Resonance Concept   70
    3.6   Resonance and Transfer Functions   72
    3.7   Notch Filter   74
    3.8   Comb Filter   75
    3.9   Transportation Lag   77
    3.10   Limitations on Use of Transfer Functions   82
           Appendix—Human Transfer Functions   85
           Problems   92

4   SPECIAL-PURPOSE NETWORKS   100
    4.1   Differentiators   100
    4.2   Integrators   108
    4.3   Step-Function Response   112
    4.4   Ladder Networks   119
    4.5   Final Comment   123
           Appendix—Identification from Impulse Response   124
           Problems   128

5   ROOT-LOCUS DESIGN   134
    5.1   Root Loci   135
    5.2   A Second Example of a Design Problem   138
    5.3   Two Forms for the Characteristic Equation   139
    5.4   Rules for Drawing Root Loci   140
    5.5   Example 1 (Root Locus)   143
    5.6   Example 2 (Root Locus)   146
    5.7   Example 3 (Root Locus)   152
    5.8   Example 4 (Root Locus)   154
    5.9   Parameter Negative   159
    5.10   Final Comment   160
           Problems   161

6   OPERATIONAL AMPLIFIERS (OP-AMPS)
     AND SYNTHESIS   165
    6.1   Op-Amps   166
    6.2   Analysis of an Op-Amp Circuit   172
    6.3   Transfer-Function Realization   177
    6.4   A Second Realization   182
    6.5   A Third Solution   184

6.6   Practical Problems in Op-Amps   188
6.7   Time Scaling   191
6.8   Amplitude Scaling   197
6.9   Final Comments   202
       Appendix—Butterworth Filters   204
       Problems   212

7   STATE MODELS   218
7.1   The State Model   219
7.2   Solutions for the Undriven System   225
7.3   Solutions for the Driven System   229
7.4   State Models for Simple Control Systems   233
7.5   Design of the Servomechanism   236
7.6   Controllability   238
7.7   Observability   243
7.8   Concluding Comment   246
       Problems   246

8   MULTIDIMENSIONAL SYSTEMS   252
8.1   Simple Multidimensional Design Problem   253
8.2   Simulation of Multi-input or Multi-output Systems   257
8.3   Multiple Inputs and Outputs   259
8.4   Controllability and Observability   268
8.5   A Final, Simple Example   271
       Problems   275

9   OP-AMP APPLICATIONS   278
9.1   An Essential Characteristic in Simulation   278
9.2   Initial Conditions   285
9.3   Op-Amps with Two Inputs   289
9.4   More General Op-Amps   292
9.5   Differentiators   297
9.6   Final Comment   299
       Appendix—Generalized Op-Amps Using a
       Difference Amplifier   299
       Problems   306

10   NONLINEAR SIMULATIONS   311
10.1   System Reduction Around a Nonlinearity   313
10.2   Linearization   316
10.3   Simulation of Nonlinearities   320
10.4   Simulation of Piecewise-Linear Input-Output
        Characteristic   322

10.5   Simulation of Even, Zero-Memory Nonlinearity   328
10.6   Simulation of a Nonlinearity Plus a Transfer Function   329
10.7   Conclusion   335
    Problems   338

11   **GAIN AND PHASE PLOTS**   343
11.1   Form of the Gain Plot   345
11.2   dB Gain for a Linear Factor   348
11.3   Asymptotic Plots for More Complex Transfer Functions   351
11.4   Second Example of Asymptotic Plots   355
11.5   Corrections for Asymptotic Gain Characteristic   356
11.6   Complex Poles and Zeros   360
11.7   Frequency Measurement of Transfer Function   363
11.8   Phase Characteristic   369
11.9   Phase for Real Poles and Zeros   372
11.10   Phase for Complex Poles and Zeros   377
11.11   Gain-Phase Relations   380
11.12   Final Comment   383
    Appendix—dB   385
    Problems   390

12   **FEEDBACK**   396
12.1   What is Feedback?   396
12.2   Why Use Feedback?   400
12.3   Sensitivity   403
12.4   Techniques of Sensitivity Control   405
12.5   Sensitivity in Active Networks   411
12.6   Disturbance Control   416
12.7   Importance of Return Difference   417
12.8   Dynamics Control   422
12.9   Radar Tracking System   429
12.10   Homeostasis   433
    Problems   440

13   **SAMPLING**   448
13.1   What is Sampling?   449
13.2   Sampling Theorem   451
13.3   Implications of the Sampling Theorem   455
13.4 z Transform   459
13.5   Discrete Data Processors   464
13.6   What Can We Deduce From $W(z)$?   470
13.7   System Design Example   473
13.8   Concluding Comments   478

Appendix—Buffalo Population Model   479
Problems   486

14   SAMPLED AND DIGITAL SYSTEMS **491**
14.1   Equivalence of Continuous and Sampled Signals   491
14.2   Hold Circuits   495
14.3   Analysis of Continuous Systems   499
14.4   Digital Simulation   504
14.5   Final Comment   508
Problems   508

15   STABILITY **513**
15.1   Stability of Linear Systems   514
15.2   Routh Test   516
15.3   More on the Routh Test   526
15.4   Stability with Two Parameters   532
15.5   Nonlinear Systems   536
15.6   Lyapunov Stability   541
15.7   Final Remarks   548
Problems   548

16   NYQUIST STABILITY CRITERION **554**
16.1   Historical Background   555
16.2   Nyquist Criterion   556
16.3   Simplifications Usually Valid   559
16.4   Examples of Nyquist Plots   561
16.5   Proof of the Nyquist Criterion   566
16.6   Example of the Nyquist Test   569
16.7   Gain and Phase Margins   572
16.8   Critical Point for Nyquist Plot   574
16.9   Simple Nonlinear Systems   576
16.10   Stability Analysis with Describing Functions   581
16.11   What Can the Nyquist Test Do?   585
Problems   585

INDEX **589**

1 SYSTEM ENGINEERING

A company manufacturing expensive cars develops and installs a device in its new models that automatically locks the doors when the car is moving (measured by wheel rotation). The "gadget" clearly represents the forward look to the future expected by the customer who invests $8,000 in a showpiece car.

Unfortunately, the best laid plans of men do not always work out as expected. New car owners quickly discover that the self-locking device is not always an advantage. When the owner leaves his vehicle as it enters an automatic car wash, he carefully closes all windows and leaves the key inside. When the clean car emerges, the doors of the empty car are securely locked.

This true episode is illustrative of the history of engineering and science during the 1960's—an unpredictable blend of astonishing successes and frustrating failures as this nation moved into an age of technology. The decade witnessed man's safe journey to the moon, the first human heart transplant, and demonstration of machines allowing the blind man to read a page of ordinary type. Parallel with these achievements, the socio-technological problems of our urban society reached crisis levels, and engineering was confronted with the challenge of the 1970's: to develop a technology both matched to *all* the people and to the society it serves, and responsive to the national goals.

1.1 SYSTEM ENGINEERING

In response to this newly emphasized challenge to engineering and science, the engineering which was so strongly concerned with performance and cost only a decade ago now must give equal attention to the interface between its products and people. This interface involves a wide variety of problems: the technology must be reasonably understood by the user, it must be capable of maintenance by individuals with reasonable training, it must not adversely affect environmental and social systems, and it must safeguard personal privacy and provide low-level risk of personal injury.

For example, because of highway and airlane congestion in the Northeast Corridor, considerable pressure exists for Federal development of a 300-mph train from Boston to Washington via New York. The engineering of such a multibillion dollar innovation encompasses conventional railroad problems:

(1) roadbed and tracks to ensure a safe and reasonably smooth ride;
(2) car design for vibration reduction;
(3) propulsion system.

In addition, acceptance of the system by the public demands equal attention to the technology—people interface, represented by such aspects as:

(1) determination of humanly acceptable acceleration signals to minimize the time spent in starting and stopping;

(2) decision on what passengers may see as they look out the windows (nearby objects passing at 300 mph are likely to cause nausea);

(3) complete design of the terminal, ticketing, reservation, and external service systems to provide an attractive mode of transportation;

(4) low-cost and convenient parking facilities and car-rental or short-distance travel facilities to provide for the fact that most intercity travel is not from the center of one city to the center of another;

(5) an absolutely safe system to avoid accidents when high-speed trains are running frequently over the same tracks (and certainly the complete elimination of grade crossings)—including scheduling and control plans to service passengers for intermediate destinations;

(6) low-cost, high-speed tunneling techniques to allow underground travel in regions where aesthetic or economic conditions require.

As soon as these factors of technology-people interplay are included, we have a set of *system engineering* problems—tasks which encompass not only the traditional concerns of the mechanical (or electrical) engineer, but also demand that the engineering effort include consideration of interaction of the particular technology with the total system in which it is to be used.

We probably should start a book such as this with a formal definition of system engineering, or at least of a system. Such definitions never seem to be very meaningful, however; perhaps a definition as good as any is that "system engineering is engineering done with a system viewpoint." In other words, system engineering is the utilization of science for the benefit of man (i.e., engineering), with emphasis on understanding and taking into account the interplay with people and with the broader social and technological environment of the technology. In our high-speed train example, system engineering may be concerned with a very specific aspect (e.g., the design of passenger seats), but this particular problem is solved with appropriate consideration of expected passenger sizes, accelerations anticipated, duration of the ride, desirability of being able to look out the windows, expected ride smoothness (with its influence on what passengers will be doing), economic considerations, and so forth.

Systems and electrical engineering

According to the above description, system engineering necessarily transcends any classical branch of engineering such as electrical, mechanical, civil, and so forth. System engineering involves not only the interconnection of components making up the artifact, but also the interplay of this device with its total environment—including the human beings within that environment. Thus, in general, system engineering must involve biological, behavioral, and social sciences.

After presentation of this broad "definition" of system engineering, we proceed to narrow radically our field of interest in this book. If we attempted to discuss those aspects of biology, physiology, psychology, and sociology, which are often important considerations in system engineering, in all probability little more than superficial coverage of any would result. While one might formulate an educational program to develop understanding in small steps across a broad front (each course or book touching lightly on a broad range of topics), it is far simpler to restrict any one book or course to more depth in a particular part of the subject.

Thus, the following chapters will focus on those analysis and design techniques within electrical engineering which have proved important in system engineering. In other words, we are concerned with *electrical system engineering*. The concepts and techniques which we consider constitute an important segment of not only system engineering, but also electrical engineering itself—particularly electronics, communication, and control.

Concern of this chapter

Before we delve into the details of various techniques and concepts, a few of the broader aspects of system engineering will be emphasized in this chapter. To accomplish this, a few familiar examples of difficult and unsolved system engineering problems will be described in the next section. Following this, in Sec. 1.3, we consider one of the outstanding successes: the cardiac pacemaker. Finally, in Sec. 1.4 we reconsider these examples and, from this overview, attempt to define some of the basic concepts underlying the system approach.

1.2 DIFFICULT SYSTEM ENGINEERING PROBLEMS

The end of the 1960s was marked by widespread public condemnation of technology for its part in the creation of all social ills, from unsafe cars to air pollution. While public awareness of the risks of automobile travel and the perils of environmental mis-management is heartening, the critics of technology tend to such extreme irrationality that they are in a position similar to those who blame the farmer for widespread obesity. (The farmer produces too much food too cheaply, hence the high mortality rate of middle-aged American men is his fault.)

On a less globally catastrophic scale, however, it is not difficult to recognize notably difficult problems of system engineering. One of the most obvious examples from the time before system engineering is the typewriter keyboard; it illustrates what system engineering could have done. The present keyboard is horribly mismatched to the human typist and to the English language (or any other language). On the typewriter, the easiest key to hit is J, the second F, and so forth, in something like the following sequence:*

J	F	K	D	H		G	U	R	M	V		N	B	Y	T	I
24	12	22	10	9		16	15	6	14	21		5	20	17	2	7

E	C	L	S	O		W	X	A	P	Q	Z
1	13	11	8	4		19	23	3	18	25	26

Under each letter, the number indicates the relative frequency of that letter in English text (that is, E is the most common, T the second most common, and so on). One can only deduce that the original designer must have had the initials JFK.

In the properly designed keyboard, the letters would be arranged approximately in order of their frequency, although with modifications to take into account the probabilities associated with various pairs or triplets of letters. For example, TH is a common pair and should be easily typed.

Experiments have shown that typing speeds can be approximately doubled if the present keyboard is replaced with an optimal arrangement. Unfortunately, the practical possibility of redesign is nil. The enormous numbers of typewriters now in use and individuals trained on the existing keyboard make even a minor change impossible. Our great grandchildren will undoubtedly be using the same keyboard.†

The rigidity imposed by existing practice is similar to problems involved in this country's switch to the metric system of weights and measures. In this area, the changeover is inevitable because of the economic and trade limitations; when the U.S. has products not conforming in size to those of the rest of the world. Even in this area, there are aspects which are unlikely to change in our lifetime. One cannot imagine the

*This sequence is merely the author's personal estimate. Others arrange the letters in somewhat different order. In all listings, however, the E, T, and A appear far down the sequence.

†The poor design of the keyboard is surprising in the light of the fact that the first commercially available typewriter did not appear until 1874 from E. Remington and Sons (long after the invention of the Morse Code, which did take into account the relative frequency of different letters). Only early in the 20th century did use of the typewriter become common, although Mark Twain was the first author to submit a typewritten book manuscript in the late 19th century.

TV announcer of a football game saying, "First down and 9.5 meters to go." We shall certainly live with a curiously mixed system for a long time, well after the official ten-year transition period.

Mass transit

The U.S. mass transit systems provide a variety of current examples of very difficult problems in system engineering. In the New York City subway system, for example, the map showing routes of the many different trains is the author's favorite example of the horror of data pollution. With several of the trains changing routes during rush hour periods, the map is so complex that a doctorate in cartography is clearly a prerequisite for successful interpretation.*

To be absolutely sure that the map is of no use to the subway rider, another disadvantage is added. The maps are placed only inside the cars. Thus, a passenger waits on the platform at the 42nd Street stop for a train; he knows his desired uptown destination. One of three possible trains appears (let us say A, D, and F as an example). In the 15 seconds the train stops to unload and load passengers, he must dash into the car, find the map, determine which train goes to his desired destination, and get back on the platform before the door closes. Actually, he is likely to be at 96th Street before he has interpreted the map.

Ridiculous as this "system" may seem, it is an order of magnitude better than the New York City bus system, where if maps do exist, they are apparently classified *Secret*. A visit to London, Paris, or Moscow will convince the person using the system that it can be designed to match his needs and capabilities. And we have not even considered the aesthetic aspects. (Moscow subway stations, for example, are well-lit, clean, and attractively decorated.)

Another example of mass transit problems occurred in 1970 when Chicago's commuter railroad line was opened: it ran down the middle of the expressway westward out of the city center. The imaginative plan was that automobile commuters, clogged in traffic, would observe the trains speeding by and be encouraged to switch to mass transit. (Unfortunately, within days after the rail system opened, there were several breakdowns and derailments—with train riders helplessly watching the cars go by—but this was merely a temporary setback.) Here we see the system design taking into account the interplay with the people—the necessity of "selling" the mass transit facility if it is to effect change in public habits.

Zip code

The postal zip code was introduced in the late 1960s to speed the sorting of mail and hopefully to allow automated machines to do the sorting and routing. After a few years' experience, it became apparent that the system problem was much more difficult than anticipated. Typically, appreciably less than half the mail was sorted correctly by the automatic equipment.

*In 1970, strip maps, showing only the stops of the particular train, began to appear inside the cars.

Several shortcomings were apparent. One was that the zip codes used the normal numbers from 0 through 9. Because of widely varying number forms in available typewriters and addressing equipment, several of the numbers were often confused (e.g., the 1 and the 7). In retrospect, zip codes should have been chosen using ten distinctly different symbols from the total of 26 available letters and ten numbers.

To sort airmail letters from first class, the equipment utilized the color of the stamps (with special dyes used to match to the characteristics of the automatic equipment).* The location of the stamp is used to orient the envelope in the sorting machine, but unfortunately the zip code may appear anywhere within a wide region.

The disappointing performance of the automation equipment led to a variety of suggestions for improvement. For example, the first three numbers of the zip code indicate the county or city. Can stamps be designed so that these three numbers, or at least the first digit, can be marked on the stamp (by blacking out a square)? The sorting machine can then accomplish the regional distribution by observation of a clearly defined signal.

Telephone system

The U.S. telephone system, not long ago a primary pride of American technology and now the target of widespread public dissatisfaction, includes some of the most impressive achievements of system engineering. One problem so far has proved beyond the capabilities of the best engineers: how to build a vandalism-proof pay telephone for an open, street location. In the central city area, it is not unusual to find more than half the public pay phones out-of-order—often with the box ripped apart. In some areas, destruction is widespread only hours after maintenance crews have completed repairs. Each time a new design is introduced to prevent theft of the coins, novel techniques are developed by the vandals.

In general, other telephone problems, sometimes acute in the larger cities, can be traced in part to the unexpectedly rapid growth in usage (for example, because of the computer) and shortages of trained maintenance personnel. Some of the publicized complaints about the system operation are of questionable justification. For example, when the area codes were introduced, there was loud bitterness over the need to dial 10 numbers. When the name of Information was changed to Directory Assistance,† there were arguments that most people would not know what the new term meant.

Role of the electrical engineer

The four examples above (the typewriter keyboard, mass transit, zip code, and telephone) perhaps begin to suggest the breadth of system engineering and, above all, the two essential characteristics:

---

*For high performance, the equipment must recognize that two eight-cent stamps can be used instead of one 11-cent airmail, if the letter weight is within limits.

†The change was made because of the tendency of some customers to call Information to ask about a broad range of subjects, unrelated to telephone numbers.

(1) the multidisciplinary aspect, with involvement of scientists, engineers, and professionals from several different fields in a single problem;

(2) the people-relatedness, with the design of each part of the system dependent on the characteristics of the human beings who interface the system.

These two characteristics are further emphasized in the next section, which discusses briefly an example of a beautifully designed system.

The electrical engineer has two motivations for his interest in systems. First, many of the important systems are primarily electrical or electronic. Automation of an industrial process, automatic control of traffic flow, guidance and navigation for a space vehicle to Mars, and an artificial arm manipulated by electrical signals from the brain are all system examples within electrical engineering. In each case, design depends on the fundamental concepts of dynamic analysis, system synthesis, feedback, and stability which we will introduce in the following chapters.

Equally important is the fact that techniques of electrical engineering are the foundations of modern system engineering, even when the systems are totally nonelectrical. The analysis of the spread of epidemics or rumors (Chap. 14) involves no electronic components, yet EE techniques are essential. Thus, the electrical engineer today is a focal individual in the modeling, analysis, and design of urban and social systems. Fortunately, the important concepts in these nonelectronic applications are those which we consider in the following chapters.

1.3 CARDIAC PACEMAKER

Electrical stimulation of human tissue was undoubtedly familiar to the coastal South Americans of ancient times through their contacts with electric eels while swimming. In 1774, there was a report in England of a three-year-old child being revived by electric shock applied to her chest after she had "died" from a fall. In 1791, Galvani published the report of his famous experiments involving electrical stimulation of frogs' legs.

In modern times, a 1929 medical report told of the restoration of heartbeat by electrical shock. It was 1952, however, before another clear demonstration was carefully reported in the literature.

During the late 1950s, the cardiac or heart pacemaker was first developed—an electronic oscillator to deliver an electrical pulse to the heart at regular intervals (for example, 70 times per minute) in order to keep the heart beating at a regular rate in spite of malfunctioning of the normal, human excitation system. Today, just a little more than a decade later, well over 100,000 pacemakers have been installed; most of the patients who received pacemakers since 1963 are alive and leading useful lives. The use of the pacemaker is today such a routine medical operation, it receives little press attention even when performed on a Supreme Court Justice.

Use of the pacemaker

The pacemaker is used in case of heart block—either total which corresponds to a stop of the heart beating, or partial, representing such a low beating rate that heart output is dangerously low. The heart block may occur because of malfunctioning of the electrical system which normally stimulates the pumping of the heart, disease of the heart, and shock from surgery or other causes.

The average adult has a heart rate about 70 pulses per minute, although athletes and others may have a much lower rate (down to 40). Under stress or physical exertion, the heart rate rises to 170 beats/minute (e.g., when an astronaut steps onto the lunar surface or a test pilot enters a particularly dangerous maneuver).

The beating of the heart can be stimulated by an electrical pulse (applied at the appropriate location) of an amplitude about 3.7 volts, and an energy in the pulse of about 100 microjoules.

A succession of pacemakers

The earliest pacemakers were external oscillators (outside the body), with wires leading from the oscillator through the patient's skin to the heart. During installation, the chest was opened to allow the doctor to implant the electrodes at the desired spot in the heart.

This early model had certain disadvantages. The patient had to carry the oscillator around, strapped to his body. Bathing was impractical. There was a tendency for infections at the point where the two wires entered the skin. A normal life was impossible.

It was fortunate for the millions of future users that during the 1950s there were extensive military research and development in the U.S. on miniaturization of components, long-life and reliable batteries, and electronic pulse circuits which require very little energy. As a direct result of this work, an implantable pacemaker became possible. The small oscillator circuit and battery (of a total size comparable to a package of cigarettes) were encased in plastic and placed under the skin on the left side at about waist level. Two wires led from the oscillator within the body up to the heart and were inserted with open chest surgery.

With a battery life of perhaps 16 months, patients had to return to the hospital regularly to have the battery replaced, but the operation was minor since it was only necessary to cut directly above the oscillator.

In 1965 the transvenous pacemaker was introduced. In this system, the oscillator (and battery) are implanted. The leads are then fed into a vein near the oscillator and pushed through the vein to the heart (just as a catheter is installed). The physician installing the pacemaker watches the wires moving through the veins toward the heart with a fluoroscope, possibly supplemented by x-ray pictures. Thus, only one operation is involved, which is at the point where the oscillator is implanted; the dangerous open chest surgery is avoided entirely.

Furthermore, the transvenous pacemaker proved especially attractive for emergency situations. The pacemaker leads can be inserted through the skin into a vein in the upper shoulder and pushed to the heart (with the oscillator-battery package left outside the body during the period of emergency care).

As always with technological development, these achievements failed to satisfy the user, in this case the physician. As soon as new technology becomes available to solve a particular problem, the customer always dreams of even greater potentialities. The next step in pacemaker evolution was in the direction of "noncompetitive" pacing: in other words, a pacemaker which pulses the heart *only* when the normal mechanism is inoperative. If the pacemaker is set for 860 milliseconds between heart beats, the oscillator emits a pulse only if there is no natural pulse by the end of 860 msec since the last beat.

There are two major advantages of such a system. First, if the patient's heart is beating normally, the pacemaker is idle and we would predict much longer life for the batteries. Actually, such devices have been used only for the last few years, and data on longevity are not yet available. It is possible that the longevity will actually be less than anticipated because of the more complex circuitry and the extra (third) lead needed to detect normal heart activity. The second advantage is of direct interest to the patient. By making the artificial pacemaker subsidiary to the normal functioning of the heart, there is possibility of a system in which heart rate increases automatically during periods of personal need—e.g., violent exercise. Thus, a more normal life is possible.*

The major disadvantage of the noncompetitive pacer is its susceptibility to electromagnetic interference. Since the electronic pulse generator operates in response to electrical signals from the heart, any stray electrical signals which pass through the skin to the oscillator and leads are apt to be detected as heart signals. In the presence of strong electromagnetic signals from external sources, the pacemaker may even stop operation.

Thus, in about 15 years, the electronic pacemaker has moved through an evolutionary cycle from a very crude device used only on critically ill patients in a few research hospitals, to a widely accepted prosthetic device, available even in small-town hospitals. As recently as 5 years ago, scientific meetings reported on the early experiments, in terms of the results achieved with 20 test patients. Today's meetings are concerned with discussions of which *type* of pacemaker is most appropriate for a particular medical problem; the pacemaker is a standard therapy for both heart block and many other irregularities in heart rhythm.

*In order to allow patient control of the heart rate, one system includes a small, portable radio transmitter with which the patient can send signals through the skin to the oscillator to command a desired change in rate. When the patient is ready for strenous exercise, he turns up the rate to perhaps 145 beats per minute; at the end of the exercise, he turns it down. The disadvantages of such a system are obvious: often we do not know in advance when we are to become excited.

Other possibilities for manual adjustment of the oscillator frequency are by an externally controlled electromagnet, or by a needle which is inserted through the skin to move the oscillator control mechanically.

Cost-benefit analysis

One of the important analytical tools widely used in business and government is cost-benefit analysis, involving a quantitiative investigation of the dollar benefits of a particular program versus the cost. While such analysis involves intensive studies of the economic impacts following a particular innovation, we can easily make an initial and superficial evaluation of the pacemaker.

When 100,000 adult males have been kept alive and leading normal lives with the pacemaker, we can estimate the saving to all levels of government as 1 billion dollars per year ($10,000 per man, representing the taxes he pays on earned income and the costs of support of his dependents had he died from heart block). One billion per year is approximately the income expected from an investment of $20 billion.

We are saying that even with these crude calculations and conservative estimates, the pacemaker justifies $20 billion of governmental investment during the last two decades in electronics research and development—a justification which includes no dollar value placed on the human lives saved and the associated improvement in the quality of life in this country.

Furthermore, the pacemaker is a byproduct of our federal research program which could not have been anticipated in the early 1950s. (We recall from the beginning of this section that it was only in 1952 that the first modern case occurred in which the heart beating was restarted by electrical pulsing.)

Technological problems

The evolution of the pacemaker over the past decade has been marked by a series of technological problems. We mention a few to illustrate the engineering contributions that have been made by researchers, particularly in companies which have been involved in manufacturing the device.

(1) Breakage of the flexible leads from the oscillator to the heart has caused some failures. The heart beats about once a second, and each beat flexes the leads.

(2) Children present a particular problem since, as they grow, the leads must extend.

(3) The plastic case containing the batteries and oscillator must remain intact in the body environment (that is, surrounded by body fluids) and must be tolerated by the body. In the early years, particularly, the plastic casing tended to crack, especially near the exit of the leads.

(4) There is a continuing search for better energy sources: batteries with life longer than the current 20 months, or implantable, nonchemical sources. In the latter category, piezoelectric sources have been tried: the mechanical motion of the heart itself or of the body leads to electric energy generation. A research program is studying nuclear energy sources.

(5) In the realm of electronic circuitry, research focuses on reduction of the energy requirements, reliability, safeguards against the oscillator running away (e.g., with pulsing rates up to 300 per minute), and effective shielding from electromagnetic interference.

In each of these problems, reasonably satisfactory solutions have been achieved, and the refinements anticipated during the 1970s will provide to the physician a more reliable and flexible device with greater longevity.

1.4 CONCEPTS UNDERLYING SYSTEM ENGINEERING

In a retrospective look at either the problems of Sec. 1.2 or the pacemaker success of Sec. 1.3, we can begin to appreciate the fundamental concepts which constitute the foundation of system engineering.

Modeling

The first concept of *modeling* is the determination of those *quantitative* features which describe the operation of the system. In the example of the typewriter keyboard design, a system-engineering approach encompasses an understanding of the frequencies of various letters and pairs of letters—in other words, a quantitative model of the language. Then the engineer must describe the appropriate characteristics of the human typist: which keys are used most easily, which pair combinations are easily struck, and so on; and what determines human accuracy and fatigue?.

There are two fundamental aspects of this modeling process. First, we restrict our model to those aspects which are important in the evaluation of system operation. In the keyboard study, we do not attempt to model the human being, but only the operation in the typing task. Second, we must be quantitative. Intelligent design demands a comparison of different possibilities and making choices. The decisions can only be made logically if we have a quantitative basis for the comparisons, even if the numbers associated with particular facets of operation are merely our best guesses or very crude estimates. In the pacemaker example, the electronic pulse-forming circuit can only be designed after we decide on the pulse voltage and energy required, the necessary reliability, the allowable energy drain from the battery, and the like.

This modeling part of system engineering is the key, starting point. The model determines which features of the system will be incorporated in the final design. The model itself is always a compromise. For the most refined picture of the system, we would like a very complex model which neglects as little as possible. For the sanity of the engineer and his chances of success in design, we would prefer the simplest possible mathematical model.

Dynamics

The second, fundamental concept of system engineering is *dynamics*: the idea that essentially all interesting system problems are dynamic in nature, with the signals changing with time and the components determining the dynamic response of the system.

In very simple terms, the system behavior depends not only on the signals at any instant, but also on the rates of change of the signals and their past values. Even in our trivial, keyboard example, the relative ease of hitting various keys cannot be meaningfully measured by having the typist hit just one key. We must be concerned with the effects of a sequence of keys and with the consequences of fatigue which build up over a period of time.

In a more complex system problem, the control of a city's air pollution, an intelligent control policy must be based on the recent history of various types of pollution and, equally, on the dynamics of the city's economics and sociology. Cleaning up the city's air is not a practical goal if the byproduct is moving all industry and all middle-class residents out of the city.

Optimization

Finally, system engineering involves *optimization*: the best possible policy or design choice among various alternatives. Every design is a sequence of compromises or choices. In the pacemaker example, the competitive model (with the pacemaker assuming complete control of the heart beating) leads to the simplest circuit, probably the most reliable device, and certainly the prosthesis which is the easiest for the surgeon to implant. The noncompetitive version has its advantages, as mentioned in Sec. 1.3, and the ultimate choice must be an optimization which assigns relative weightings to the importance of each aspect.

In system engineering, optimization is often exceedingly difficult. Very rarely are we able merely to list all possible designs and determine a value for each, so that we can simply choose the best. Often, indeed, which design is optimal depends critically on the subjective decisions we make in regard to the relative importance of different features. For example, if a city mayor decides economic growth is more crucial than cleaner air, he may logically reach the conclusion that he cannot impose severe restraints on electric energy generation (and thereby increase the cost of electricity used industrially).

Optimization is frequently complicated further by the fact that complex systems often respond in a manner which Forrester* calls counter-intuitive—i.e., the response tends to be just the opposite of what one would think. Even simple electronic systems may exhibit such behavior. For example, Fig. 1.1 shows the beginning of the response

*In "Urban Dynamics," M.I.T. Press, Cambridge, Mass., 1969, Jay W. Forrester emphasizes (with computer simulations) the counter-intuitive behavior of many urban system problems.

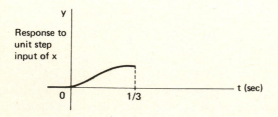

y

Response to
unit step
input of x

0   1/3   t (sec)

FIG. 1.1 Initial response of the system
$y^{iv} + 6.5y''' + 12.25y'' + 6y' + 2.25y$
$= 0.375x' - 1.75x'' - 3.125x$.

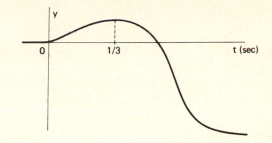

FIG. 1.2 Total response of system of Fig. 1.1. (Exactly this sort of behavior often occurs in the vertical control of an airplane; motion of the pilot's stick to cause an increase in altitude initially causes the plane to drop slightly.)

described by the given differential equation. If we continue the response, however, it eventually changes sign completely (Fig. 1.2). In other words, the initial response gives a false indication of the future behavior. If we were trying to position a system with this characteristic and adjusted the input signal every 1/3 second, the human control task would be impossible.

1.5 FINAL COMMENT

In this introductory chapter, we have discussed system engineering in broad terms. As we become engrossed in the following chapters with the details of modeling, analysis, and design of electrical systems, the broader aspects of system engineering tend to become obscured by the specific minutiae. Even the key concept of optimization assumes a very secondary role, essentially because intelligent optimization is more easily considered after we have a background in analysis and design.

The importance of electrical system engineering derives, however, from its applicability to an awesome range of problems, many of which have no electrical elements at all. The ideas of dynamic performance, simulation, feedback and stability, with which we shall be concerned in the chapters that follow, are fundamental concepts within the modern system approach.

2 SYSTEM DIAGRAMS

It was suggested in Chap. 1 that system studies require quantitative models describing the significant elements of the system and the way these elements are interconnected. Only after such a model is developed can we hope to understand system behavior or design a system to meet performance specifications.

The most familiar models in electrical engineering are circuit diagrams, displaying the interrelationship of the component R's, C's, L's, sources, and so forth. In system engineering (or the systems part of EE), we are interested not only in electric circuits, but also in systems involving human beings, economic or social elements, and physical elements from all fields of science and engineering. Consequently, circuit diagrams are of limited use, requiring that we seek more general, pictorial models. In this chapter, two of the most important types are discussed: *block diagrams* and *signal flow diagrams*. These two types of *system diagrams* are very nearly equivalent; the one that is used is largely a matter of the engineer's personal preference.*

---

*In later chapters we use the two forms interchangeably and rather arbitrarily. In Sec. 2.8 we compare the two in more detail.

Before defining the block diagram, however, a very simple system problem (totally nonelectrical) is considered in order to understand better the type of problem we want the system diagrams to describe.

2.1 THE SYSTEM OF URBAN APARTMENTS*

An interesting (although oversimplified) system governs the number of vacant apartments in a particular city each year. We focus on apartments in a specific, upper price-range where, when the demand is high, there is an economic incentive for entrepreneurs to construct new buildings. The system is interesting because there is a time lag between the decision to build new apartments and the availability: a time lag typically of three years.

Before we do any modeling, we can try to guess what the system will do. If there is a shortage of apartments for a year or two (vacancies are few, profits for the owners are high), new construction will be extensive. Years later, when the new apartments become available, there may be excessive vacancies, causing a sharp drop in new construction starts. With the three-year time lag, the system may tend to oscillate. Will this happen? Will the oscillations be severe? Can the city government adopt a regulatory policy to smoothe out these oscillations? How should an individual entrepreneur improve his position in the field (e.g., should he start new construction when vacancies are high, but falling, to try to anticipate a shortage of apartments)? These are questions we might logically ask of the model. These are answers widely sought throughout government and business today as people increasingly turn to "the systems approach."

To construct a model of the urban-apartments system, we need to *quantify* the relationships which describe system behavior. Can we associate numerical values with the various factors we have described above?

The first question is: What are the signals (variables) which describe the state of the system at any time? In our case, we might decide to look at the system once a year, or take an average of any signal over that year (conditions probably do not change too greatly from day to day, and perhaps we can get by with a year-by-year variation). Assuming this to be valid, we are certainly interested in the number of vacancies each year—how many apartments are unrented? We call this signal

$$V_n$$

the number of vacancies in the year n. V_n is the measure of the condition of the total system in year n. In a sense, V_n is the *output signal* of the system: the effect of the system or the response of the system.

---

*This system example is one of many described in "Problems in Industrial Dynamics," edited by W. E. Jarmain, M.I.T. Press, Cambridge, Mass., 1963—an offshoot of the classic modeling text, "Industrial Dynamics," by Jay W. Forrester, M.I.T. Press, 1961. That book focused on computer modeling of management systems; a later work "Urban Dynamics," (by Forrester, 1969) centered on urban systems.

What is the *input signal,* causing the system to change each year? It is the number of new rentals (called R_n). There may be new rentals because of people moving into the city, people moving into this cost-range of apartments, or young people leaving parents and setting up their own home. R_n is the net increase, and also must include the reductions due to people leaving apartments in our price class.

In its simplest form, the system has one input (R_n) and one output (V_n). Now we have to move inside the system: how is V_n determined from R_n? Again, a quantitative model is desired.

The number of vacancies in the year n depends on the number of vacancies the year before minus the new rentals during the preceding year.

$$V_n = V_{n-1} - R_{n-1} \tag{2-1}$$

This equation must be modified, however, to take into account the new construction started three years ago. If C_n is the number of construction starts in the year n, the complete equation for V_n is

$$V_n = V_{n-1} - R_{n-1} + C_{n-3} \tag{2-2}$$

In other words, C_{n-3} is the number of additional apartments newly available in the year n. (Actually, some apartments may be abandoned or may move out of the price-range we are considering, but this merely adds another term to the equation. For simplicity, we neglect this factor).

We now have one equation in our model, but we have also introduced a second unknown, C. Consequently, an equation for C_n is required. How many new construction starts are there in the year n? Obviously, the number depends on the policy of the entrepreneurs who build apartment houses. To construct a model, we would have to observe past behavior and see if there is a relation between C_n and the vacancies V_n during that same year. Let us assume such observations indicate that

$$C_n = \alpha(1000 - V_n) \tag{2-3}$$

Each year the number of new starts is a constant α times the deviation of V_n from the normal value of 1000.

Equations (2-2) and (2-3) now constitute the quantitative model for the urban-apartments system. To predict the vacancies each year in the future, we can calculate one year at a time from the unknown data for this year and the past.

For example, if we are interested in V_{1973}, this is

$$\left. \begin{aligned} V_{1973} &= V_{1972} - R_{1972} + C_{1970} \\ C_{1970} &= \alpha(1000 - V_{1970}) \end{aligned} \right\} \tag{2-4}$$

V_{1973} can be found from V_{1972}, R_{1972}, and V_{1970} (data for earlier years).

Cause-effect relationships

The two Eqs., (2-2) and (2-3), are *cause-effect* relationships within the system. Equation (2-2) states the effect V_n which results from causes V_{n-1}, R_{n-1}, and C_{n-3}. C_{n-3} is itself also an effect [Eq. (2-3)] resulting from V_{n-3}.

Thus, the equations of the model are representations of the cause-effect pairs which, in total, constitute the system.

Final comments on this model

Problem 2.1 involves calculation of the response of this simple model when $\alpha = 0.8$. There we find that, starting with 1000 vacancies, the system is actually unstable: V_n oscillates with growing amplitude and finally goes negative (an absurd situation, of course, meaning only that the model no longer is valid).

Clearly, this is an exceedingly simple system. We could easily make it somewhat more realistic by making the rents charged depend also on the number of vacancies (with many vacancies, rents would be reduced to try to attract more customers or increase R_n). The model is also more complex if the number of new construction starts depends on not only the current vacancies V_n, but also the vacancies the preceding year V_{n-1}. Finally, governmental policy might limit or discourage the number of new construction starts.

Figure 2.1 shows the actual data for new housing (in all price ranges) in New York City over the past 25 years. Attempts to explain the decrease in the late 1960s uncover the important dependence of new construction on mortgage interest rates—a factor which we have not considered in our model. The wide signal variations show the general tendency of the system to instability.

The purpose of the example is to emphasize the nature of the modeling step: the transaction from an understanding of how the system works in general terms to a quantitative description of the ways the signals are interrelated. We shall find in the following sections that a system diagram is often of great help in visualizing the significance of the model relationships.

2.2 THE BLOCK DIAGRAM

A block diagram shows the interplay of cause-effect relations in a system. The diagram consists in most cases of the two types of elements indicated in Fig. 2.2 (addition or subtraction can be shown with either the circle or the block; usually we use the circle). In every case the input signals (the causes or drives for the element) are shown by arrows entering the circle or block.

A *block diagram* is the interconnection of these building blocks to show the relations among the signals within the system. For example, Fig. 2.3 shows a possible block diagram. To find the equations corresponding to this system, it is necessary to define each signal at the output of a block or circle. These are called $y_1, y_2, \ldots, y_8$

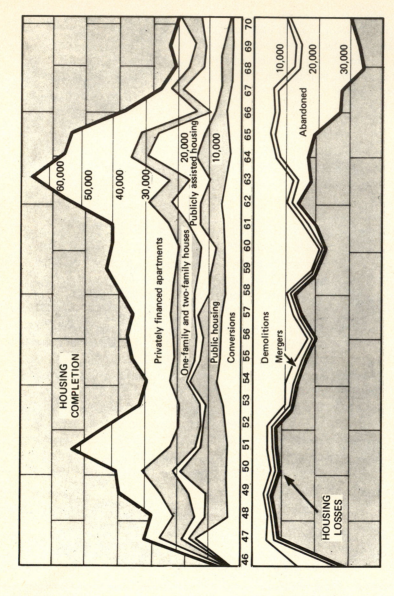

FIG. 2.1 New York City housing situation during the last 25 years. The term "conversion" means splitting existing housing into more units; "merger" is a loss by change to a non-housing use. The curves show emphatically the bases for the alarm about 1963 over excessive housing availability and the concern only seven years later about the severe shortage (with the dire predictions that by 1975 thousands of people would have to be housed in armories and public buildings). © 1971 by The New York Times Company. Reprinted by permission. From the *New York Times*, Sec. 8, p. 1, March 14, 1971.

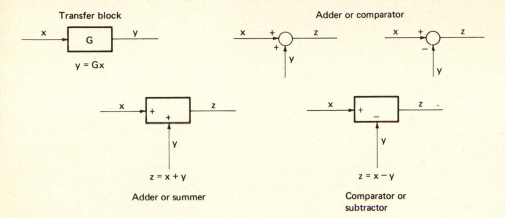

FIG. 2.2 Elements used in drawing a block diagram.

in Fig. 2.4. The equations describing the system then, are:

$$
\left.
\begin{array}{ll}
y_1 = G_1 x & y_6 = G_3 y_5 \\[2mm]
y_2 = y_1 + y_4 - y_8 & y_7 = H_2 y \\[2mm]
y_3 = G_2 y_2 & y_8 = H_3 y \\[2mm]
y_4 = H_1 y_6 & y = G_4 y_6 \\[2mm]
y_5 = y_3 - y_7 &
\end{array}
\right\}
\qquad (2\text{-}5)
$$

An equation is written for each of the *dependent* variables (x is an independent variable or drive signal, which must be given if the other signals are to be determined).

Obviously the set of nine equations is unnecessarily cumbersome; several can be combined immediately to yield a more compact mathematical model, but this expanded form does show how we can always go from a block diagram to a set of equations. *The same information is contained in either form of the model: the equations or the block diagram.*

Why use a block diagram?

Why use a block diagram if the equations are equivalent? The block diagram is most useful when it portrays the flow of signals through the system. For example, Fig. 2.5 shows an elementary model for the system by which a man controls the temperature of the water during his shower. The system input signal is the desired temperature; the output, the actual temperature. There is an input comparator representing his measurement of the error. He uses any knowledge he has of the shower to decide, on

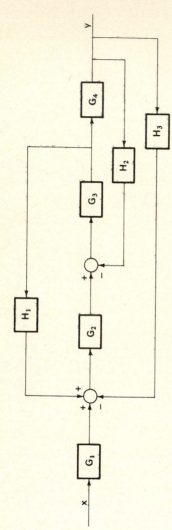

FIG. 2.3 A block diagram of moderate complexity.

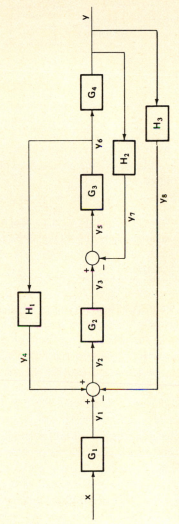

FIG. 2.4 All different signals indicated.

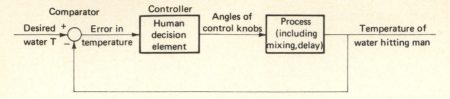

FIG. 2.5 Basic system for controlling shower temperature.

the basis of this error, how much to turn the knobs controlling hot and cold water flow. This signal (the angles the knobs are turned) eventually causes a change in water temperature.

The block diagram shows three distinct parts of the system:

Comparator (or the error measurement)
Controller (here the human decision element)
Process (the water-mixing system, the time delay until this temperature change reaches the man)

If we wished to study the system operation either mathematically or with a computer, the characteristics of each of these components would have to be determined in detail.

Actually, the human decision element is quite complex in this example: the man normally makes a change in knob angles, then waits the amount of time he considers reasonable to see how the temperature changes before he makes a new adjustment. In other words, he disregards the error for an interval of time. If he is impatient (or underestimates the time required for the temperature to change), he inserts an additional correction, then the first temperature change occurs, then he adds another correction, and so on. The system is very likely to be unstable. The detailed description of the controller block must include the man's ability to estimate accurately the length of time required before the process responds to an input control signal.

Thus, the block-diagram model portrays the cause-effect relationships in the system. It shows pictorially the various factors which influence system operation. Figure 2.6 is a more typical electrical engineering example: the block diagram for an electronic amplifier with three stages, represented by the three transfer functions $K_1(s)$, $K_2(s)$, and $K_3(s)$. In addition, there are networks feeding signals backward (leftward) into the

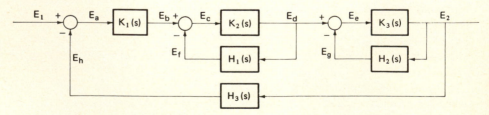

FIG. 2.6 A three-stage, electronic amplifier.

comparators. In later chapters, we will see how $H_1(s)$, $H_2(s)$, and $H_3(s)$ can be used to improve the performance of the total system.

In this system, the block diagram is equivalent to the set of equations:

$$
\left.
\begin{aligned}
E_a &= E_1 - E_b & E_e &= E_d - E_g \\
E_b &= K_1 E_a & E_f &= H_1 E_d \\
E_c &= E_b - E_f & E_g &= H_2 E_2 \\
E_d &= K_2 E_c & E_h &= H_3 E_2 \\
& & E_2 &= K_3 E_e
\end{aligned}
\right\} \tag{2-6}
$$

Most systems engineers find the picture of Fig. 2.6 much easier to interpret than the Eqs. (2-6); this is the merit of the block diagram as a form of the system model.

Constructing the block diagram

In many cases, the block diagram can be drawn directly from an understanding of how the system operates: what are the various cause-effect relationships which combine to determine the output from the input. In the shower-temperature system of Fig. 2.5, we developed the block diagram in this way.

There are occasional situations when we want to go from a set of equations to a block diagram. This transition is always possible and the key is evident upon inspection of Eqs. (2-5) or (2-6): we simply write the equations in a form in which each dependent variable appears *alone* on the left side of one equation.

As an example, we can consider the simple system described by the three equations:

$$
\begin{cases}
G_1 Y + G_2 Z + G_3 W = X & \tag{2-7} \\
G_4 Y - G_5 W = 0 & \tag{2-8} \\
G_6 Z + G_7 W = G_8 X & \tag{2-9}
\end{cases}
$$

The input (or independent variable or drive signal) is X; the dependent variables (or responses) are Y, Z, and W—three in number since there are three simultaneous equations. The G's are constants (or, more generally, depend on the complex frequency s or the real frequency ω). The three equations may come from circuit analysis (they might be node equations) or from an analysis of operation of the separate parts of the system.

To begin, we want to rewrite the equations in the desired form: each equation solved for a separate dependent variable. If we arbitrarily choose to solve the first for Y, the second for W, and the last for Z,

$$Y = \frac{1}{G_1}(X - G_2Z - G_3W) \tag{2-10}$$

$$W = \frac{1}{G_5}(G_4Y) \tag{2-11}$$

$$Z = \frac{1}{G_6}(G_8X - G_7W) \tag{2-12}$$

Now the block diagram is drawn in the steps shown in Fig. 2.7.

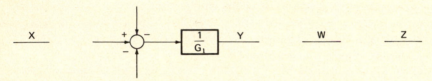

(a) The variables are indicated. X, the input, is shown at the
left. The three equations are in a form indicating a flow from
X to Y to W to Z.

(b) Equation (2-10) says that Y is $1/G_1$ times the sum of three terms.

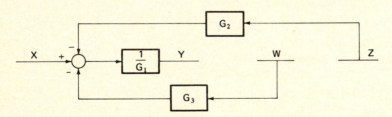

(c) Equation (2-10) is realized, term by term.

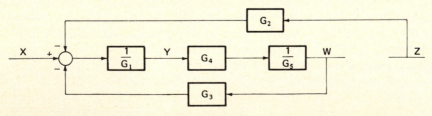

(d) Equation (2-11) for W in terms of Y is added.

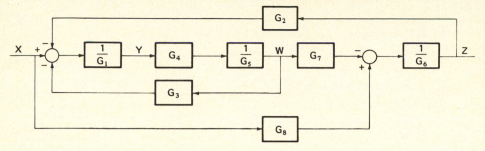

(e) Final realization of all three equations.

FIG. 2.7 Steps in construction of the block diagram representing Eqs. (2-10), (2-11), and (2-12).

This block diagram is only one of many which can be drawn to represent the original set of system equations, (2-7)–(2-9). For example, we could have solved the three equations for Z, Y, W, respectively (rather than Y, W, Z), to obtain a very different block diagram. We could form a new set of three equations by using linear combinations of Eqs. (2-7)–(2-9): e.g., Eqs. (2-7), (2-8), and the sum of Eqs. (2-8) and (2-9).

Usually, in drawing a block diagram, we try to maintain the greatest amount of correspondence possible with the manner in which we intuitively believe the signals flow through the system. This is an important concept, but also difficult to describe precisely. In the above example, for instance, we would use Fig. 2.7 if it seemed natural to picture the

$$X \rightarrow Y \rightarrow W \rightarrow Z$$

flow as the primary set of cause-effect relationships.

The block diagram is the most important model of the systems engineer. There is a slightly different form, called the signal flow diagram, which we consider next.

2.3 SIGNAL FLOW DIAGRAMS

A signal flow graph is a pictorial representation of a set of simultaneous algebraic equations in which the variables are represented by a graphical symbol called a *node*, and the dependencies between pairs of variables are represented by *directed branches* drawn between pairs of nodes. Signal flow graphs are particularly appealing when used to portray the equations describing dynamical systems because the nodes can then be interpreted as representing the signals, or time functions, of the system, and the branches can be visualized as the transmission paths, or transfer functions, which perform specified operations on the signals flowing along them.

A signal flow graph differs in a minor way from a set of conventional equations in that an explicit causality is indicated in the graph. As an elementary example, Newton's

FIG. 2.8 Elementary signal flow graph.

FIG. 2.9 Alternate interpretation.

second law of motion is expressed:

$$\text{Force} = \frac{d}{dt}(\text{momentum}) \; ; \quad F = \frac{d}{dt}p \tag{2-13}$$

where $p = Mv$ is the momentum. To graph this relation, it is necessary first to decide which variable is regarded as independent (cause), and which as dependent (effect). With momentum taken as the cause, the graph appears as in Fig. 2.8 where s symbolizes differentiation. For a body of constant mass, M, the graph might be drawn as in Fig. 2.9.

To reverse the cause-and-effect relation, new graphs are needed such as those in Fig. 2.10, where $1/s$ symbolizes intergration. (Because of the integration process involved, it is necessary to think about an initial condition when the graph portrays a physical system).

Nodes

In addition to their role as symbols representing signals or variables, the nodes possess two interesting operational properties:

(1) they are summing devices each of which sums all signals arriving by way of incoming branches;

(2) they also act as signal repeaters: from each node, the node signal is dispatched or propagated equally along all branches leaving that node.

Branches

The branches operate upon the signals entering the branch in the arrow direction and deliver a new branch signal to the node on which they terminate. The operations are indicated by the branch labels which may be simple multiplication by a constant, or more generally by a transfer function. The direction of signal flow is always in the arrow direction, irrespective of the sign of the branch label.

As a consequence of these basic properties, certain aspects of flow graphs become obvious. Thus, for the two graphs of Fig. 2.11, it is a simple matter to evaluate the overall transmittance or transfer function in each case.

FIG. 2.10 Causality inverted.

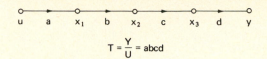

$$T = \frac{Y}{U} = abcd$$

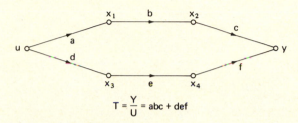

$$T = \frac{Y}{U} = abc + def$$

FIG. 2.11   Basic operations.

It is noteworthy that the nodes of Fig. 2.11 labeled u have only outgoing branches, while those labeled y have only incoming ones. In these special circumstances where the nodes perform but one of their operational functions, they are termed respectively *sources* and *sinks*. The sources, which are free of incoming branches, are clearly representative of the independent variables inasmuch as they have no dependence upon any other variables of the system.

As a further consequence of the basic node properties, it is a simple matter to write the set of equations portrayed by the graph. Each node signal represents a summation of incoming branch signals; therefore, the node signals represent separate equations, each of which can be read directly from the graph. If the graph of Fig. 2.12 is used as an example, the equations are:

$$\left.\begin{aligned}
x_1 &= au + hx_2 \\
x_2 &= bx_1 + eu + jx_2 + ky \\
x_3 &= fx_1 + cx_2 \\
y &= dx_3 + gx_3 = (d + g)x_3
\end{aligned}\right\} \tag{2-14}$$

This is exactly the same equation format we found from the block diagram in the preceding section: each *dependent* variable is a linear function of other variables. Thus, the signal flow and block diagrams are really equivalent models:

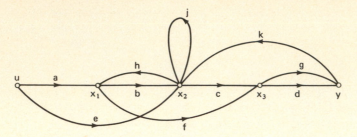

FIG. 2.12 Example—loops, nodes, and paths.

Block diagram	*Signal flow diagram*
Line carrying a signal	Node
Block G_1	Branch with "gain" G_1
Comparator or adder	(Addition and subtraction done at the node)

It is noteworthy that this graph displays several *closed* signal paths or *feedback loops*. The branch j which begins and terminates on x_2 is called a self-loop and is characterized by containing only a single node. In general, loops contain several nodes, as for example the loop with gain $fgkh$, which also illustrates the fact that the recognition of a loop may not be obvious.

This is an appropriate place to call attention to the fact that a proper loop is one in which no node is encountered more than once per cycle. The closed path $fgkjh$ is improper for the reason that when the self loop, j, is included as part of the path, the node x_2 is encountered twice. A similar situation exists in recognizing and identifying open paths; again no node is encountered more than once in a proper path. In Fig. 2.12 there are several open paths from the input u to y, with the following path transmittances or gains

$$\left.\begin{aligned}
P_1 &= abcd \\
P_2 &= abcg \\
P_3 &= ehfd \\
P_4 &= ehfg \\
P_5 &= afd \\
P_6 &= afg \\
P_7 &= ecd \\
P_8 &= ecg
\end{aligned}\right\} \tag{2-15}$$

A path such as *ehbcd* is improper because x_2 is encountered twice. The multiplicity of paths could have been halved, of course, by first combining the branches d and g into a single branch $(d + g)$.

Formulating a graph from a set of equations

The nature of the node equations suggests a procedure for formulating a graph from a given set of equations. In the process, it is necessary to select a *different* dependent variable from each equation and to rewrite the equation so that the selected variable equals a sum of terms. The right-hand side of the equation then effectively specifies the branches required to portray the equation. For the set of equations given below, it is assumed that four dependent variables have been selected arbitrarily and the equations written in the required form. It is also assumed that it is known that u is the independent variable.

$$
\left.
\begin{aligned}
x_1 &= au + jy \\
x_2 &= bx_1 + hy \\
x_3 &= eu + cx_2 + fx_3 + gy \\
y &= dx_3
\end{aligned}
\right\}
\tag{2-16}
$$

To construct the corresponding signal flow graph, we first sketch the nodes in any, arbitrary position [usually the independent variables—here only u—are placed at the left and the output of primary interest (here we assume y) at the right]. Figure 2.13 (a) results, and we are now ready to add the branches for each equation, one at a time. Figure 2.13 (b) shows the equation for x_3 alone graphed, and part (c) shows the completed graph.

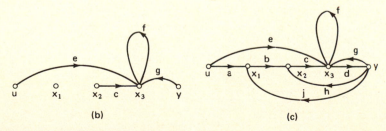

FIG. 2.13 Graph formulation.

Second example

The following set of equations is given:

$$\left.\begin{array}{l} 4x_1 - 2x_2 - x_3 = u_1 + 2u_2 \\ 2x_1 + 3x_2 = 3u_1 \\ -x_2 + 2x_3 = 3u_2 \end{array}\right\} \qquad (2\text{-}17)$$

and it is known that u_1 and u_2 are excitation signals or drives. Each equation is then solved for a different dependent variable:

$$\left.\begin{array}{l} x_1 = \frac{1}{4}u_1 + \frac{1}{2}u_2 + \frac{1}{2}x_2 + \frac{1}{4}x_3 \\ x_2 = u_1 - \frac{2}{3}x_1 \\ x_3 = \frac{3}{2}u_2 + \frac{1}{2}x_2 \end{array}\right\} \qquad (2\text{-}18)$$

The graph for this selection of solutions appears in Fig. 2.14. Clearly if we had decided to solve the equations for x_3, x_1, and x_2, respectively, we would have obtained a very different signal flow diagram. The diagram is not unique; finding the simplest signal flow diagram for a given system often involves considerable trial and error.

Graph for an electrical network

In developing a signal flow graph from an electric circuit, the first step is to write a set of circuit equations. There is considerable flexibility in this procedure since the equations may be developed in several different ways. For example, a set of node equations, or a set of mesh equations, or superposition may be used.

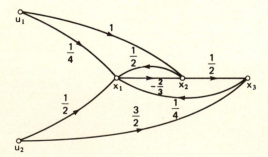

FIG. 2.14 Second example—graph formulation.

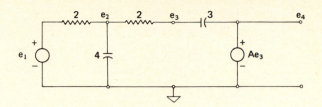

FIG. 2.15 Electric circuit model.

The circuit of this example is drawn in Fig. 2.15 and includes a controlled source Ae_3. The node* equations for the dependent or free junctions are

$$4se_2 + \frac{1}{2}(e_2 - e_1) + \frac{1}{2}(e_2 - e_3) = 0 \qquad (2\text{-}19)$$

$$\frac{1}{2}(e_3 - e_2) + 3s(e_3 - e_4) = 0 \qquad (2\text{-}20)$$

and the controlled source yields

$$e_4 = Ae_3 \qquad (2\text{-}21)$$

Solving for the dependent variables in an arbitrary order gives

$$e_2 = \frac{1}{8s + 2} e_1 + \frac{1}{8s + 2} e_3$$

$$e_3 = \frac{1}{6s + 1} e_2 + \frac{6s}{6s + 1} e_4 \qquad (2\text{-}22)$$

$$e_4 = Ae_3$$

The corresponding flow graph of Fig. 2.16 results.

*The *node* of an electric network has distinctly different properties from a node of a flow graph.

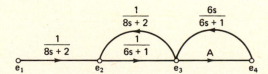

FIG. 2.16 One possible flow graph.

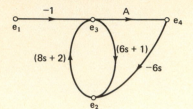

FIG. 2.17 A second version of the graph.

With a different choice in the order of solving for the dependent variables, the equations appear as:

$$e_3 = (8s + 2)e_2 - e_1$$

$$e_2 = (6s + 1)e_3 - 6se_4 \qquad\qquad (2\text{-}23)$$

$$e_4 = Ae_3$$

The corresponding graph of Fig. 2.17 presents an entirely different appearance, yet it depicts the same interrelations of the variables and the overall transmittance E_4/E_1 remains unchanged.

One final variation is shown, particularly to underline the versatility of form which is at the analyst's disposal. The network is redrawn in Fig. 2.18 with three independent variables: e_2 and e_5 (the capacitor voltages) and the output e_4. (When e_5 is defined, the e_3 in the original Fig. 2.15 is simply $e_4 + e_5$, and the controlled source can be labeled correspondingly.) Now the circuit equations can be written in terms of these variables:

$$\begin{cases} 4se_2 + \frac{1}{2}(e_2 - e_1) + \frac{1}{2}(e_2 - e_5 - e_4) = 0 \\[2mm] 3se_5 + \frac{1}{2}(e_5 - e_2) = 0 \\[2mm] e_4 = A(e_4 + e_5) \end{cases} \qquad\qquad (2\text{-}24)$$

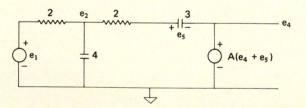

FIG. 2.18 Circuit redrawn with capacitor voltages (e_2 and e_5) and output e_4 as the dependent variables.

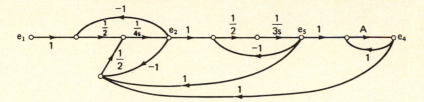

FIG. 2.19 Signal flow diagram for Eqs. (2-25). The interesting feature of this model is that s appears only in the form $1/s$ (integration) and in only two branches for this second-order system.

These are rewritten in the form

$$
\left.
\begin{aligned}
e_2 &= \frac{1}{4s}\left[\frac{1}{2}(e_1 - e_2) + \frac{1}{2}(e_4 + e_5 - e_2)\right] \\[2mm]
e_5 &= \frac{1}{3s}\left[\frac{1}{2}(e_2 - e_5)\right] \\[2mm]
e_4 &= A(e_4 + e_5)
\end{aligned}
\right\}
\qquad (2\text{-}25)
$$

and the signal flow diagram is shown in Fig. 2.19.

As the figure caption states, this is a very particular form (called a state model) which we will study in more detail in Chap. 6. At this point, we are only interested in emphasizing that a particular circuit or system can be represented by many different signal flow diagrams (or many different block diagrams).

2.4 MASON'S THEOREM FOR REDUCTION OF SYSTEMS

Given the block diagram or signal flow diagram of a linear system with various feedback loops, we often wish to write the transfer function from a particular input to a particular response signal. A straightforward approach would entail writing the simultaneous algebraic equations described by the system diagram, elimination of all variables (signals) except the particular input and output, and rewriting in terms of a transfer function.

Mason's theorem allows us to write the transfer function directly from the block diagram or signal flow diagram. The theorem expresses the transfer function in terms of the various loop gains and the parallel transmittances from input to output.

The theorem

Mason's theorem states that the transfer function from input X to response Y is

$$T = \frac{Y}{X} = \frac{\sum_i P_i \Delta_i}{\Delta} \tag{2-26}$$

where the terms are defined as:

(1) Δ is the determinant of the feedback configuration and is calculated from the equation

$$\Delta = 1 - \sum L_j + \sum{}' L_k L_l - \sum{}' L_m L_n L_o + \cdots \tag{2-27}$$

Here the L_j (or L_k, etc.) are loop gains—transmittances all the way around a feedback loop in the system. Thus ΣL_j means the sum of *all* loop gains. The next term $\Sigma' L_k L_l$ is the sum of all products of pairs of different loop gains—e.g., $L_1 L_3$ and so forth. The prime on the summation means we use only the products for pairs of gains of *non-touching* loops. In other words, $L_1 L_3$ is included only if loop 1 does not touch loop 3.* Likewise, $\Sigma' L_m L_n L_o$ is the sum of all products three at a time, where again *each* of the three loops does not touch the other two.

(2) P_i is a direct transmittance from input X to output Y. P_1, for example, is the gain of one path from X to Y (a path which contains no loops); P_2 the gain of a second parallel path (if there is a second).

(3) Δ_i is the system determinant (Δ), after we have excluded all loops which touch the P_i path.

Since the formula depends on the various closed or feedback loops in the diagram and the various direct paths from input to output, we list first all the loops, then all the direct paths. Then Δ can be determined. Finally, the Δ_i are found, and the desired transfer function is written.

Example 1 in a signal flow graph

Figure 2.20 shows the signal flow graph for a system in which we wish to determine the transfer function Y/X. The various branch transfer functions are labeled a, b, c, through i, as shown. The overall transfer function is determined in the following steps:

(1) First we list all loop gains. There are three closed loops (three separate paths along which it is possible to return to the starting point while always following arrow directions):

$$L_1 = bg \qquad L_2 = df \qquad L_3 = bcdh$$

---

*Two loops "touch" if they have at least one node in common.

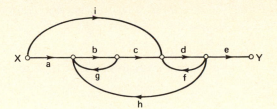

FIG. 2.20 First example of Mason's theorem.

Hence

$$\sum L_j = bg + df + bcdh$$

(2) Next we determine the non-touching pairs of loops. Here L_3 touches both L_1 and L_2 (i.e., L_3 and L_1 have at least one node in common), but L_1 and L_2 are non-touching. Hence

$$\sum' L_k L_l = bgdf$$

(3) There are no non-touching triplets.
(4) Now we can write the determinant Δ:

$$\Delta = 1 - \sum L_j + \sum' L_k L_l$$

or

$$\Delta = 1 - (bg + df + bcdh) + bgdf$$

(5) We now enumerate all the P_i—all the direct paths from X to Y:

$$P_1 = abcde$$

$$P_2 = ide$$

There is no other way to move from X to Y without forming a closed loop in the process.

(6) Next we calculate the Δ_i. Path P_1 touches all loops (L_1, L_2, and L_3); hence Δ_1 is calculated with no loops involved and thus is unity (*not zero*):

$$\Delta_1 = 1$$

Path P_2 touches loops L_2 and L_3, but not L_1. Hence, Δ_2 is calculated as

$$\Delta_2 = 1 - L_1 = 1 - bg$$

(7) Now we are ready to substitute in Mason's formula:

$$T = \frac{Y}{X} = \frac{P_1\Delta_1 + P_2\Delta_2}{\Delta} \tag{2-28}$$

$$T = \frac{abcde + ide(1 - bg)}{1 - (bg + df + bcdh) + bgdf} \tag{2-29}$$

Actually, the seven steps above have been described in much more detail than necessary when we gain familiarity with the calculations. After working only a few examples, we should be able to bypass all detailed steps and write the T of Eq. (2-29) directly from inspection of the diagram of Fig. 2.20.

Example 2 in a signal flow graph

Figure 2.24 illustrates a problem in which there are three non-touching loops. In this case, the loop gains are

$$L_1 = af$$

$$L_2 = be$$

$$L_3 = cd$$

$$L_4 = abcg$$

Here L_4 touches all three others, so the non-touching pairs are

$$L_1L_2 \qquad L_1L_3 \qquad L_2L_3$$

and the non-touching triplet is

$$L_1L_2L_3$$

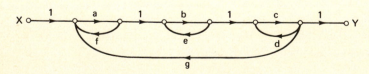

FIG. 2.21 Second example—three non-touching loops.

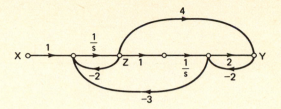

FIG. 2.22 Determination of an internal variable z.

Hence,

$$\Delta = 1 - (L_1 + L_2 + L_3 + L_4) + (L_1 L_2 + L_1 L_3 + L_2 L_3) - L_1 L_2 L_3 \tag{2-30}$$

and

$$T = \frac{Y}{X} = \frac{abc}{1 - (af + be + cd + abcg) + (afbe + afcd + becd) - (afbecd)} \tag{2-31}$$

Example 3 in a signal flow diagram

Only one other aspect is noteworthy in the application of Mason's theorem. Often we wish to find the transfer function from the drive x to an internal variable z. In this case, we merely treat Z as the response, but we must include loops which originate at Z in the determination of Δ and Δ_i. For example, Fig. 2.22 gives:

$$L_1 = -\frac{2}{s} \qquad L_2 = -4 \qquad L_3 = -\frac{3}{s^2} \qquad L_4 = \frac{24}{s}$$

$$\Delta = 1 + \frac{2}{s} + 4 + \frac{3}{s^2} - \frac{24}{s} + \frac{8}{s} = 5 - \frac{14}{s} + \frac{3}{s^2}$$

$$P_1 = \frac{1}{s} \qquad \Delta_1 = 1 + 4 = 5 \quad \text{(all loops except } L_2 \text{ touch } P_1\text{)}$$

$$T = \frac{Z}{X} = \frac{5/s}{5 - 14/s + 3/s^2} = \frac{5s}{5s^2 - 14s + 3} \tag{2-32}$$

Actually, there is no complication here. If preferred, we can add a new output node (called Z_0) with a gain of unity from Z, as shown in Fig. 2.23, where we would find Z_0/X. Occasionally, this "dummy" output clarifies the analysis, but it is not really necessary.

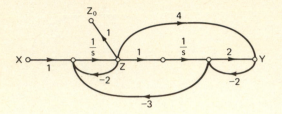

FIG. 2.23 Redrawn version of Fig. 2.22, showing Z_0 as an output variable.

Example 4 in a block diagram

When we use block diagrams, rather than signal flow diagrams, Mason's theorem is unchanged. The configuration of Fig. 2.24 illustrates the approach.

Now the loop gains are determined (we must remember to include the minus signs introduced by the comparators):

$$L_1 = -\frac{6}{s} \qquad L_2 = -\frac{11}{s^2} \qquad L_3 = \frac{-5}{s^3}$$

and all three loops touch (all include the Y-line output).* The two forward paths are

$$P_1 = \frac{4}{s^3} \qquad P_2 = \frac{2}{s^2}$$

Since each touches all loops, $\Delta_1 = \Delta_2 = 1$, and

$$\frac{Y}{X} = \frac{4/s^3 + 2/s^2}{1 + 6/s + 11/s^2 + 5/s^3} = \frac{4 + 2s}{s^3 + 6s^2 + 11s + 5} \qquad (2\text{-}33)$$

*In a block diagram, two loops touch if any one signal within the system appears in both loops.

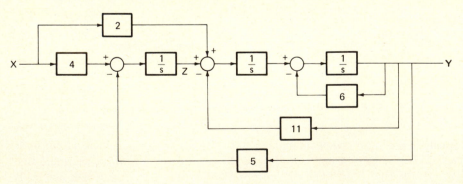

FIG. 2.24 Final example, in terms of a block diagram.

In this example, it is interesting also to calculate Z/X, where Z is the output of the leftmost integrator. Since Δ depends only on the loops (not on the output chosen), Δ is the same as in the Y/X calculation:

$$\Delta = 1 + \frac{6}{s} + \frac{11}{s^2} + \frac{5}{s^3}$$

There are now two "forward" paths from X to Z: one through 4 and $1/s$; the other through 2, $1/s$, $1/s$, 5, (-1), $1/s$.* Hence

$$P_1 = \frac{4}{s} \qquad P_2 = \frac{-10}{s^3}$$

The corresponding Δ_1 and Δ_2 are

$$\Delta_1 = 1 - \left(-\frac{6}{s} - \frac{11}{s^2}\right) \qquad \Delta_2 = 1$$

(Δ_1 includes the $-11/s^2$ loop since the direct path from x to z does not touch this loop). Thus,

$$\frac{Z}{X} = \frac{(4/s)(1 + 6/s + 11/s^2) - 10/s^3}{1 + 6/s + 11/s^2 + 5/s^3} = \frac{34 + 24s + 4s^2}{s^3 + 6s^2 + 11s + 5} \tag{2-34}$$

In such a calculation, errors are apt to arise in the determination of the numerator because of omission of the P_2 path (which winds circuitously from x to z), or through overlooking some of the loops involved in Δ_1.

Final comment
Two points should be emphasized:

(1) Mason's theorem gives a simple, fast procedure for writing a desired transfer function directly from the signal flow diagram or block diagram.
(2) For all transfer functions found from a single signal flow or block diagram, Δ is the same. Hence, the denominators of all transfer functions are identical.†

---

*We should particularly note that Z is the output of the leftmost integrator, *not* the output of the comparator. Consequently, to go from X to Z, we can travel this circuitous route.

†The numerators, of course, differ. We may find that in some transfer functions certain factors of the denominator are cancelled by like numerator factors. If the transfer functions are reduced to simplest form, the denominators may not all be the same.

2.5 COMPLEX SYSTEMS

If the system is very complex (possibly involving dozens of feedback loops), we run the risk of inadvertently omitting some of the loops in our enumeration of L_1, L_2, ... If such a case is unfortunately encountered, we can proceed in a systematic manner. First, a node in the signal flow diagram or a signal line in the block diagram is arbitrarily selected; we then find all loops passing through this point by following all possible paths leaving this point and subsequent branching points. As soon as all loops passing through this point have been listed, we can break the system there and proceed to the remaining system to find other loops in the same way.

Figure 2.25 is a horrendous example which illustrates the procedure.* Let us first focus on node z and try to find all loops originating here. There are three paths by which we can leave z: through i, d, or k. In each case (one at a time), we list all the possible ways to return to z without looping on ourselves:

$$i - c$$
$$k - n - a - b - c$$
$$ - j - e - l - c \;†$$
$$d - e - 1 - c$$
$$ - f - g - n - a - b - c$$

Thus, there are five loops starting from z and returning to z.

$$L_1 = ic \qquad\qquad L_3 = knajelc$$
$$L_2 = knabc \qquad\quad L_4 = delc$$
$$\qquad L_5 = defgnabc$$

*It is of interest to attempt to list all loops before reading further.

†Starting with k, we have no choice but to move through n and a. Then if we choose j, we must pass through e. Next we can go through h, but this loops back. l passes then to c and back to z so this is admissible. If we take f rather than l, we must go through g and then we are back at y, where we were when we originally passed through k.

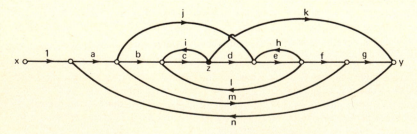

FIG. 2.25 Example illustrating attack on a complex system.

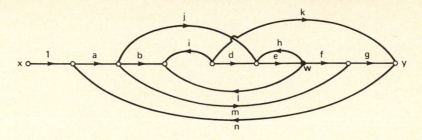

FIG. 2.26 Simplified diagram with node z split (omitting all loops through z).

Since *no* other loops pass through z, it is necessary to continue our search for loops omitting z. The only way to enter z is by c; hence, we redraw the diagram omitting branch c (Fig. 2.26).* We now ask: what are the loops in this diagram?

The node w seems central, so all loops through w should be found. We can leave w through h or f or l. To return to w, we can follow these paths:

$$h - e$$
$$f - g - n - a - j - e$$

Then there are two loops through w:

$$L_6 = he \qquad L_7 = fgnaje$$

Redrawing the diagram with w split results in Fig. 2.27. Upon inspection, we find only one loop left:

*Actually, we break the system at node z—in Mason's terminology, we split node z. All incoming branches are separated from the outgoing branches.

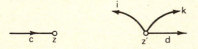

This is equivalent to just omitting all branches entering z.

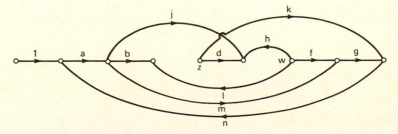

FIG. 2.27 Diagram with both z and w split.

$$L_8 = mgna$$

Hence, the original system has eight loops, which we have now enumerated.

The above paragraphs outline a systematic approach for enumerating all loops, but should not obscure the inherent simplicity of Mason's theorem. In the majority of system-analysis problems, the theorem permits the engineer to write a desired transfer function directly by inspection of the block diagram or signal flow diagram. In only the most complex cases must we resort to a systematic procedure to be sure all the loops are enumerated.

2.6   ALGEBRA OF SIGNAL FLOW GRAPHS*

The signal flow graph provides pictorial representation of the interrelationships of variables in a linear system. In order to amplify understanding, the system may be simplified by eliminating one or more internal variables. The signal flow graph merely represents graphically the set of algebraic equations. We could go back to these equations, eliminate the undesired variable, and redraw the graph. Actually, it is possible to work directly from the graph according to the basic rules by which the graph is defined.

Elimination of a node

To eliminate one node (say x_3), we simply construct the new diagram:

(1)   All nodes except x_3 are drawn.

(2)   All original branches *not entering or leaving* x_3 are inserted.

(3)   We add branches representing *every* possible path (in the original diagram) through x_3.

The last step is, of course, the tricky one; the meaning of these words is illustrated by the following example.

Figure 2.28 shows a signal flow diagram in which we wish to eliminate x_4—i.e., the aim is to draw an equivalent diagram with only x_1, x_2, x_3, and x_5. We follow these steps:

*Section 2.6 is written in terms of signal flow graphs, but the same techniques are applicable to block diagrams (as illustrated in the Problems at the end of this chapter). Actually, this section is the least important part of this chapter, but on rare occasions the techniques are very useful.

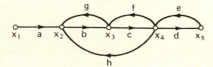

FIG. 2.28 Original signal flow diagram.

FIG. 2.29 Desired nodes.

FIG. 2.30 All original branches not contacting x_4.

(1) First we draw all nodes except x_4 (Fig. 2.29).

(2) We insert all original (Fig. 2.28) branches not touching x_4 (Fig. 2.30).

(3) Now we add all possible paths through the original x_4. We can enter x_4 from x_3 through c;

Then we can go from

x_3 to x_3 through cf

x_3 to x_5 through cd

x_3 to x_2 through ch

We can also enter x_4 from x_5 through e; therefore we can go from

x_5 to x_3 through ef

x_5 to x_2 through eh

x_5 to x_5 through ed

Hence, the final diagram is as drawn in Fig. 2.31. (This figure could be simplified by combining the g and ch branches into one of transmittance $g + ch$).

The six paths through x_4 in the original diagram are enumerated by entering x_4 along one path at a time—in each case combining with each possible departure path.

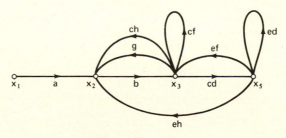

FIG. 2.31 Final equivalent of original signal flow diagram.

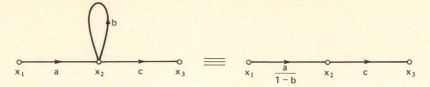

FIG. 2.32 Removal of a self-loop.

$$\begin{cases} x_2 = ax_1 + bx_2 \\ x_3 = cx_2 \end{cases} \quad \text{equivalent to} \quad \begin{cases} x_2 = \dfrac{a}{1-b}x_1 \\ x_3 = cx_2 \end{cases}$$

Elimination of a self-loop

A self-loop of gain T_a is eliminated at a node by dividing all *incoming* transmittances by $(1 - T_a)$. The equivalence (which can be proved by reference to the algebraic equations describing the flow graph) is indicated by Figs. 2.32 and 2.33. Any self-loop at a node should be removed before that node is eliminated.

Repeated node elimination

One node after another can be eliminated until the diagram is simplified to any desired extent. Figure 2.32 illustrates the elimination of x_2 and x_3 in Fig. 2.31 to derive a single branch yielding the x_5/x_1 of the original system of Fig. 2.28.

Actually, if we want a single transfer function or overall transmittance (as x_5/x_1 in Fig. 2.34), Mason's theorem is obviously preferable to the step-by-step elimination of nodes. The algebraic techniques of node elimination and self-loop removal are primarily useful to simplify or slightly modify the signal flow diagram.

2.7 ISOLATION OF ONE ELEMENT

In system analysis and design, we are frequently interested in a study of the effects of variation of a particular parameter (the gain of a controlled source or an amplifier, an R or C, the compressibility of the oil in a hydraulic amplifier, and so on). How do system characteristics depend on the value of this parameter?

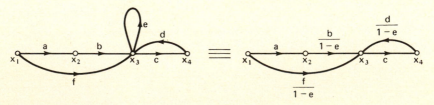

FIG. 2.33 Another example of removal of a self-loop.

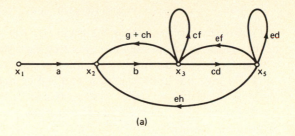

(a)

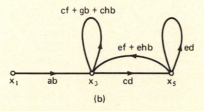

(b)

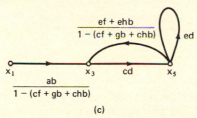

(c)

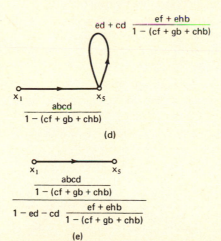

(d)

$$\frac{abcd}{1-(cf+gb+chb)}$$

$$1-ed-cd\ \frac{ef+ehb}{1-(cf+gb+chb)}$$

(e)

FIG. 2.34 Repeated node elimination in system of Fig. 2.31: (a) Redrawn version (with two parallel branches combined); (b) x_2 eliminated; (c) Self-loop at x_3 removed; (d) x_3 eliminated; (e) Single branch realized.

There are two common reasons for such an interest:

(1) We may anticipate that the parameter will vary significantly during operation of the system. In an automobile power steering system, the oil compressibility changes with temperature and with the amount of air captured in the oil.

(2) We may not know the parameter and be unable to measure it easily with any precision. Indeed, in complex systems (particularly those concerned with human or social problems), we must often guess at many parameters since experiments are too expensive or impractical.

In either case, understanding of system operation requires a *sensitivity analysis*: an evaluation of the sensitivity of the system performance to changes in the parameter value. This analysis may be carried out either analytically or experimentally (in the latter case, for example, by building a simulation in which the parameter can be varied by changing a potentiometer or gain constant).

Chapter 12 on *Feedback** explains how to define and calculate sensitivity mathematically. Here we are interested only in the simpler question: Can we redraw the signal flow or block diagram in such a way that the parameter under study appears *alone as the gain of a branch or block*?

Parameter initially a gain multiplier

If the parameter (called c for convenience) initially appears only once in the system diagram, and then only as the multiplier of a transfer function, the solution is trivial. Figure 2.35 shows an example of simply factoring out the c and breaking the corresponding branch into two tandem branches.

Parameter embedded in a transfer function

When c appears in one coefficient of a transfer function, we have to decompose that branch or block into a subsystem configuration with c representing the gain of one branch. An example is shown in Fig. 2.36. First we use straightforward factoring to simplify as much as possible the transfer function containing c; in this example,

---

*The principal technique available to reduce sensitivity to a particular parameter is to design the system with feedback around that parameter.

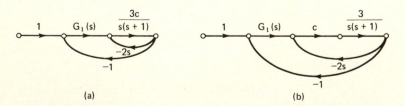

(a)   (b)

FIG. 2.35 System with a multiplicative parameter c: (a) Initial system diagram; (b) Diagram redrawn with c alone in a branch.

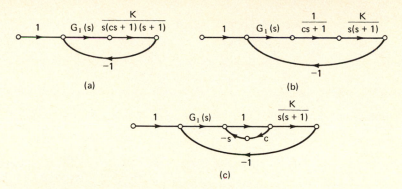

(a)

(b)

(c)

FIG. 2.36 Example with c a coefficient within a transfer function: (a) Original system diagram; (b) First step: function containing c simplified by factoring; (c) One possible solution.

consideration can be focused on

$$\frac{1}{cs + 1}$$

This transfer function must then be viewed as the input-output relation for a subsystem in which c appears alone. We know that a single-loop system (Fig. 2.37) is described by

$$\frac{\text{Out}}{\text{In}} = \frac{a}{1 - ba} \tag{2-35}$$

Direct comparison of Eq. (2-35) with our desired $1/(cs + 1)$ suggests the diagram of Fig. 2.36(c). (There are, of course, other possible configurations.)

In retrospect, we took the transfer function involving c as a parameter and rewrote this in a form which could be recognized as representing a subsystem configuration in which c is a simple gain. In general, the transfer function involving c may have either of the two forms shown in Fig. 2.38.* In either case, the decomposition is straightforward; we can redraw the total system diagram to display c alone in a single branch.

Figure 2.39 shows one final example in terms of a block diagram.

*We might also have a transfer function $n(s)/c$. In such a case, we treat $1/c$ as the parameter rather than c.

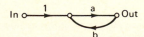

FIG. 2.37 General single-loop system.

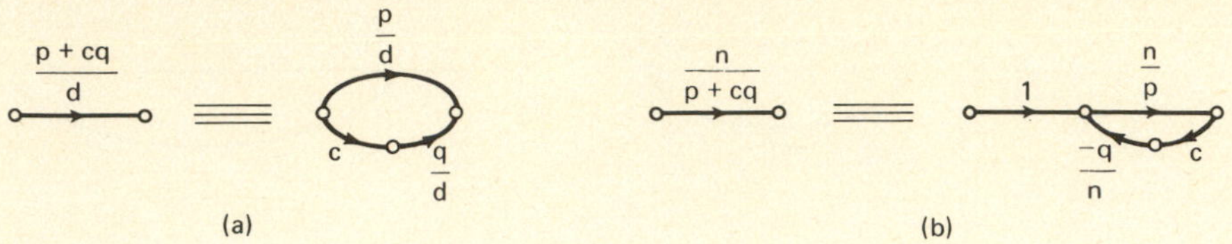

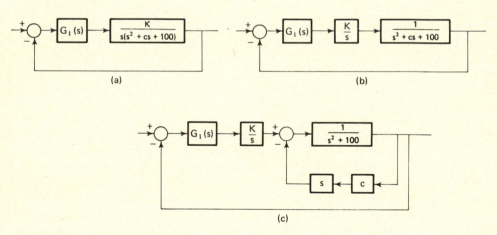

FIG. 2.38 c can appear either in the numerator or the denominator of the transfer-function. (Here p, q, d, n are in general polynomials in s.)

(a) $\dfrac{p + cq}{d} = \dfrac{p}{d} + c\,\dfrac{q}{d}$ Numerator

(b) $\dfrac{n}{p + cq} = \dfrac{n/p}{1 + cq/p}$ Denominator

Parameter appears more than once

Sometimes when we formulate the system equations and then the signal flow or block diagram, the parameter c appears in two or more places. Figure 2.40 shows one possible signal flow diagram for the simple network with R the parameter of interest. Here R appears in two different branches; it is not obvious how to modify the diagram so that R is alone only once.

FIG. 2.39 Decomposition when the parameter is a damping coefficient: (a) Original system with c the parameter of interest; (b) Simplification of the transfer function encompassing c. Now we view

$$\frac{1}{s^2 + cs + 100}$$

as

$$\frac{1/(s^2 + 100)}{1 + c[s/(s^2 + 100)]}$$

(We divide numerator and denominator by the part of the denominator *not* multiplying c.) (c) The final system diagram.

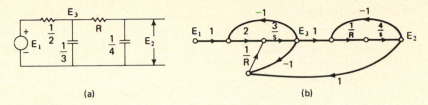

(a)   (b)

FIG. 2.40 The circuit, and the system model with the parameter R appearing twice: (a) Original circuit; (b) Signal flow diagram drawn by writing two node equations and solving for sE_3 and sE_2.

If R is a single parameter in the actual system, the appearance twice (or more) in the signal flow diagram means that we have not written the system equations in an appropriate form. We must return to the system and rewrite the equations with R only appearing once. For example, in the analysis of Fig. 2.40(a), we can replace R by a voltage source (Fig. 2.41); in the equations we use either I_R or E_R (as needed), and R only appears in the relation

$$E_R = I_R R \tag{2-36}$$

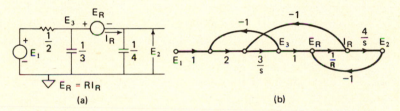

(a)   (b)

FIG. 2.41 Example of rewriting circuit equations so that a parameter appears only once: (a) Redrawing the circuit of Fig. 2.40 described by the equations

$$\begin{cases} 2(E_3 - E_1) + \frac{1}{3}sE_3 + I_R = 0 \\ \frac{1}{4}sE_2 - I_R = 0 \\ E_R = RI_R \\ E_3 = E_2 + E_R \end{cases}$$

(b) Signal flow diagram for (a) and the equations

$$\begin{cases} E_3 = \frac{3}{s}[2(E_1 - E_3) - I_R] \\ E_2 = \frac{4}{s}I_R \\ I_R = \frac{1}{R}E_R \\ E_R = E_3 - E_2 \end{cases}$$

2.8 FINAL COMMENT

Block diagrams or signal flow diagrams are the basic model for the systems engineer—a model which is applicable to the study of economic and social problems as well as to strictly engineering systems. The problems are designed to illustrate some of the breadth of applicability of such system diagrams.

In the 1950s, just after Mason introduced the signal flow graph, there was considerable discussion in the EE literature about differences between signal flow and block diagrams—the advantages and disadvantages of each form. We have tried to point out in this chapter that the differences are largely a matter of artistic form, rather than of technical substance.

It can be argued that block diagrams are used to represent the interconnection of *non-interacting* parts of the system; when we have two blocks in tandem, we mean two distinct parts of the system, with the output of the first being the input to the second (the *effect* of the first is the *cause* for the second).

According to this argument, the signal flow diagram is, in contrast, a graphical representation of a set of equations. The different branches need not represent distinct, noninteracting parts of the system. In other words, the signal flow diagram is a more detailed model of the system.

This argument is answered only in terms of the viewpoint of the engineer. He can always convert back and forth between the two forms at will. In the literature today, we find both forms used, and the engineer must be able to work from either. In subsequent chapters both forms are used, with no particular logic underlying the choice. However, some effort is made to use the form most common in the literature on that particular topic.

PROBLEMS

2.1 In Sec. 2.1, we derive a simple model for the number of vacancies annually in a city's apartments. We assume $\alpha = 0.7$, and we start with 1000 vacancies in each of the first three years; each year we rent 700 apartments. Determine the way the number of vacancies varies from the fourth year on. If $\alpha > 0.7$, what kind of system behavior results? Explain with an α of 0.8.

2.2 Determine a block diagram to represent the model of urban-apartment vacancies described in Sec. 2.1. What are the inputs or independent variables (including the initial conditions necessary to permit solution)? What are the possible outputs or responses of interest?

2.3 Write the equations for the signal flow graph shown below. Which are the independent variables? Which the dependent variables?

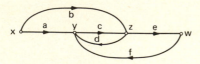

2.4 For the signal flow graph shown, write the transfer function Y/X.

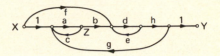

2.5 The block diagram shown represents a radar tracking system we will consider in more detail in Chap. 12. Each block is described by a transfer function. Determine the overall transfer function Y/X in two ways: (a) by algebraic solution of the simultaneous equations represented by the block diagram; (b) by use of Mason's reduction theorem.

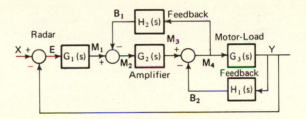

2.6 The operation of a system is described by the set of equations

$$3x - 2y = 5u$$
$$x + y + z = 3u$$
$$x + 2y + 1.6z = 4u$$

Here u is the input and the primary response is z. x and y are subsidiary, dependent variables.

Construct a signal flow diagram which represents this set of equations. Determine the overall transmission from u to z. Discuss your result.

2.7 In the signal flow graph of Prob. 2.3, find a new graph involving only the nodes x, y, and w. In other words, eliminate z.

2.8 For the signal flow graph of Prob. 2.4, determine the transfer function Z/X.

2.9 Calculate the overall transmission, y/x, for the system described by the signal flow graph shown. In the course of this solution, list the various loop gains, those which are non-touching pairs, those which comprise non-touching triplets, and so forth.

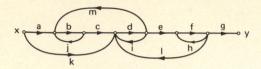

2.10 For the system shown, calculate the overall transmission y/x by direct application of Mason's theorem.

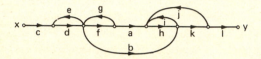

The diagram reveals that the system can be partitioned by a cut through branches a and b. In other words, a vertical line through these branches cuts only paths which represent transmittances from left to right. As a result, the overall transmission can be expressed as a product of transmissions in the left half and in the right half of this cut. Demonstrate that the two answers obtained are the same.

Show that the simplification is considerably greater if there is only one branch from left to right across the cut. (This can be demonstrated by assuming that $b = 0$ above.)

2.11 For the rather complicated system portrayed in Prob. 2.9, construct the block diagram. Comparison of these two system diagrams indicates how the electrical engineer often chooses between the two possibilities. The signal flow diagram is used to yield a simple picture when the system description is very detailed and complex. The block diagram is often used to show the interrelation of a few major subsystems, as in Prob. 2.5, where each block itself may represent a system of some complexity.

2.12 Signal flow diagrams have been used to describe the characteristics of various organizational arrangements. For example, the diagram shown represents the flow of memoranda in an organization. Each man in the chain of command sends a memoranda back up to his boss for every one received, perhaps to indicate he is working hard and has taken action. (a is obviously less than unity.)

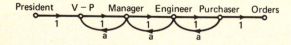

The diagram shows the flow of information when the President makes a decision (perhaps ordering the purchase of staples, clips, or the like). The President sends out a series of signals, and memoranda result down the line.

The interesting output is the number of orders actually resulting from the President's signals. Determine the system transfer function. For what minimum value of a does the transfer function become infinite, meaning an uncontrolled flood of orders? How would this answer be changed if we added additional loops in the chain of command?

The signal flow diagram is, of course, oversimplified, since no time delays are shown. In general, such delays tend to make the system even more unstable.

2.13 The signal flow graph can also be used to portray the way a system moves from one state to another. For example, in a coin-flipping exercise, we can consider the problem of finding the probability of throwing the third consecutive head for the first time on the sixth toss of the coin. Here we are interested in three consecutive heads; we reach this state by going from one head to two in a row, then to three in a row. These three states are called H_1, H_2, and H_3, and having just thrown a tail is called the state T. Then, the probabilities of transitions among the states are shown by the signal flow graph given.

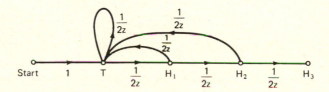

For example, when we are at state H_1 (we have just thrown a head preceded by a tail), we can move to H_2 with probability $1/2$ or move back to T with probability $1/2$. The $1/z$ term symbolically represents a delay of one toss.

(a) Determine the overall transfer function as the ratio of polynomials in z.

(b) Divide the denominator into the numerator to find the transfer function as an infinite series in $1/z$.

(c) The coefficient of $1/z^6$ is the probability of reaching H_3 after exactly six tosses (or on the sixth toss). What then, is the answer to our original problem?

(d) What is the most likely toss on which the third consecutive head will appear?

3 TRANSFER FUNCTIONS

The transfer function is the most important tool used by the system engineer to describe the behavior of a system. The basic analysis problem, in terms of Fig. 3.1, is: Given the input or excitation signal x and the system, what is the output or response signal y? We might also have the related problems:

Given the system and the desired y, what should x be?
Given the input x and output y, what is the system?

All of these problems need a convenient mathematical way to describe or define the system. In this chapter, we consider the transfer function* for this purpose; in later chapters, alternative descriptions are discussed (state models, analog or digital simulations, and so forth).

---

*The term "system transfer function" is also used; naturally, we then tend to abbreviate this to "system function."

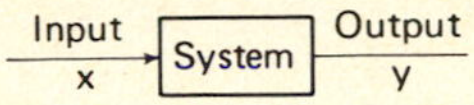

FIG. 3.1   A two-port system (the input port and the output port).

3.1   WHAT IS THE TRANSFER FUNCTION?

The transfer function for a two-port system (with a single input or excitation and one response or output) is defined as the ratio of the Laplace transforms of the output and the input. In other words, in Fig. 3.1, the transfer function $T(s)$ is given by

$$T(s) = \frac{Y(s)}{X(s)} \tag{3-1}$$

The transfer function is a useful concept only when the system is *inert*,* *linear*, and *time-invariant*; under these circumstances, $T(s)$ describes the response of the system to any input signal $x(t)$, and the resulting $y(t)$ can be calculated by Laplace transform methods. If the system is not inert (i.e., if there is initial energy storage), $y(t)$ depends on the initial conditions; if the system is nonlinear or time-varying, a transfer function cannot be defined in the ordinary way. The remainder of this chapter, then, emphasizes linear, time-invariant systems with no initial energy storage.

Furthermore, our primary interest is in *lumped* systems, where the components are R, L, C, M, and controlled sources for electrical networks or the analogous parameters in nonelectrical systems. Under this restriction, the transfer function is a ratio of polynomials in the complex frequency s. This situation occurs when the system is the interconnection of elements of impedance [a, bs, $1/(cs)$, ds^2, $1/(es^2)$, etc.], where impedance is defined as (across variable/through variable) [(voltage/current) in the electrical case]. Indeed, in most systems, the impedances are of the form a or bs or $1/(cs)$, and such elements are combined with simple controlled sources.

Thus, we are most often concerned with transfer functions which are ratios of polynomials in s. For example, we might have

$$T_1(s) = \frac{2}{s^2 + 2s + 2} \qquad T_2(s) = \frac{8(s^2 + s + 4)}{s^5 + 2s^4 + 6s^3 + 12s^2 + 4s + 1}$$

$$\tag{3-2}$$

Relation to differential equations

The system can also be described by a differential equation relating the response y to the excitation x. The $T_2(s)$ above, for instance, states

$$\frac{Y(s)}{X(s)} = \frac{8s^2 + 8s + 32}{s^5 + 2s^4 + 6s^3 + 12s^2 + 4s + 1} \tag{3-3}$$

*By "inert," we mean that no energy is stored in the network at the time $x(t)$ begins (normally at $t = 0$).

If we simply cross-multiply,

$$(s^5 + 2s^4 + 6s^3 + 12s^2 + 4s + 1)Y = (8s^2 + 8s + 32)X \qquad (3\text{-}4)$$

For an inert system (zero initial energy storage), multiplication by s corresponds to differentiation. Hence, Eq. (3-4) can also be written

$$y^v + 2y^{iv} + 6y''' + 12y'' + 4y' + y = 8x'' + 8x' + 32x \qquad (3\text{-}5)$$

The system is described by a single, fifth-order differential equation.

We can convert back and forth between the differential equation and the transfer function by recognizing s as a representation of differentiation. In this sense, s is an operator (sometimes called D or p in texts).

If the system is described by a set of differential equations, a particular transfer function can be found algebraically. For example, the equations

$$y' + 2y - z = 2x \qquad (3\text{-}6)$$

$$z' + 3z - y = x \qquad (3\text{-}7)$$

define a system with one input (x) and two possible outputs (y and z). If we are interested in the transfer function Y/X, we first transform the equations to obtain

$$(s + 2)Y - Z = 2X$$
$$(s + 3)Z - Y = X \qquad (3\text{-}8)$$

Now Z is eliminated algebraically

$$Z = (s + 2)Y - 2X$$
$$(s + 3)(s + 2)Y - 2(s + 3)X - Y = X$$

Simplification yields the transfer function

$$\frac{Y}{X} = \frac{2s + 7}{s^2 + 5s + 5} \qquad (3\text{-}9)$$

Alternatively, the x-y relationship can be defined by the single differential equation

$$y'' + 5y' + 5y = 2x' + 7x \qquad (3\text{-}10)$$

which is equivalent to the two equations, (3-6) and (3-7).

Thus, the transfer function Y/X is entirely equivalent to the differential equation relating y to x.

Relation to forced response

There is one other, possible definition of the transfer function. If a system is described by the transfer function

$$\frac{Y}{X} = T(s) \tag{3-11}$$

and if the input x is a single exponential

$$x = Ae^{s_1 t} \tag{3-12}$$

the *forced* part of the response is

$$y_f = AT(s_1)e^{s_1 t} \tag{3-13}$$

In other words, $T(s)$ *is the gain of the system at the frequency s.* When an input e^{st} is applied, the output contains a component (the forced component) at the same frequency. The amplitude of this output component is just the input amplitude multiplied by the transfer function evaluated at that frequency.

To show the validity of this interpretation of the transfer function as the system gain, we consider a specific example

$$\frac{Y}{X} = \frac{as + b}{s^2 + cs + d} \tag{3-14}$$

The differential equation is

$$s^2 y + csy + dy = asx + bx \tag{3-15}$$

where s represents differentiation. If x is $Ae^{s_1 t}$, an exponential at the frequency s_1, the forced component of y has the form

$$Be^{s_1 t}$$

where B is the amplitude to be determined. Now differentiation of x gives $s_1 Ae^{s_1 t}$; similarly with y. Hence Eq. (3-15) becomes

$$(s_1^2 + cs_1 + d)Be^{s_1 t} = (as_1 + b)Ae^{s_1 t} \tag{3-16}$$

Then

$$B = \frac{as_1 + b}{s_1^2 + cs_1 + d} A \tag{3-17}$$

The amplitude of the forced component of y is just the transfer function evaluated at s_1 multiplied by the amplitude of the input. The transfer function is the gain at s_1.

Summary

The transfer function $T(s)$ or (Y/X) is

(1) The Laplace transform of y divided by the Laplace transform of x;
(2) The representation of the differential equation relating y to x (with s meaning differentiation);
(3) The system gain for the exponential drive of frequency s.

To a large extent, the power of the transfer function concept results from the availability of all three interpretations. We can go back and forth at will among these.

3.2 CRITICAL FREQUENCIES

The transfer function

$$T(s) = \frac{p(s)}{q(s)} \tag{3-18}$$

is the ratio of the polynomials p and q—the numerator and denominator, respectively. The denominator $q(s)$ is called the *characteristic polynomial*; the *characteristic equation* is formed by equating q to zero:

$$q(s) = 0 \tag{3-19}$$

The zeros of the characteristic polynomial (or the roots of the characteristic equation)* are the *poles of the transfer function.*

The zeros of the numerator polynomial p are the *zeros of the transfer function.* Taken together, the poles and zeros of $T(s)$ are called the *critical frequencies*—the values of s at which T is either infinite or zero.

For example,

$$T(s) = \frac{3(s + 2)(s - 2)}{s(s + 1 - j2)(s + 1 + j2)(s + 4)^2} \tag{3-20}$$

---

*An equation has *roots*; a polynomial has *zeros*. Some authors use the terms interchangeably.

has the following critical frequencies

Poles	Zeros
0	−2
−1 + j2	+2
−1 − j2	
−4	
−4	

The $(s + 4)^2$ factor in the denominator can be described as two poles at −4 or, usually, as a *double* pole at −4.

Importance of poles and zeros

The importance of the poles and zeros of T is apparent from our third interpretation of the preceding section: the idea that $T(s)$ is the gain at the frequency s. When T has a zero at s_1, it means the gain is zero at this frequency. Even if an input $Ae^{s_1 t}$ is applied, there is *no* output at this frequency. The network blocks transmission of this frequency.

In contrast, if T has a pole at s_2, the gain at that frequency is "infinite." There is an output even with no input. In other words, the system output contains a term $Be^{s_2 t}$ even when there is no similar input. Thus, the poles are the *natural frequencies* of the system: the frequency components present in the output as a result of the system, rather than the drive.

For example, in the problem of Fig. 3.2, we see at once that the output y contains the three components

$$y = \underbrace{Ae^{-2t}}_{\text{Forced}} + \underbrace{Be^{-t} + Ce^{-3t}}_{\text{Free}} \tag{3-21}$$

The first is the forced component (at the same frequency as the input); the last two are the free or natural components at the natural frequencies of the system (the poles of the system function: −1 and −3).

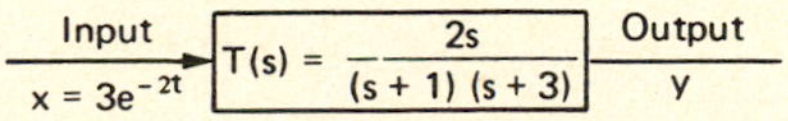

FIG. 3.2 Particular system and drive signal.

$$\text{Input} \quad x = 3e^{-2t} \quad \rightarrow \quad T(s) = \frac{2s}{(s+1)(s+3)} \quad \rightarrow \quad \text{Output} \quad y$$

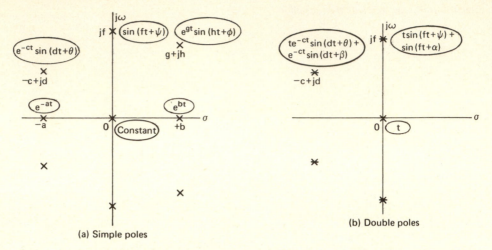

FIG. 3.3   Response for different pole locations: (a) Simple poles; (b) Double poles.

Stability

The importance of the poles of the transfer function is demonstrated in Fig. 3.3, which shows the type of natural response anticipated from poles located in various regions of the s plane. (When a complex pole appears, it always occurs in a conjugate pair.) The figure reveals that the poles determine system stability (that is, whether the free response eventually decays or whether it grows without bound). A system is unstable if there are any poles inside the right half of the s plane or any multiple poles on the $j\omega$ axis. Simple poles on the $j\omega$ axis represent the borderline between stability and instability.

Alternatively, we can say that a system is stable if, and only if, all system-function poles are inside the left half s plane, or, if on the $j\omega$ axis, are simple.

Different drive and response signals

Figure 3.4 shows an electric network with two different inputs (the voltage source e_1 and the current source i_2); there are also two outputs we might wish to measure: the current i_3 and the voltage e_4. Since the system is linear and superposition applies,

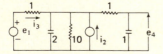

FIG. 3.4   Electric network with two sources and two outputs (values in ohms and farads).

the responses are determined from the four transfer functions

$$I_3 = T_{13}E_1 + T_{23}I_2$$

$$E_4 = T_{14}E_1 + T_{24}I_2$$

(3-22)

Each transfer function is a ratio of polynomials. Are all these polynomials different?

All the characteristic polynomials are the same. In other words, T_{13}, T_{23}, T_{14}, and T_{24} all have the same denominator, which happens to be

$$(s + 0.32)(s + 1.73)$$

The natural frequencies of the system are the same, regardless of how the system is excited or where the response is measured.

Is this a general phenomenon? The answer is yes, although with restrictions. First, the sources inserted as drive signals must not change the structure of the network (a voltage source must be in series with another element, a current source in parallel). Second, we may find that a certain transfer function possesses a numerator which cancels part of the characteristic polynomial. For example, T_{14} above might have the factor $(s + 0.32)$ in the numerator. Then -0.32 would not be a pole of T_{14}, even though it is a pole of the other transfer functions. Figure 3.5 illustrates a somewhat different form of this situation. Here

$$E_4 = \underbrace{\frac{6}{(2s + 1)(8s + 1)}}_{T_{14}} E_1 + \underbrace{\frac{1}{8s + 1}}_{T_{24}} I_2$$

(3-23)

The pole at $-1/2$ never appears in T_{24} since the 1-ohm resistor and 2-farad capacitor are totally isolated from the E_4/I_2 network. In a sense, we can view this situation as a cancellation of one term of the characteristic polynomial for T_{24}; in other words, we might write

$$T_{24} = \frac{2s + 1}{(2s + 1)(8s + 1)}$$

(3-24)

FIG. 3.5 A system with different characteristic polynomials for E_4/I_2 and E_4/E_1.

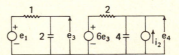

Thus, when a system is described by several transfer functions, we can anticipate that the various characteristic polynomials will be generally the same, although some may have certain factors missing.*

3.3 TOTAL RESPONSE

The transfer function can be used to find the total system response resulting from a given input signal. The procedure (the Laplace transform analysis) is illustrated by the specific example depicted in Fig. 3.6. Here the input is applied at $t = 0$ to the inert system; we desire to find the total output $y(t)$.

In this section, we present the solution, then discuss specific techniques which have been used.

Solution

The Laplace transform $Y(s)$ of the output is

$$Y(s) = T(s) X(s)$$

In this case,

$$Y(s) = \frac{26s}{(s + 1)^2 (s^2 + 4s + 13)} \left(\frac{1}{s} - \frac{1}{s + 2} \right)$$

or

$$Y(s) = \frac{52s}{(s + 1)^2 (s^2 + 4s + 13)(s + 2)s}$$

A partial fraction expansion of $Y(s)$ yields the equivalent form

$$Y(s) = \frac{5.2}{(s + 1)^2} + \frac{-6.24}{s + 1} + \frac{(0.578 \underline{/53°})/2j}{s + 2 - j3} + \frac{(0.578 \underline{/-53°})/-2j}{s + 2 + j3}$$

$$+ \frac{5.78}{s + 2} + \frac{0}{s} \qquad (3\text{-}25)$$

*We discuss this problem of missing factors again when we consider the concepts of controllability and observability.

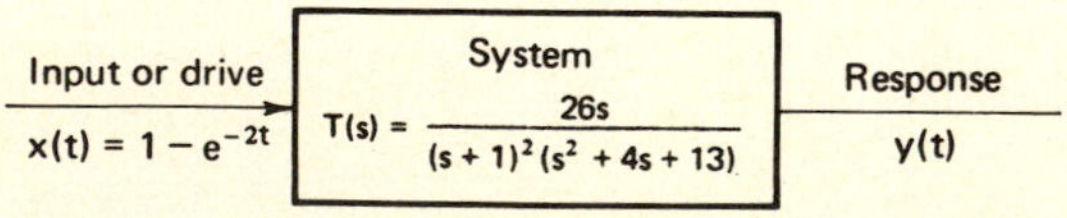

Input or drive $x(t) = 1 - e^{-2t}$

System $T(s) = \dfrac{26s}{(s + 1)^2 (s^2 + 4s + 13)}$

Response $y(t)$

FIG. 3.6 Analysis example illustrating determination of total response.

Now we can find the time function term-by-term:

$$y(t) = 5.2te^{-t} - 6.24e^{-t} + 0.578e^{-2t} \sin(3t + 53°) + 5.78e^{-2t} \qquad (3\text{-}26)$$

As a partial check on this result, we know that $y(0)$ should be zero [at very high frequencies, $Y(s)$ behaves as $52/s^5$ so that $y(t)$ starts off as $52t^4/4!$]. Substitution of $t = 0$ into Eq. (3-26) gives

$$\begin{aligned} y(0) &= -6.24 + 0.578 \sin 53° + 5.78 \\ &= -6.24 + 0.46 + 5.78 = 0 \end{aligned} \qquad (3\text{-}27)$$

In retrospect, the solution involves three steps:

(1) Finding $X(s)$;
(2) Partial fraction expansion of $Y(s)$;
(3) Writing $y(t)$.

If $x(t)$ is given as a sum of generalized exponentials, step (1) is straightforward (though possibly requiring that we consult a table of transforms).* Likewise, step (3), the reverse of (1), again involves only use of the transform table. The only tedious part of the analysis is the partial fraction expansion.

Partial fraction expansion

The procedures for making a partial fraction expansion can be illustrated by the preceding example. Here

$$Y(s) = \frac{52s}{(s + 1)^2 (s^2 + 4s + 13)(s + 2)s} \qquad (3\text{-}28)$$

We proceed in the following steps.

(a) We first identify the probable poles of $Y(s)$ by inspection of the denominator polynomial. The poles of Y are the values of s where the denominator goes to zero.† In this example,

---

*If $x(t)$ is given graphically from measured data, we may approximate with exponentials or use numerical analysis techniques (Chap. 14).

†In rare cases, we may have a $Y(s)$ with the degree of the numerator (d_N) equal to or greater than the degree of the denominator (d_D). Then the partial fraction expansion includes a polynomial of degree ($d_N - d_D$); this polynomial is found by dividing the denominator into the numerator by ordinary long division. For example,

$$Y(s) = \frac{s^3}{(s + 1)(s + 2)} = s - 3 + \frac{-1}{s + 1} + \frac{8}{s + 2}$$

Poles of $Y(s)$

-1 A double pole because of $(s + 1)^2$ factor

$\left.\begin{array}{l}-2 + j3 \\ -2 - j3\end{array}\right\}$ A conjugate complex pair coming from the factor $(s^2 + 4s + 13)$ To see the poles from the factor, we can complete the square started by the first two terms

$$s^2 + 4s + 13 = \underbrace{s^2 + 4s + 4}_{\substack{\text{Complete} \\ \text{square}}} + \underbrace{9}_{\substack{\text{Rest of constant} \\ \text{term to give 13}}}$$

$$= (s + 2)^2 + 3^2$$

Once the term is written as the sum of two squares, we can factor

$$(s + 2 - j3)(s + 2 + j3)$$

-2 From factor $(s + 2)$

0 From factor (s). We might cancel this factor with the same factor in the numerator. In other cases, the cancellation possibility may not be obvious. Even if we do not cancel, the residue will turn out to be zero, since there really is no pole at $s = 0$.

(b) Now we can write the form of the desired partial fraction expansion

$$Y(s) = \frac{A_{-1}}{(s + 1)^2} + \frac{k_{-1}}{s + 1} + \frac{k_{-2 + j3}}{s + 2 - j3} + \frac{k_{-2 - j3}}{s + 2 + j3} + \frac{k_{-2}}{s + 2} + \frac{k_0}{s} \tag{3-29}$$

The symbol k_α is used for the residue at a pole α [the residue is the coefficient of the $1/(s - \alpha)$ term in mathematics; hence A_{-1} is not a residue above, but merely a coefficient]. If there were a pole of order three, there would be three corresponding terms in the partial fraction expansion.

(c) Residues at simple, real poles: There are two apparent poles which are simple (first order) and real, at -2 and 0. The residue k_{-2} is found from Eq. (3-28) by omitting (covering up) the $(s + 2)$ factor, and letting $s = -2$ in the rest of $Y(s)$:

$$k_{-2} = \left. \frac{52s}{(s + 1)^2(s^2 + 4s + 13)s} \right|_{s = -2} = \frac{52(-2)}{(-2 + 1)^2(4 - 8 + 13)(-2)}$$

$$= \frac{52}{9} = 5.78 \tag{3-30}$$

Similarly

$$k_0 = \left. \frac{52s}{(s + 1)^2 (s^2 + 4s + 13)(s + 2)} \right|_{s = 0} = 0 \tag{3-31}$$

(d) Residues at simple, complex poles: In Eq. (3-29), we wish to find

$$k_{-2 + j3} \quad \text{and} \quad k_{-2 - j3}$$

Actually, since the coefficients of $Y(s)$ are real, these two residues are conjugates, and we only need to find the former. If we again omit (or cover up) the term representing the pole at $-2 + j3$, we have

$$k_{-2 + j3} = \left. \frac{52s}{(s + 1)^2 (s + 2 + j3)(s + 2)s} \right|_{s = -2 + j3} \tag{3-32}$$

Direct substitution of the pole value of s gives

$$k_{-2 + j3} = \frac{52(-2 + j3)}{(-2 + j3 + 1)^2 (2j3)(-2 + j3 + 2)(-2 + j3)} \tag{3-33}$$

It turns out to be convenient to find $(2jk_{-2 + j3})$ rather than the residue; hence

$$2jk_{-2 + j3} = \frac{52}{(-1 + j3)^2 3(j3)} = \frac{52}{10 \times 9} \angle{-90°} - 2 \tan^{-1} \frac{3}{-1}$$

$$= 0.578 \angle{53°} \tag{3-34}*$$

Then

$$k_{-2 + j3} = \frac{0.578 \angle{53°}}{2j} \tag{3-35}$$

The calculation of this residue is more tedious than the case for a real pole, but we have really found two residues at once (for the two conjugate poles).

  Why have we left the $2j$ term in the denominator? In other words, why not simply write

$$k_{-2 + j3} = 0.289 \angle{-37°} \tag{3-36}$$

*We must note here that $\tan^{-1}(3/-1)$ is not the same as $\tan^{-1}(-3/1)$; the former is $108.5°$, the latter $-71.5°$.

When we later write the time function for $y(t)$, the complex poles

$$\frac{k_{-2+j3}}{s+2-j3} + \frac{k_{-2-j3}}{s+2+j3}$$

give a time function

$$|2jk_{-2+j3}| e^{-2t} \sin(3t + \angle 2jk_{-2+j3}) \tag{3-37}$$

The amplitude and phase depend on $(2jk_{-2+j3})$ so it is convenient to find the residue in the form of a complex number divided by $2j$. This is perhaps merely an idiosyncrasy of the author.

(e)   Terms for multiple poles: Finally in Eq. (3-29) we must find

$$\frac{A_{-1}}{(s+1)^2} + \frac{k_{-1}}{(s+1)}$$

the terms corresponding to the double pole at -1. The coefficient A_{-1} is found by omitting the $(s+1)^2$ factor and then letting $s = -1$:

$$A_{-1} = \frac{52s}{(s^2+4s+13)(s+2)(s)}\bigg|_{s=-1} = \frac{52(-1)}{(1-4+13)(-1+2)(-1)} = 5.2 \tag{3-38}$$

The residue k_{-1} is evaluated by differentiating $[(s+1)^2 Y(s)]$ once, then letting $s = -1$:

$$k_{-1} = \frac{d}{ds}\left[\frac{52}{(s^2+4s+13)(s+2)}\right]_{s=-1}$$

$$= \left[\frac{-52(s^2+4s+13) + (s+2)(2s+4)}{(s^2+4s+13)^2(s+2)^2}\right]_{s=-1}$$

$$= \frac{-52(10+2)}{100} = -6.24 \tag{3-39}$$

If we have a third order pole at -1, the third coefficient is $1/2!$ times the second derivative of $(s+1)^3 Y(s)$ evaluated at $s = -1$, and so forth for higher order poles. The tedium of differentiation is usually not necessary, however, since we rarely have multiple poles—certainly seldom more than double poles.

Comment

In this section, we have evaluated the total response (forced plus free components) for a system described by a transfer function and driven by a sum of generalized exponentials. The Laplace transform approach requires that the denominator of the transfer function be factored [so that we can determine the poles of $Y(s)$ and effect the partial fraction expansion]. In the study of reasonably complex systems (e.g., fifth or higher order), this factoring may be the most tedious part of the analysis. As we become interested in systems of greater complexity, we often find that the simulation techniques discussed in later chapters are much more attractive than the transform analysis.

Furthermore, even in reasonably simple systems, often we are not interested in the complete response to a particular input signal, but rather in the general characteristics of system behavior (e.g., does the system exhibit resonance, how long does the transient last, and so on?). In the next few sections, we consider how such information can be deduced from $T(s)$ without the detailed evaluation of the total response.

3.4 STEADY STATE RESPONSE

The total system response can be viewed as having two components: forced and free. The forced component includes the complex frequencies of the drive signal, the free component the complex or natural frequencies of the system. Actually, the distinction is rather artificial. As Fig. 3.7 indicates, we can assume that the signal $x(t)$ is generated by a hypothetical system with the transfer function $X(s)$. Then $y(t)$ is just the response of the overall system

$$X(s)\ T(s)$$

to a unit impulse. The roles of $X(s)$ and $T(s)$ are really interchangeable in the determination of $y(t)$.

Furthermore, the partial fraction expansion of $Y(s)$ decomposes the overall, equivalent system of Fig. 3.7(b) into a parallel combination of simple systems. The specific

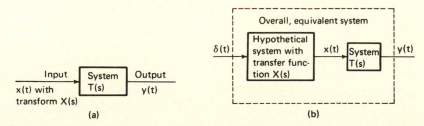

FIG. 3.7 Dualism of $X(s)$ and $T(s)$: (a) System excited by $x(t)$; (b) Equivalent system excited by a unit impulse.

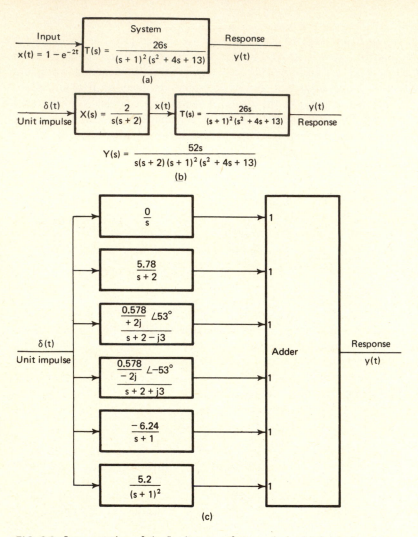

FIG. 3.8 Interpretation of the Laplace transform analysis: (a) Original problem; (b) Laplace transform formulation; (c) Decomposition into simpler systems (by partial fraction expansion).

example of the preceding section is really analyzed as shown in Fig. 3.8. In this decomposition, the roles of $x(t)$ and $T(s)$ become interrelated. Thus, when we talk about the forced and free components of $y(t)$, we must recognize that the forced component depends on $T(s)$, the free component depends on $x(t)$. Each of the coefficients in Fig. 3.8(c) depends on both $x(t)$ and $T(s)$.

There is an alternate way to break the output $y(t)$ into two parts. We can talk about the *transient* and the *steady-state* components. The term transient refers to those parts which tend to decay as time passes. If we wait long enough, the transient components are negligibly small, and the response is entirely the steady state part. For example, transient components might have the forms

$$
\begin{cases}
Ae^{-\alpha t} & \text{(from a simple pole on the negative real axis)} \\
Bte^{-\alpha t} & \text{(from a double pole on the negative real axis)} \\
Ce^{-\alpha t}\sin(\beta t + \theta) & \text{(from a conjugate pair of simple poles inside the left half plane)} \\
Dte^{-\alpha t}\sin(\beta t + \psi) & \text{(from a conjugate pair of double poles inside the left half plane)}
\end{cases}
$$

The steady state component of the response consists of one or more terms of the form

$$
\begin{aligned}
& E\sin(\beta t + \phi) && \text{(from a conjugate pair of simple poles on the } j\omega \text{ axis)} \\
& F && \text{(from a simple pole at } s = 0\text{)}
\end{aligned}
$$

These terms neither decay nor increase as time progresses.

Clearly, we might also have a response with unstable terms, corresponding to poles of $Y(s)$ in the right half plane or multiple-order poles on the imaginary axis. If the system is unstable, discussion of steady state and transient components is meaningless, since with the passage of time the response is dominated by the unstable terms. Consequently, if we use the terms "transient" and "steady state," we are assuming a stable, overall system (in other words, a bounded output).

Determining steady state response

In most cases, the steady state response results from driving the system with an input signal which has either a constant or sinusoidal component.* When such a signal is applied, we often want to know the steady state output only—the output after the transient has disappeared. The evaluation is simple if we recognize that the transfer function is the system *gain* at the frequency s.

Figure 3.9 is an illustrative problem. Here we want to determine the steady state component of the response y. We reason that:

*The duality of signals and systems shows that the steady state response component may also come from poles of $T(s)$ on the $j\omega$ axis (in other words, the system behaves as an oscillator).

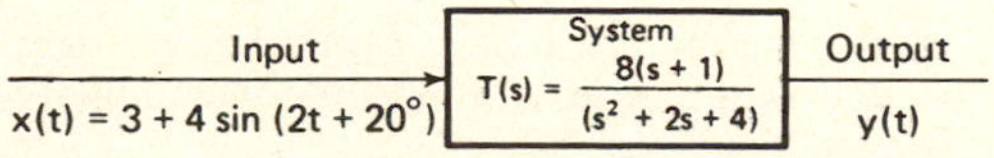

FIG. 3.9 Example for calculation of steady state response.

Input
$x(t) = 3 + 4\sin(2t + 20°)$

System
$T(s) = \dfrac{8(s+1)}{s^2 + 2s + 4}$

Output
$y(t)$

(1) The system is stable, and $T(s)$ has no natural frequencies which are undamped (no poles on the $j\omega$ axis).

(2) The input contains two components which cause steady state output:

$s = 0$ Component of amplitude 3

$s = \pm j2$ Component which is a sine of amplitude 4, phase $+20°$ ($s = \pm j2$ because $\sin 2t$ is the sum of terms e^{j2t} and e^{-j2t})

(3) The gains at the particular values of s are

$$T(0) = \left.\frac{8(s + 1)}{s^2 + 2s + 4}\right|_{s = 0} = \frac{8}{4} = 2 \tag{3-40}$$

$$T(j2) = \left.\frac{8(s + 1)}{s^2 + 2s + 4}\right|_{s = j2} = \frac{8(1 + j2)}{-4 + j4 + 4} = 2\frac{1 + j2}{j}$$

$$= 2\sqrt{5}\ \underline{/63.5°} - 90° = 4.48\ \underline{/-26.5°} \tag{3-41}$$

(4) Then the steady state output is

$$y_{ss}(t) = T(0)3 + |T(j2)|4 \sin[2t + 20° + \underline{/T(j2)}]$$

$$= 6 + 17.92 \sin(2t - 6.5°) \tag{3-42}$$

The sinusoidal component of the input is amplified by 4.48 and shifted in phase by $-26.5°$ as it "passes through" the system.*

Thus, the steady state component of the response is found in this example without the necessity of transforming $x(t)$, or making the partial fraction expansion of $Y(s)$. We need only the concept of the transfer function as the complex gain at the frequency s.

3.5 RESONANCE CONCEPT

Resonance is a concept which appears again and again not only throughout engineering, but also in the study of behavior in other biological, natural, and social systems. Resonance means that a system responds with unusual strength when excited by signals of a particular frequency (called the resonant frequency).

---

*If the input is described as a sine or cosine signal, the output is written in the same way. For example, here if x were $4 \cos(2t + 70°)$, the corresponding otuput term would be 17.92 cos $(2t + 43.5°)$.

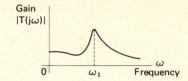

FIG. 3.10 Gain versus frequency for a resonant system.

Figure 3.10 shows a possible curve for the gain of a system as a function of the sinusoidal driving frequency. The graph might be determined analytically from a known $T(s)$. With $s = j\omega$, various values of ω are used in turn; for each, $|T(j\omega)|$ is calculated exactly the same as in finding the steady state output. [Chapter 11 is devoted to a simpler procedure for plotting this gain characteristic from the $T(s)$.]

Alternatively, the gain curve might be measured experimentally on the actual system. A sinusoidal signal at ω_a is applied as the input; the amplitude ratio output/input is just $|T(j\omega_a)|$. Then ω_a is changed and the measurement repeated, with this process continuing until the complete curve of Fig. 3.10 is obtained.

The system described by Fig. 3.10 exhibits resonance. The resonant frequency is ω_1 rad/sec. The gain has a sharp, "resonant" peak at this frequency. The gain is comparatively large at this frequency, compared to values obtained when the frequency is well away from ω_1.

Some of the earliest engineering investigations of resonance were made by mechanical engineers around 1930. As a motor is brought from rest up to normal operation, the speed increases gradually from zero to its normal value. If the rotor is slightly asymmetrical, force is applied regularly at each revolution to the mounting through the bearings. As speed increases, the frequency of this applied force increases proportionally, from zero to its value in normal motor operation. If the mounting system for the motor happens to resonate at a frequency this force passes through during motor startup, the entire motor can begin to vibrate with very large amplitudes as we excite at the resonant frequency. The motor vibrations may become destructive. This situation encouraged mechanical engineers to study how to control the resonant frequency and the magnitude of the gain peak at resonance.

We meet resonance in a host of other situations. As a car drives on a modern concrete highway, the vertical suspension system is excited by the slab dividers. As the speed increases, the excitation frequency rises correspondingly. In many of today's cars, we hit resonance at about 55 mph, when the large-amplitude vertical vibrations (the output) make riding uncomfortable.

Resonance is exhibited by the vibration of tall buildings excited by the wind, by aircraft wing vibrations in many different modes, by pendulums and tuning forks, and of course by piezoelectric crystals and RLC networks.

3.6 RESONANCE AND TRANSFER FUNCTIONS

Can we tell by inspection of the system function $T(s)$ whether resonance occurs? The answer is obviously yes (otherwise we would not raise the question). We can show that, for any system described by a transfer function (that is, any linear, lumped, time-invariant system), resonance is always associated with a pair of conjugate complex poles. The denominator factor is

$$s^2 + Bs + \omega_r^2$$

When the coefficient of s^2 is made unity, the constant term is the square of the resonant frequency. Furthermore, the coefficient of s is the *bandwidth* (the frequency separation between half-power points, or points at which the gain is 0.707 times the peak gain).

As soon as these relations are accepted, we can look at the factored transfer function and immediately describe all resonances. For example, the system described by

$$T(s) = \frac{36(s + 2)(s + 3)}{(s + 1)(s^2 + 0.1s + 4)(s^2 + 0.2s + 9)} \tag{3-43}$$

has two resonances:

$$\begin{cases} \text{Resonant frequency} = \sqrt{4} = 2 \text{ rad/sec} \\ \text{Bandwidth} = 0.1 \text{ rad/sec} \end{cases}$$

$$\begin{cases} \text{Resonant frequency} = \sqrt{9} = 3 \text{ rad/sec} \\ \text{Bandwidth} = 0.2 \text{ rad/sec} \end{cases}$$

No plotting of the gain-versus-frequency curve is required.

The proof of the validity of these relations between the coefficients of the denominator quadratic factor and the resonance features depends on the fact that, for frequencies very close to the resonant frequency, all other terms in $T(s)$ change very little in value. Hence, right around the resonant frequency (say the one at 2 rad/sec), $T(s)$ in Eq. (3-43) is

$$T(s) \longrightarrow \frac{K}{s^2 + 0.1s + 4} \tag{3-44}$$

where K is a complex constant. Then

$$|T(j\omega)| \longrightarrow \frac{|K|}{\sqrt{(4 - \omega^2)^2 + (0.1\omega)^2}}$$

If we plot this gain versus frequency around $\omega = 2$ [or, alternatively, if we let $\omega = 2(1 + \delta)$ with δ small], we find that the maximum value is very close to $\omega = 2$ and that the half-power points are approximately 2 ± 0.05 (i.e., the bandwidth is 0.1).

In other words, if the transfer function contains the quadratic factor

$$(s^2 + as + b)$$

we can say that the system exhibits resonance with

$$\begin{cases} \text{Resonant frequency} = \sqrt{b} \text{ rad/sec} \\ \text{Bandwidth} = a \text{ rad/sec} \end{cases} \tag{3-45}$$

When we investigate the validity of Eq. (3-45) in more detail, we find that the frequency at which the gain is a maximum is not quite $\sqrt{b}$. The actual value depends on:

(a) The sharpness of the resonance, usually measured by the *quality factor* or Q, defined as

$$Q = \frac{\text{Resonant frequency}}{\text{Bandwidth}} \tag{3-46}$$

The higher the Q, the sharper the peak of the gain characteristic.

(b) The location of other poles and zeros of $T(s)$. As an extreme example,

$$T(s) = \frac{s^2 + 0.1s + 16}{s^2 + 0.5s + 16} \tag{3-47}$$

seems to have a resonance at 4 rad/sec with a bandwidth of 0.5 rad/sec. Actually the complex zero at the same ω_n yields an even sharper dip, and $|T(j\omega)|$ has a minimum (not a maximum) at $\omega = 4$.

Thus, Eq. (3-45) is valid only when Q is reasonably large and only if there are no other $T(s)$ factors with similar resonances or antiresonances (the latter referring to numerator factors).

What do we mean by a reasonably large Q? Actually there really is no gain peak unless Q is larger than unity, and the peak is not really significant unless Q is 4 or greater. Consequently, we are not normally interested in resonances with Q less than 4. For larger Qs, the approximations of Eq. (3-45) are excellent except in the very rare, odd cases symbolized by Eq. (3-47). Thus, when resonance is clear and pronounced, the resonance can be described by the resonant frequency, bandwidth, and Q found directly from the quadratic denominator factor of the system function.

3.7 NOTCH FILTER

In a low-frequency instrumentation amplifier, we find that objectionable distortion of the output is caused by 60-Hz pickup. In the instrument, in spite of the best attempts at shielding, we have a large 60-Hz signal coming originally from the power line, a signal so large that it masks the signal we are trying to measure.

To decrease the 60-Hz component, we would like to insert a network with the gain characteristic shown in Fig. 3.11. The gain is unity except in the immediate vicinity of 377 rad/sec (60 Hz). At $\omega = 377$, the gain is 1/20.

Can we find a transfer function with this gain characteristic?

Clearly, we desire an antiresonant system, one with a numerator of the form

$$s^2 + BS + 377^2$$

This is not a suitable transfer function, however, since the gain at low and high frequencies is not unity. (Also, networks are vastly easier to build if the transfer function has a denominator of a degree at least as high as that of the numerator.) Both factors suggest

$$T(s) = \frac{s^2 + Bs + 377^2}{s^2 + Cs + 377^2} \tag{3-48}$$

This T gives the required low-frequency and high-frequency behavior, and we now can select B and C to give the desired behavior near $\omega = 377$.

At $\omega = 377$, the gain is just

$$T(j377) = \frac{B}{C} \tag{3-49}$$

since s^2 is then -377^2. Hence, to realize the gain of 1/20 at 377, we can select any B or C arbitrarily, with $B/C = 1/20$.

$$T(s) = \frac{s^2 + Cs/20 + 377^2}{s^2 + Cs + 377^2} \tag{3-50}$$

Chapter 6 tells how to build networks from a specified transfer function. Actually, it turns out that, if we select C large enough so that the denominator has real poles, the

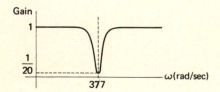

FIG. 3.11 Desired gain characteristic.

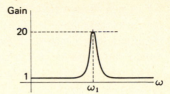

FIG. 3.12 Gain for $G(s)$ of Eq. (3-51).

$T(s)$ can be realized by a network of resistors and capacitors; this might be one guide to the choice of C. Alternatively, we can choose C to control to some extent the width of the gain dip in Fig. 3.11 (by calculating the shape of the dip for different values of C).

Because of the form of the gain characteristic (Fig. 3.11), this system is called a *notch filter*. In communication work, we also find the term band-rejection filter or band-elimination filter, to emphasize the fact that the system rejects or eliminates the narrow band of frequencies around 377.

3.8 COMB FILTER

If we simply invert the transfer function of Eq. (3-50), we obtain

$$G(s) = \frac{s^2 + Cs + \omega_1{}^2}{s^2 + Cs/20 + \omega_1{}^2} \tag{3-51}$$

Figure 3.12 shows that $G(s)$ is a sharply resonant circuit or, in communication terms, a band-pass filter.*

One of the interesting extensions of this concept is the comb filter shown in Fig. 3.13. Here the gain is high near ω_1, $2\omega_1$, $3\omega_1$, and so on, but low elsewhere. Such a filter (called a *comb filter* because of the appearance of the gain characteristic) can be used to select a periodic signal of fundamental angular frequency ω_1 from a total signal (such as noise) with components spread over the ω domain. Each of the Fourier components of the signal is amplified, but *most* of the noise is not.

*The term filter refers to any system in which the gain characteristic favors certain frequency components, discriminates against others.

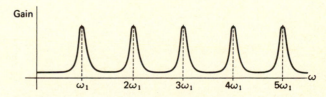

FIG. 3.13 Gain characteristic for a comb filter.

This gain characteristic is used, for example, to detect radar echoes from Venus when the radar signal has the appearance of pure noise. The noise may be several orders of magnitude larger than the signal; if we know the periodic frequency accurately, we can build a filter with extremely narrow "teeth" for the comb.

To find the transfer function of Fig. 3.13, clearly we can just add on terms similar to Eq. (3-51):

$$H(s) = \frac{s^2 + Cs + \omega_1^2}{s^2 + Cs/20 + \omega_1^2} \; \frac{s^2 + Ds + (2\omega_1)^2}{s^2 + Ds/20 + (2\omega_1)^2} \cdots \tag{3-52}$$

We need as many biquadratics as the significant number of Fourier components in the periodic signal we are measuring. The system rapidly becomes much too complicated.

There is a far simpler way to build a comb filter if we allow exponentials in our transfer function. For example, the transfer function

$$L(s) = \frac{1}{1 - e^{-\alpha s}} \tag{3-53}$$

has poles when

$$e^{-\alpha s} = 1 \quad \text{or} \quad -\alpha s = 0, \; \pm j2\pi, \; \pm j4\pi, \ldots$$

or

$$s = 0, \; \pm j\frac{2\pi}{\alpha}, \; \pm j\frac{4\pi}{\alpha}, \; \pm j\frac{6\pi}{\alpha}, \ldots \tag{3-54}$$

To obtain poles (or resonances) at $\omega_1, 2\omega_1, \ldots$, we need only choose

$$\frac{2\pi}{\alpha} = \omega_1 \quad \text{or} \quad \alpha = \frac{2\pi}{\omega_1} \tag{3-55}$$

But ω_1 for the periodic signal is $2\pi/T$ where T is the period. Hence, α is made equal to the period T, and the desired transfer function is

$$L(s) = \frac{1}{1 - e^{-Ts}} \tag{3-56}$$

We can rewrite this $L(s)$ as an infinite series

$$L(s) = 1 + e^{-Ts} + e^{-2Ts} + e^{-3Ts} + \cdots \tag{3-57}$$

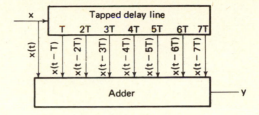

FIG. 3.14 System for realizing $L(s)$. Practically, a finite number of terms must be added. Using eight as shown,

$$L(s) = \frac{1 - e^{-8Ts}}{1 - e^{-Ts}}.$$

This relation states that the comb filter can be viewed as performing the following function: the output is the input, plus the input delayed one period, plus the input delayed two periods, plus . . . .* In other words, $L(s)$ can be realized by the system shown in Fig. 3.14.

The example illustrates that often the very elaborate devices developed for particularly exotic applications are in actuality rather simple conceptually—simple, at least, once they have been invented.

3.9 TRANSPORTATION LAG

The driver of an automobile moving at 60 miles/hour suddenly spots an obstacle in the road ahead. From that instant until the driver starts to respond by moving his foot toward the brake, a time of perhaps 0.3 second passes (during which the car moves 25 feet). We are witnessing the *human reaction time*: the minimum time required before a man can response to the unexpected need for a control signal.

The same *dead time* often appears in production or chemical processes. A valve controls the input flow of a particular ingredient, but this substance has to travel to a mixing tank some distance away before the effects of valve motion are noticeable in the output. The time to move the substance, called the *transportation lag*, is a dead time, a time of absolutely no response.

In both these examples, the system fails to respond immediately. For any system, the response tends to be slower in starting, the greater the degree of the denominator polynomial of $T(s)$ compared to the numerator. In a case such as

$$T(s) = \frac{1}{s^7 + 7s^6 + \cdots + 7s + 1} \tag{3-58}$$

the response does not really get started (Fig. 3.15) until t_a seconds have elapsed. If the denominator were of even higher degree, the response would start even more slowly.

Systems such as described by Eq. (3-58) or Fig. 3.15 do not, however, exhibit a transportation lag. There is some response immediately after $t = 0$, although the response may be very small. For the $T(s)$ of Eq. (3-58), the unit-step response y starts

*e^{-3Ts} means a pure delay of $3T$ seconds.

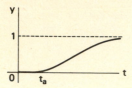

FIG. 3.15   Step response for $T(s)$ of Eq. (3-58).

off as the inverse transform of $1/s^7$ since at high frequencies (small t) $T(s)$ behaves in this way. Thus,

$$y \xrightarrow[t \to 0]{} \frac{t^6}{6!} \qquad (3\text{-}59)$$

When $t = 0.01$, y is about 1.4×10^{-15}—certainly too small to measure, but not strictly zero as we would have with a transportation lag. How, then, do we represent a transportation lag in a transfer function?

Figure 3.16 shows the pure transportation lag. For this system

$$y(t) = x(t - \alpha) \qquad (3\text{-}60)$$

The fact that the response is $x(t)$ delayed by α seconds is represented mathematically by replacing t by $t - \alpha$ in the argument of x.

If we take the Laplace transform of Eq. (3-60), we find

$$Y(s) = \int_0^\infty x(t - \alpha) e^{-st} dt \qquad (3\text{-}61)$$

In this relation, we replace $(t - \alpha)$ by u under the integral sign. Then

$$Y(s) = \int_{-\alpha}^\infty x(u) e^{-s(\alpha + u)} du \qquad (3\text{-}62)$$

Since $x(u)$ is zero for u negative, we can change the lower limit to 0. Furthermore, we

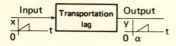

FIG. 3.16   A transportation lag.

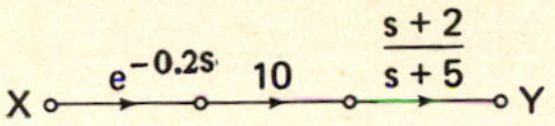

FIG. 3.17 Model for human being responding in a control task.

$$X \circ \xrightarrow{e^{-0.2s}} \circ \xrightarrow{10} \circ \xrightarrow{\frac{s+2}{s+5}} \circ Y$$

can take $e^{-\alpha s}$ outside the integral to obtain

$$Y(s) = e^{-\alpha s} \int_0^\infty x(u)\, e^{-su}\, du \qquad (3\text{-}63)$$

The integral is just the Laplace transform of $x(t)$. Hence,

$$Y(s) = e^{-\alpha s} X(s) \qquad (3\text{-}64)$$

The transfer function for a pure transportation lag of α seconds is

$$\boxed{T(s) = e^{-\alpha s}} \qquad (3\text{-}65)$$

Response calculation

Figure 3.17 shows one model which has been found to describe the response of a man in a control task.* We wish to calculate the response of this system to a square pulse of duration 0.2 sec. (Fig. 3.18.)

First, we find $X(s)$. The signal $x(t)$ is a step of amplitude 2 occurring at $t = 0$, plus a step of amplitude -2 occurring 0.2 sec later. Thus,

$$X(s) = \frac{2}{s} - \frac{2}{s} e^{-0.2s} = \frac{2}{s}(1 - e^{-0.2s}) \qquad (3\text{-}66)$$

*We know from experience that human response depends on the particular task (driving a car, flying an airplane, dodging cars as a pedestrian, and so on). In addition, the transfer function is influenced by training, fatigue, distraction, mental attitude, and so forth. These terms are discussed briefly in the appendix to this chapter.

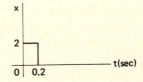

FIG. 3.18 Input signal for the example.

Then

$$Y(s) = T(s) X(s) = 10 \frac{s + 2}{s + 5} e^{-0.2s} \frac{2}{s} (1 - e^{-0.2s})$$

$$Y(s) = 20 \frac{s + 2}{s(s + 5)} (e^{-0.2s} - e^{-0.4s})$$

(3-67)

The exponentials inside the parentheses are merely delay factors. We find the time function corresponding to the rest of $Y(s)$—let us call this Y_1:

$$Y_1(s) = 20 \frac{s + 2}{s(s + 5)} = \frac{8}{s} + \frac{12}{s + 5}$$

(3-68)

Then

$$y_1(t) = 8 + 12e^{-5t}$$

(3-69)

Equation (3-67) states that the desired output $y(t)$ is just this y_1 delayed by 0.2 sec minus the same signal delayed 0.4 sec. In other words

$$y(t) = \begin{cases} 0 & 0 < t < 0.2 \\ 8 + 12e^{-5(t - 0.2)} & 0.2 < t < 0.4 \\ [8 + 12e^{-5(t - 0.2)}] - [8 + 12e^{-5(t - 0.4)}] & 0.4 < t \end{cases}$$

(3-70)

These terms are written by noticing that: to delay $y_1(t)$ by α seconds, we replace t by $t - \alpha$. Finally, this relation can be slightly simplified algebraically

$$y(t) = \begin{cases} 0 & 0 < t < 0.2 \\ 8 + 33e^{-5t} & 0.2 < t < 0.4 \\ -56e^{-5t} & 0.4 < t \end{cases}$$

(3-71)

The waveform is shown in Fig. 3.19. [Actually, it is most easily plotted from Eq. (3-71).]

It is interesting to try to interpret this response in terms of what we know about human behavior in control tasks. The large rise after the transportation lag (human reaction time) and the subsequent decay show an overreaction to the large stimulus

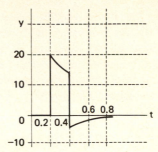

FIG. 3.19 Response of man described by Fig. 3.17.

suddenly applied. After the input drops to zero, the man again overcompensates. We can visualize situations where this response might approximate the behavior of a man (for example, when a sideways gust of wind hits the car he is steering).

Transportation lag in a feedback loop

Figure 3.20 shows a particularly difficult system to analyze. In this case the overall system function is

$$T(s) = \frac{G(s)e^{-\alpha s}}{1 + G(s)e^{-\alpha s}} \tag{3-72}$$

The $e^{-\alpha s}$ factor in the numerator causes no trouble; it merely means the output is delayed by α seconds.

The exponential in the denominator, however, usually makes straightforward analysis impossible. The poles of $T(s)$ are the roots of the transcendental equation

$$1 + G(s)e^{-\alpha s} = 0 \tag{3-73}$$

Even with relatively simple $G(s)$ functions, this equation may have many roots or even an infinite number.*

---

*For example, if $G(s) = 1$, the roots are where $e^{-\alpha s}$ is −1, or

$$s = \pm j\frac{\pi}{\alpha}, \; \pm j\frac{3\pi}{\alpha}, \; \pm j\frac{5\pi}{\alpha}, \ldots$$

(The impulse response is an infinite train of impulses in this case, and the system is oscillatory.)

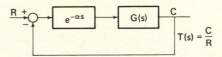

FIG. 3.20 Feedback system with transportation lag.

The easiest way to analyze such a problem is to modify the system by inserting a sampler; we postpone discussion of this until Chap. 14, when we consider feedback systems with sampling. The problem is mentioned here merely to emphasize that there are common linear systems which are not easily studied or designed. These systems with transportation lag are particularly important because they occur whenever a man is an element of a feedback system.*

3.10 LIMITATIONS ON USE OF TRANSFER FUNCTIONS

We have suggested some of the power of the transfer function concept in this chapter. In the next chapter, particularly, but also to some extent through the remainder of the book, we will expand the range of problems in which transfer functions serve as a useful tool. Before becoming totally enamored by the concept, however, we should keep in mind clearly the important system limitations implied when we use a transfer function.

(1) First, the system must be *linear*. No real-world system is linear over the entire possible range of input signals. For example, normally a system eventually saturates as the input signal is made larger and larger (amplifiers saturate or give the maximum possible output, components fail, and so on).

We are interested in linearity over the normal operating range—linearity for the signals we anticipate in system operation. Even with this restriction, important systems are nonlinear and transfer functions are not useful. An example is the water temperature-control system in a home shower. In many cases, when the man tries to adjust the temperature of the water hitting him, he turns either the hot or cold valve. If this valve-knob system has backlash,† the system is exceedingly difficult to regulate. The backlash nonlinearity is a critical element in determining system operation (for instance, system stability).

Fortunately, many very important systems can be considered linear, or at least can be adequately represented by a linear model over significant regions of operation. When the input signal changes radically, we may have to use a different linear model;

---

*We can hope that the transportation lag in such cases is sufficiently small compared to the time constants of the system that we can represent $e^{-\alpha s}$ by $1/(\alpha s + 1)$. The approximation is derived from the first two terms of the power series for $e^{\alpha s}$:

$$e^{-\alpha s} = \frac{1}{e^{\alpha s}} = \frac{1}{1 + \alpha s + \alpha^2 s^2/2! + \cdots} \approx \frac{1}{1 + \alpha s}$$

Hence, it is valid if all frequencies of interest satisfy

$\alpha\omega \ll 1$.

†Backlash means: The cold knob is turned clockwise to decrease the amount of cold water. When direction is reversed to counterclockwise, we have to turn through perhaps 25° before the valve starts to move and the amount of cold water starts to increase.

for example, aircraft transfer functions are quite different at low speeds (during landing) and at cruising speed.

(2) The system must be *time-invariant*. In terms of the differential equations modeling the system, this constraint means that the coefficients should not depend on time. The equation

$$\frac{d^2y}{dt^2} + 3(1 - t^2)\frac{dy}{dt} + 2y = x \tag{3-74}$$

describes a linear system (superposition applies), but the system is time-varying. The coefficient of dy/dt changes with time. In terms of the system or network, one or more of the parameters (R, L, or C in passive electrical networks) is time-dependent.

If the parameter variation is much slower than the system response, we can analyze the system with transfer functions, merely by changing the transfer function in discrete steps. In the interesting time-varying system, parameters change significantly during a single response. As a missile consumes fuel during launch, the mass and moment of inertia change while the vehicle is responding to a wind gust or other cause of oscillation.

Some of the most familiar time-varying systems are those in which queues, or waiting lines, form. Here, an exceedingly important parameter is the average arrival rate of customers. In most queueing problems, this parameter changes long before steady state (or equilibrium) operation is reached—long before the transient dies out. In a city barbershop, for example, arrival rates commonly peak in the early morning, during lunch hours, and in the late afternoon. At toll gates, commuting traffic peaks in the morning and evening.

(3) The system must be *lumped*, except for the few problems we can handle involving transportation lags. In other words, the system must be describable in terms of simple elements, each defined by a relation between through and across variables. The interconnections are described by the Kirchhoff equations, or their analogs for nonelectrical systems.

Room or concert hall acoustics serve as subjects where the system is not usefully described by lumped models. The sound waves reflected from room structures result in standing waves, destructive and constructive interference, and then diffuse sound persisting throughout the room after the source has stopped. The extreme difficulties involved in analysis and design are illustrated by the unsatisfactory nature of New York's Lincoln Center Philharmonic Hall, which became apparent after its opening in the early 1960s, in spite of the high cost of elaborate design work by acoustical engineers.

(4) The system must be *inert*; the transfer function describes the response resulting from a specified input signal, and does not indicate the effects of initial energy storage at the time the drive signal is applied. If initial energy storage is significant, we can, of course, represent each initial condition by an independent source. The system then has a number of separate inputs and is described by a family or matrix of transfer functions. While this approach is straightforward, it tends to complicate the problem so greatly

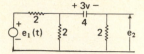

Values in ohms and farads

$$e_1 = \begin{cases} 0 & t < 0 \\ 1 & t > 0 \end{cases}$$

FIG. 3.21 Circuit analysis problem.

that we lose the real power of the transfer-function concept—the possibility of deducing important system properties merely by inspection of $T(s)$.

(5)   The system and signals must be described analytically: the system in terms of $T(s)$ and the signals as sums of generalized exponentials. If the input signal and/or the system impulse response are given graphically (for example, from experimental measurements), we must first approximate these time functions by a sum of exponentials. In most cases, this approximation is likely to be too tedious to be attractive.

(6)   Finally, the system must be of *reasonable complexity*. If the system is first or second order, analysis and design are usually straightforward without any powerful, mathematical tools. For example, in the circuit of Fig. 3.21, we are asked to determine the output voltage $e_2(t)$ with e_1 a unit step and the initial voltage on C equal to 3v as shown.

The system is first order (one capacitor), so the response is described by a single time constant which is C times the R faced by the capacitor. By inspection

$$\tau = 4\left(2 + \frac{2 \times 2}{2 + 2}\right) = 12 \text{ sec} \tag{3-75}$$

The initial value of e_2 (due to initial capacitor voltage) is -2; the component due to e_1 is $+1/3$; hence at $t = 0+$, e_2 is $-5/3$.

The final value (due solely to the drive e_1) is zero. Hence

$$e_2 = -\frac{5}{3}e^{-t/12} \tag{3-76}$$

and the transfer function concept is really not needed.

On the other hand, if the system is too complex, any analytical approach is decidedly limited in value. For example, if we are studying the heating system in the author's house, the detailed model should represent 17 windows, three doors to the outside, 11 radiators, various internal sources of heat, and nine rooms of varying shapes. As we attempt to include all the details of heat flow in the model, we move rapidly toward a transfer function of order 20 or greater.*

---

*This is a relatively simple system compared to models of a city's traffic flow or of an aircraft wing flutter (where hundreds of different modes of oscillation may be involved).

A primary goal of modeling is to bring the system complexity within the realm of human circumspection. A model should include only the primary and particularly significant factors. Once the "simplified" system is studied, we can start to add secondary factors which might be important.

Part of the motivation for simplification is to work with a model for which the transfer function can be interpreted—preferably a model of the order less than six (although a higher order is admissible if most of the poles are so far from the $j\omega$ axis that they contribute very little to the dynamics).

With all the above constraints on the system, it might be logically concluded that the transfer function concept is of very limited value. Fortunately, the engineer is primarily interested in designing systems (or in analyzing systems designed by other people). Furthermore, we can often realize the performance characteristics we seek by using a system which is indeed linear, time-invariant, lumped, and so forth. Consequently, because we work with the artifacts of the man-made world, engineering systems are often vastly simpler to design or analyze than the systems the scientist finds in the physical or biological world.

APPENDIX—HUMAN TRANSFER FUNCTIONS

As technology is developed which has strong impacts on so many aspects of our daily lives, much greater emphasis must be placed on matching the machines to the human beings with which they are to be used. Only a few decades ago, trains were the only convenient method for travel from city to city, for example from Chicago to Denver. Without competition in economy, speed, or attractiveness, the railroads had little concern for matching the service provided to the desires of the people.

Today, with the elaborate interstate highway system and an extensive airline industry, there are public demands for a high-speed, passenger railroad system similar to those being developed in Japan and western Europe. The success of such a system in this country depends, however, on making the system so attractive that it changes the travel habits of the people. This means that the system must provide convenient facilities for making reservations, expeditious ticketing at the terminal, very frequent and reliable service, car rental facilities at every terminal, cleanliness, and so on—all characteristics seemingly foreign to the American passenger railroad service of the recent past. Precisely because of these questions of what is really required for public acceptance of train service, the Federal government (through the Department of Transportation) has fallen well behind other nations in this direction technologically. We are not sufficiently knowledgeable about how to match the technology to the U.S. public.

Rather than consider such broad questions further, we turn to a much narrower problem: matching machines to men in relatively simple control tasks. For example, in piloting an airplane, what are the human characteristics which combine with the airplane characteristics to determine the behavior of the total system? How then, can we design the airplane so the overall system behaves properly?

History of research

While man has always been interested in man's characteristics, perhaps the first major engineering modeling was carried out by the Germans during the 1920s. Under terms of the Versailles Treaty after World War I, the Germans were forbidden to build motored airplanes. As a consequence, they carried out extensive research in gliders and in the possibility of using human power to extend the glider's flight and improve its maneuverability.

In this connection, research was done to determine how to draw the maximum possible power from the human being. It was found that maximum power was achieved when the man pedaled with his feet and legs (as in an amusement park paddle boat). Man could deliver more than one horsepower for intervals of about a second, but only about 0.5 horsepower over many seconds (Fig. A3.1).*

After the Germans disregarded the Versailles Treaty, interest in gliders almost disappeared. The next major motivation for intensive study of man's capabilities arose during World War II, primarily because of the increases achieved in aircraft speed. Man's sluggishness in responding to control signals was critical in landing an airplane (particularly on the deck of an aircraft carrier), in aiming antiaircraft guns at planes moving at close range, and in maneuvering the plane during combat. In all three situations, evaluation of the total-system performance required characterizing the dynamics of the man who was a part of the system.

The simplest control system involving a human operator is shown in Fig. A3.2. The man performs two functions. He observes or senses the input and the output to determine the error. He then generates a control signal which drives the process, hopefully in the direction to reduce the error toward zero. In the aiming of antiaircraft guns, the output is the gun direction, the input is the target position predicted ahead in time to give the lead required to compensate for the time required for the bullet or projectile to reach the target. The process is the gun mount with its inertia, friction, and other dynamic characteristics.

*There is an interesting corollary to Fig. A3.1. Calculations of the power required for self-powered flight indicate that man should just be able to fly short distances (a mile or so) if the equipment could be designed to utilize most of the human power. This fact has encouraged men to try to build self-powered aircraft.

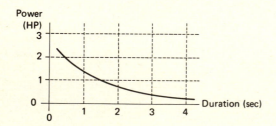

FIG. A3.1 Power that can be drawn from a man.

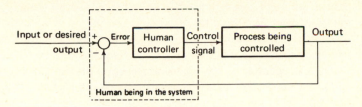

FIG. A3.2 A man-machine control system.

Starting with research on how to model the human being, one would hope that man could be described by an error measurement followed by a transfer function for the controller block in the figure. In an attempt to determine this human transfer function, the experiment depicted in Fig. A3.3 was used. The man sits in front of a cathode-ray tube on which he can see two dots which are free to move horizontally. One dot (the input) is driven by a random signal generator, so that its motion is unpredictable. The man controls the other dot by motion of a joystick. In the simplest case, the controlled-dot position follows the joystick command exactly (the process in Fig. A3.2 is just a constant);* in more complex cases, the system output is determined from the control signal (the joystick position) by a transfer function for the process.

Extensive experiments over the past 25 years with this type of test system have revealed several important features:

(1) The input signal must be random or at least unpredictable. Early tests used a sinusoidal input, just as we might measure the transfer function of a circuit by applying a sinusoidal signal with the frequency changed over the range of interest. The human controller, observing a sinusoidal variation of the reference spot, rapidly learns to

*The system is similar to the amusement park driving game, where the man tries to keep his "car" on the road which is varied randomly.

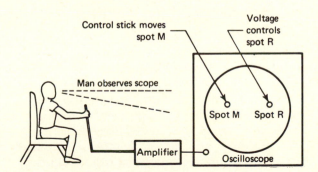

FIG. A3.3 Experiment for evaluating human transfer function.

predict the input signal and is able to follow with high accuracy. This ability is given the fancy name "precognitive tracking."

Actually, a truly random signal is not needed. We can simply use the sum of perhaps five sinusoids (*not* harmonics of a single frequency, so that the sum is not periodic). The human operator is then unable to recognize the waveform pattern.

(2)   Human performance depends strongly on motivation, fatigue, training, boredom, and the magnitude of distractions. Furthermore, there is, of course, variation from individual to individual. The best we can hope for is an average transfer function which approximately represents typical subjects in the task.

(3)   Human performance deteriorates rapidly when the task complexity is increased beyond the level where the man is able to do a reasonable job. For example, as the bandwidth of the input signal is increased, the man can track reasonably well as long as the input has no significant frequency components above 1 to 1.2 Hz. When the bandwidth is raised to 2 Hz, the man simply gives up.

(4)   The man is remarkably *adaptive*: he changes his transfer function according to the transfer function of the process being controlled. He automatically adjusts his transfer function to give a stable total system which has a response time of about 2 secs. He is able to adapt over quite a wide range of process transfer functions (as we know from the observed fact that a man can drive a monocycle, bicycle, car, car in reverse, truck, airplane, and helicopter—processes with very different dynamics).

(5)   Finally, the man cannot be described accurately by a simple transfer function for two other reasons. First, he samples the signals rather than observing them continuously. In other words, he observes the signals at discrete instants of time, takes corrective action, and then later observes again. This concept of sampling is discussed in more detail in Chap. 13. Second, he generates a control signal (Fig. A3.2) which has a component related to the error by a transfer function, but which also has a second component (called a *remnant*) which is random in nature—for example, a higher frequency variation which keeps this spot moving slightly in anticipation of major control signals needed in the immediate future. Both of these effects (sampling and remnant generation) are often negligible.

The transfer function

In spite of these difficulties, we are able to use a transfer function to describe the human controller at least approximately, and we can estimate the dynamic performance to be anticipated from simple man-machine systems. It is not at all surprising that many different transfer functions have been suggested; indeed, the dynamic performance of a system is often startlingly insensitive to major changes in the poles and zeros of the transfer function. In other words, for any device (mechanical or human), we can always find quite different transfer functions which yield essentially the same performance.

The most common transfer function for the human controller is

$$\frac{\text{Control signal}}{\text{Error}} = H(s) = K \frac{T_0 s + 1}{(T_1 s + 1)(T_2 s + 1)} e^{-Ts} \qquad \text{(A3-1)}$$

The following parameters are included:

(1) T—the "transportation lag," or the *human reaction time*: the time required for the man to react regardless of the urgency of the stimulus. The minimum value of T is about 0.2 sec., but the value may be much greater if the man is fatigued, or under the influence of alcohol.

(2) K—a gain constant which, in the control of simple processes, is automatically adjusted by the man to give a stable system with a response time of about 2 sec.

(3) T_0—a lead or prediction time constant. If the transfer function were

$$K(T_0 s + 1)$$

the man would generate a control signal $m(t)$ which is related to the error $e(t)$ by

$$m(t) = K[e(t) + T_0 e'(t)] \qquad \text{(A3-2)}$$

T_0 measures the amount of "derivative control," the extent to which the man tries to predict the future error.

(4) T_1 and T_2—lag time constants, represent delays in the man's processing of the signals. From the time the man senses the error, there are delays due to the need to transmit the error to the brain, make a decision on action, transmit the electrical signal to the appropriate muscles, and then start the muscle action.

When the control task is very simple (e.g., when the process is described by the simple transfer function K_p/s and the input signal has only very low frequencies), the man is relaxed in his control task. Under these conditions

$$T_0 = 0 \qquad T_1 = T_2 = \tfrac{1}{3} \qquad \text{(A3-3)}$$

The man does not bother to predict and the response is sluggish.

As the control task becomes more difficult $\{$(e.g., a process increasingly complicated toward $K_p/[s^2(s+1)]$ or $K_p/[s(s^2 + \alpha s + \beta)]\}$, the man first introduces prediction. T_0 rises toward a value as high as unity. With still more complicated processes, the man speeds up his action: the lag constants T_1 and T_2 are reduced to as low as 1/20 sec. Finally, when the process is so complicated and unstable that it cannot be adequately controlled even with these limits of $T_0, T_1,$ and T_2, the man gives up.

From a system design viewpoint, the engineer must recognize that the level of effort required of the man increases as T_0 is increased and T_1 and T_2 are decreased. If the man is to continue for any length of time in the task, the process must be designed so that it can be controlled with T_0 appreciably less than unity, T_1 and T_2 near 1/3 sec. Thus, understanding of the dynamics of the man leads to design criteria for the process.

Other control problems

The single-loop system of Fig. A3.2 is, of course, only one way that a man can be incorporated in a control system. Often the system designer would like to use the man in parallel with a mechanical component. The "device" performs the control most of the time; the man is only used when a specific task arises for which he is particularly skillful (e.g., an emergency situation, when the man recognizes what is happening and can make a decision as to required action).

In another example, a man is asked to control two systems concurrently (Fig. A3.4). When the pilot tries to control altitude h, he directly controls the pitch angle θ, which in turn ultimately controls h. In this case, the inner loop is relatively fast (wide bandwidth), and the transfer function described above is appropriate. The outer, very slow loop can involve a very different pilot transfer function. Typically, in aircraft control problems,

$$H_o(s) = K_0(1 + 6s) \tag{A3-4}$$

There are no significant lags and a very large amount of prediction.

Thus, the appropriate transfer function depends critically on the particular task. This field of research is relatively young, and studies have been concentrated primarily on the aircraft piloting problem. Only in the very recent past has analogous research been initiated for automobile driving and other control functions.

Future problems

In addition to our need for knowledge about human dynamics in various control tasks, to match machines to men, we also must learn much more about the way man's

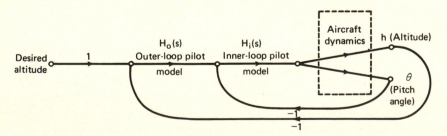

FIG. A3.4 Human pilot controlling two loops.

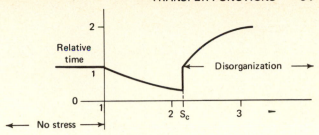

FIG. A3.5 Time required by man for specific tasks.

performance varies with the task. For example, one of the areas of interest in recent years has been the optimization of the planning and scheduling of a series of work tasks involving both men and machines. How should the system be organized to minimize cost by improving both machine utilization and human productivity?

Studies of human performance in a sequence of specific tasks have shown that the time required to complete the tasks varies as indicated in Fig. A3.5. The independent variable is stress, which is defined as

$$\text{Stress} = \frac{\text{Time to complete under leisurely performance}}{\text{Time remaining}} = \frac{T_l}{T_r} \qquad \text{(A3-5)}$$

When there is no urgency, the man uses a certain length of time T_l to complete the series of tasks. If only T_r remains before he must be done, the stress on him increases as T_l/T_r increases.

With no stress (no need to hurry), the time required is known (and called 1 on the vertical scale). As the stress increases beyond 1, the man works faster and the required time falls. When the stress exceeds a certain critical value S_c, the man is unable to finish on time and he becomes disorganized.

Clearly, optimization requires that we operate at a stress level below S_c but appreciably above unity. To design an optimum total system, we need to study the component tasks, evaluate man's performance in each, and model the tasks performed by machines. We then choose the number of machines and men and schedule the overall operation so that cost is minimized, or performance maximized, within the constraints imposed by human fatigue, machine reliability, and so on.

In such a problem, the matching of machines to men goes beyond merely finding human transfer functions to the broader problem of measuring quantitatively human capabilities and attitudes. The transfer function research described briefly in this appendix is merely one facet of man-machine system design.

PROBLEMS

3.1   A network is linear with constant coefficients and has a transfer function

$$\frac{Y}{X} = \frac{1}{s + 2}$$

(a) Find the differential equation of the network.

(b) Find the complete response to $x(t) = 3\cos 2t$ when the system is inert and x is applied at $t = 0$. Evaluate all coefficients.

3.2   The flow diagram represents a system to which the input is $2\sin(10t + \theta)$. Determine the value of θ which will result in a transient which lasts less than 0.5 sec. Assume that the system is inert when the signal is applied at $t = 0$.

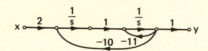

3.3   A system is assumed to be linear with constant coefficients and zero initial conditions; when tested with a unit step, the response is as shown.

(a) Find the transfer function of the system.

(b) Determine the response to a unit impulse.

(c) How would you test to determine if the above assumptions are fulfilled?

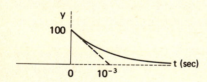

3.4   A system is described by the transfer function

$$\frac{Y}{X} = 3000 \frac{s + 2}{(s + 4)(s^2 + 6s + 100)(s + 10)}$$

(a) When $x(t)$ is a unit step function, about how long does the transient last?

(b) With x a unit step, what is the steady state response?

(c) If the system exhibits resonance, what are the resonant frequency and bandwidth in Hz?

3.5   What is the transfer function of the system described by the signal flow graph shown? Show that this system does not exhibit oscillatory behavior. Show further that a feedback path from z to w can be used to produce such oscillatory behavior. If the gain of the added path is −1, how must you choose a and b in order to have a resonant frequency of 3 and a Q of 1/2?

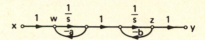

3.6   A system is described by the model shown, with x the input and y the output.

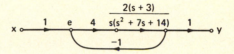

(a)   Determine $T(s) = Y/X$ and $U(s) = E/X$.
(b)   What is the order of the system?
(c)   What time constants govern the response?
(d)   Approximately how long does the transient last?
(e)   The system operates ideally when $y = x$. At what values of complex frequency does ideal operation occur?
(f)   What is the steady state response to an x which is a step of amplitude 3?
(g)   What is the steady state response to an x which is a ramp of slope 3?
(h)   Over what frequency range does the system behave nearly ideally in the steady state if x is a sinusoid?
(i)   What is the s - s response to an x of $4 \sin(2t + 60°)$?
(j)   Is the system resonant? If so, what are the resonant frequency and Q?

3.7   A system is described by the signal flow diagram shown. The inert system is driven by a unit impulse as $x(t)$. The values of the three parameters a, b, and c are to be used as shown in the diagram.

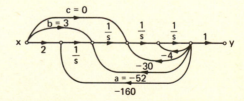

(a)   What is the transfer function, $Y(s)/X(s)$?
(b)   What is the Laplace transform of the output $y(t)$?
(c)   What are the initial values $y(0)$, $y'(0)$, and $y''(0)$?
(d)   What is the initial value of the third derivative of $y(t)$?

(e) We now change the input signal to $4 \cos(2t + 20°)$. We wish to choose b and c so that the steady state output is zero. Select b and c.

(f) The transfer function has a denominator which can be factored into two quadratics, one of which is $s^2 + 2s + 16$. What resonant frequencies are present in the system?

(g) What is the Q for each resonance?

(h) What is the relative damping ratio associated with each pair of poles?

(i) Show the poles and zeros of Y/X in the s plane when the parameters $a, b,$ and c have the values given at the start of the problem.

3.8   This problem refers to the transfer function

$$T(s) = \frac{250s(s + 20)}{(s + 1)(s + 10)(s^2 + 20s + 500)}$$

(a) What is the steady state output when a unit ramp function is applied?

(b) The drive signal is $4 \sin(10t + \theta)$. What should θ be if the transient is to be over in less than 0.5 sec?

(c) What is the behavior of the system for small values of t? Determine two terms in the series.

(d) A signal $4 \sin 10t$ is applied. What is the s - s output?

(e) The system is resonant. Find the resonant frequency, Q, bandwidth, and approximate half-power frequencies.

3.9   The system with the transfer function $T(s)$ is unstable, where

$$T(s) = \frac{s^2 + 3}{(3s^2 - 2s + 3)(s^2 + 2s + 2)}$$

Show that this is so, and also that the input

$$x(t) = \cos t - 2 \sin(\sqrt{3}\, t + 30°)$$

suppresses the increasing portion of the free response.

3.10 Construct a flow diagram for the transfer function

$$\frac{Y}{U} = \frac{4}{s^2 - 4}$$

by determining the partial fraction expansion and representing the separate terms by parallel, simple systems. In each case, the dynamics should be represented only by branches with transmittances of $1/s$. In this diagram, show the initial conditions for the two integrator outputs, with the values being the impulses entering the inputs of the integrators. We now wish to choose

$U = X + HY$. That is, the input u is the sum of a total system input x and an output of a network H in which the input is y. If the total system is to be stable, what must H be? Is there any choice? What would be a suitable value of H? This is a rather simple example of the capability of feedback to control inherently unstable systems.

3.11 To evaluate the polynomial $p(s)$ at a particular value s_1 (a calculation required in making a partial-fraction expansion), we can divide $p(s)$ by $(s - s_1)$. The remainder is $p(s_1)$. The validity of this fact is demonstrated by the equation for the division:

$$\frac{p(s)}{s - s_1} = q(s) + \frac{R}{s - s_1}$$

Multiplication by $(s - s_1)$ gives

$$p(s) = (s - s_1)q(s) + R$$

When $s = s_1$, we have

$$p(s_1) = R$$

Use this approach to evaluate the polynomial $s^5 + 2s^4 + 3s^3 + 6s^2 + 4s + 3$ at $s = -1.25$.

3.12 The procedure of Prob. 3.11 is cumbersome when s_1 is complex. Then it is simpler to divide by $(s - s_1)(s - \bar{s}_1)$. We substitute s_1 into the remainder $R_1 s + R_0$. Use this procedure to evaluate the polynomial $3s^6 + 8s^5 + 12s^4 + 20s^2 + 6s + 1$ at $s = -1 + j$.

3.13 We can measure the up-and-down resonant frequency of a car body on the shock absorbers by jumping off the back bumper and observing the period of the damped oscillation. Suppose we find this period is 1.8 sec. As we drive along a concrete highway with slab dividers, we seem to excite the resonance of the vertical vibration at a speed of 55 mph. How far apart are the slab dividers?

(This is actually an interesting experiment to perform if a car is available—we can compare the two measurements of the resonant frequency. It is interesting how often a highway seems to be designed exactly so the most popular cars hit resonance when traveling just at the speed limit—which encourages the driver to go at a significantly lower or higher speed.

3.14 In various texts, we find the assertion that the poles of a transform or a transfer function are the *really important features*. The implication is that, if we know $F(s)$ has a pole at -2, this is a truly significant fact.

In a sense, this sort of argument is utter nonsense, which we shall show in two ways. Since the transfer function and the transform of the input signal are analogous functions in analysis, we focus our attention on the transform of the input.

(A) *Real-life system analysis*

A realistic analysis problem can be described in terms of the notation shown here. The signal $x(t)$ is applied at $t = 0$ and continues thereafter. We want to find the output $y(t)$ at all times from $t = 0$ to $t = T$. (In an actual problem, we are interested in the output for a finite period of time; no one really cares about the state of the world indefinitely into the future.)

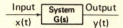

For simplicity, let's assume the input is Ae^{-2t}. Since we are interested in the system response y only until $t = T$, the input for $t > T$ is obviously immaterial. (The system is causal; the input in the future has no effect on the present output.)

The above is a description of a real analysis problem. Now let us consider its solution.

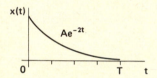

(1) *An elementary solution.* Here we would agree that x for $t > T$ is immaterial. So, we choose the simplest mathematical form for $x(t)$ beyond T: a continuation of the exponential decay forever. Then

$$X_1(s) = \frac{A}{s + 2} \tag{1}$$

and we say that the input signal is described by the fact *its transform has a pole at* -2.

(2) *An alternate approach.* For some inexplicable reason, we decide that the input signal decays somewhat faster after $t = T$—e.g., as e^{-10t}. In other words,

$$x_2(t) = \begin{cases} 0 & t < 0 \\ Ae^{-2t} & 0 < t < T \\ (Ae^{-2T})e^{-10t} & t > T \end{cases} \tag{2}$$

This equation is a perfectly valid description of $x(t)$. The corresponding $y(t)$ is exactly what we are looking for over the entire interval 0 to T. Furthermore, if

T is greater than 3, the signal is so small by T that a plot doesn't show whether we have Eq. (2) or an $x(t)$ which is just Ae^{-2t}.

Where does $X(s)$ for Eq. (2) have poles? Here

$$X_2(s) = \frac{A}{s+2}(1 - e^{-2T}e^{-Ts}) + \frac{Ae^{-2T}}{s+10}e^{-Ts} \tag{3}$$

This $X_2(s)$ has only one pole: at $s = -10$. *Our input signal has a transform with a pole at* -10.

(3) *Other approaches.* By varying the form assumed for $x(t)$ when $t > T$, we can obtain *any poles we want* for $X(s)$. Indeed, if we just make $x(t) = 0$ for $t > T$, $X(s)$ has no poles at all. Only one conclusion can be drawn: the poles of $X(s)$ really describe the $x(t)$ when t is very, very large—indeed, far into the future beyond the point where anyone cares.

(B) *Approximation of $x(t)$ by exponentials*

We can attempt to understand the importance of $X(s)$ in another way. Suppose we measure an $x(t)$ over the time span of interest (0 to T). During this interval we find that

$$x_m(t) = 3e^{-2t} \qquad 0 < t < T \tag{4}$$

where the subscript m means this is an observed or measured value. Now, we represent this signal approximately by a single exponential

$$y(t) = Ae^{-s_1 t} \tag{5}$$

where the two parameters A and s_1 are to be chosen to minimize the error between the approximation $y(t)$ and the actual signal $x_m(t)$.

First, it must be decided what we mean by "minimize the error." What error? Here we can obtain a mathematical solution by minimizing the integral of the square of the error. That is, we minimize

$$E = \int_0^T [y(t) - x_m(t)]^2 \, dt \tag{6}$$

Substitution of Eqs. (4) and (5) gives

$$E = \int_0^T [Ae^{-s_1 t} - 3e^{-2t}]^2 \, dt$$

$$E = \int_0^T A^2 e^{-2s_1 t} dt - 6A \int_0^T e^{-s_1 t} e^{-2t} \, dt + 9 \int_0^T e^{-4t} \, dt$$

$$E = \frac{A^2 (1 - e^{-2s_1 T})}{2s_1} - \frac{6A (1 - e^{-(s_1 + 2)T})}{s_1 + 2} + \frac{9 (1 - e^{-4T})}{4} \tag{7}$$

Now, to minimize by selection of A and s_1, we want to set $\partial E/\partial A$ and $\partial E/\partial s_1$ equal to zero. There is no need to do this algebra. If we did, however, we would find that the two conditions for minimun error are (when T is large)

$$\left. \begin{aligned} Y(2) &= X_m(2) \\ Y'(2) &= X'_m(2) \end{aligned} \right\} \tag{8*}$$

In other words, the transform of our approximation must match the transform of the measured signal, not at the pole (-2) but at the *negative of the pole* (which lies in the right half plane). Indeed, the derivatives with respect to s must also match at this point within the right half plane. Thus, it is not the behavior of $X(s)$ at its poles which is important, but rather the behavior of $X(s)$ at appropriate points inside the right half plane. The poles are merely an analytical convenience.

Concluding comment

The above discussion is intended to emphasize that the poles of a signal transform or a transfer function are truly significant *only* when the function is a ratio of polynomials in s. Only then can we infer that the poles do describe important characteristics of the significant part of the signal.

For example, if we know that $X(s)$ is $a/(s^2 + bs + c)$, we can state that b and c determine the shape of the signal during all positive time. If, on the other hand, we merely know that a signal transform has a pole at -10, we can only deduce that *eventually* the signal has a component of the form Ae^{-10t}. This may

*These relations were derived by P. R. Aigrain and E. M. Williams, "Synthesis of n-reactive networks for desired transient response," *J. Appl. Physics,* Vol. 20, pp. 597–600, June 1949. The material in these pages is based on a paper by W. H. Huggins, "Poles Are Like Prince Rupert's Drops," June 18, 1970.

not appear, however, until long after the time interval of interest, or until the term Ae^{-10t} is so small that it is unmeasurable. Under such circumstances, the specific values of the poles are irrelevant.

Questions

(1) What is the transform of a signal $x(t)$ which has the following characteristics: it behaves as $3e^{-t}\sin(10\pi t)$ from $t = 0$ to $t = 4$. The transform of the signal has only a single pole at $s = -1$.

(2) Sketch the signal described in (1).

(3) Is there a unique answer to question (1)? If not, give at least one additional, valid answer.

(4) If we were to approximate the signal of (1) by a pair of exponentials and wished to minimize the integral of the square of the error, at what values of s would we be particularly interested in $X(s)$?

4 SPECIAL-PURPOSE NETWORKS

This chapter should not be taken too seriously. In any branch of engineering, there are certain "tricks" of the trade, certain "rules of thumb" which are very useful in design. They form part of the "art" of engineering. Network design is no exception. In this chapter we want to describe just a few of these tricks which the experienced system engineer has at his command.

Consequently, there really is no continuity in the chapter. Instead, each section discusses a different, special-purpose system. Thus, the chapter is actually a potpourri of small topics.

4.1 DIFFERENTIATORS

A differentiator or *differentiating system* yields an output which is proportional to the first derivative of the input. In terms of Fig. 4.1,

$$Y = AsX \quad \text{or} \quad y = A\frac{dx}{dt} \tag{4-1}$$

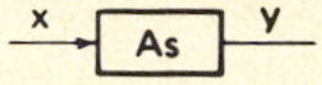

FIG. 4.1 Differentiator. $\xrightarrow{x}$ $\boxed{\text{As}}$ $\xrightarrow{y}$

Why differentiation?

A differentiator is an important element of a system designed to predict the future value of a signal from its present and past values. In Fig. 4.2, the signal up to time t_1 can be observed, and we want to predict (at t_1) the value of the signal at t_2.

The simplest prediction assumes that the derivative at t_1 will remain constant in the interval from t_1 to t_2: in other words, the signal will change along a straight line from its present value to the future, predicted value. Under this assumption

$$x^*(t_2) = x(t_1) + \frac{dx}{dt}\bigg|_{t=t_1} (t_2 - t_1) \tag{4-2}$$

where the asterisk signifies a predicted value.

In other words, prediction is accomplished by a transfer function

$$1 + s(\Delta t)$$

where (Δt) is the time interval over which the system is to predict. To realize this simple predictor, we must be able to construct a differentiator.

Importance of prediction

Prediction is frequently necessary because of the time delays inherent in operation of the total system. For example, in an antiaircraft fire-control system, the guns must be aimed at the location of the target aircraft when the bullets reach that point, not at the location when the gun is fired. In other words, the system must *predict* target motion in order to insert the required lead.

It was this fire-control problem which led to Norbert Wiener's classic work during World War II.* In this study of optimum predictors, Wiener worked from the random

*Norbert Wiener (1894-1964) probably ranks as the principal founder of modern systems engineering. His years as a child genius (he received his Harvard doctorate at the age of 18) are fascinatingly described in his book "Ex-Prodigy," published in 1953 by the M.I.T. Press. Two of his other books of primary interest are "Cybernetics" (1948) and "The Human Use of Human Beings" (1950).

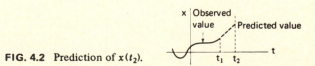

FIG. 4.2 Prediction of $x(t_2)$.

characteristics of the signals: the fact that the signals were not precisely predictable, but instead had to be treated in terms of *probability distributions.* Wiener's work led to a fundamental revolution in system engineering, which previously had been concerned with sinusoidal signals.

Prediction is an essential element of all system engineering. If we are designing a system for solid waste disposal in New York City, for example, seven years are required before a new incinerator is in operation. Clearly, system innovations must be designed for the future, not for the past or present.

Desirable gain characteristic for differentiators

If prediction is to be accomplished on the basis of the measured signal and its first derivative, the ideal differentiator transfer function $D(s)$ is

$$D(s) = 1 + Ts \tag{4-3}$$

where T is the prediction interval. Actually, there are two problems in such an approach:

(1) The first problem is not insurmountable. It arises because a transfer function with a numerator of higher degree than the denominator is somewhat difficult to build. If we try to use R, L, and C components only, the derivative part of Eq. (4-3) requires a network such as shown in Fig. 4.3: a voltage-to-current conversion and an inductor. If α is perhaps 1 ma/v and T is 0.1 sec, the required inductance is $100\,h$—a size which may be undesirably large, heavy, and expensive. Furthermore, the parasitic elements associated with a practical L tend to distort the transfer function.

(2) The second problem is fundamental: we usually do not really want this "ideal" transfer function of Eq. (4-3). In most cases, the total input consists not only of the signal e_{in}, but also of undesired noise. The signal typically is of limited bandwidth (for example, it might have most of its energy from 0 to 10 rad/sec in a control problem). The noise, on the other hand, commonly has a much wider bandwidth (in the same control problem, perhaps to 100 rad/sec).

We are only trying to differentiate the signal; the transfer function Ts is to be realized only from 0 to 10 rad/sec in our example. Beyond 10 rad/sec, we would like the system gain to fall off; we certainly do not want to continue the differentiation on out in frequency indefinitely. Ts has a gain at 100 rad/sec ten times as great as at 10 rad/sec. *The ideal differentiator tends to accentuate the high-frequency noise*—indeed to such an extent that the desired output (the derivative of the signal part of the input) may be masked entirely by the derivative of the noise.

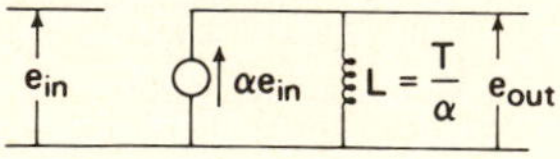

FIG. 4.3 Realization of $\dfrac{E_{out}}{E_{in}} = Ts$.

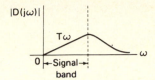

FIG. 4.4 Desired gain characteristic of an actual differentiator.

What we really want is a filter-differentiator: a network which behaves as Ts in the signal frequency band, but which has a small gain outside this band.

A possible transfer function

Figure 4.4 is the gain of a possible transfer function, behaving as Ts at low frequencies and falling off toward zero at high frequencies. The latter requirement means that the denominator should be of higher degree than the numerator. Thus, we might try

$$D(s) = \frac{As^2 + Bs + C}{(s + a)(s + 2a)(s + 3a)} \tag{4-4}$$

We arbitrarily pick the poles at $-a, -2a$, and $-3a$.* The numerator is selected of degree 2 to be sure that D tends to zero as $s \to \infty$. Now A, B, and C are to be chosen to give the desired low-frequency behavior.

To determine how D behaves at very low frequencies, we simply divide the denominator into the numerator. Since $s \to 0$ are the frequencies of interest, the polynomials are first arranged in *ascending* powers of s

$$D(s) = \frac{C + Bs + As^2}{6a^3 + 11a^2s + 6as^2 + s^3} \tag{4-5}$$

Before starting the long division, we observe that C must be zero if $D(s)$ is to behave as Ts at low frequencies:

$$D(s) = \frac{Bs + As^2}{6a^3 + 11a^2s + 6as^2 + s^3} \tag{4-6}$$

---

*It is shown in network synthesis courses that, if the poles are simple and on the negative real axis, the network can be built with Rs and Cs only—no inductor is required. In Chap. 6 we note that poles anywhere in the left half plane can be chosen if we are willing to use amplifiers as well as resistors and capacitors.

Ordinary long division then gives

$$6a^3 + 11a^2s + 6as^2 + s^3 \overline{\left)\begin{array}{l} \cfrac{B}{6a^3}s + \cfrac{A - \cfrac{11B}{6a}}{6a^3}s^2 + \cfrac{-\cfrac{B}{a^2} - \cfrac{11\left(A - \cfrac{11B}{6a}\right)}{6a}}{6a^3}s^3 \\[2em] Bs \quad + \quad As^2 \\[1em] Bs \quad + \quad \cfrac{11B}{6a}s^2 \quad + \quad \cfrac{B}{a^2}s^3 \qquad\qquad + \cdots \\[2em] \hline \left(A - \cfrac{11B}{6a}\right)s^2 - \cfrac{B}{a^2}s^3 \qquad\qquad - \cdots \\[2em] \left(A - \cfrac{11B}{6a}\right)s^2 + \cfrac{11\left(A - \cfrac{11B}{6a}\right)}{6a}s^3 \qquad - \cdots \\[2em] \hline \left[-\cfrac{B}{a^2} - \cfrac{11\left(A - \cfrac{11B}{6a}\right)}{6a}\right]s^3 - \cdots \end{array}\right.}$$

Hence

$$D(s) = \frac{B}{6a^3}s + \frac{A - \dfrac{11B}{6a}}{6a^3}s^2 + \frac{-\dfrac{B}{a^2} - \dfrac{11\left(A - \dfrac{11B}{6a}\right)}{6a}}{6a^3}s^3 + \cdots \qquad (4\text{-}7)$$

Now B and A can be chosen. Since D is to behave as Ts at low frequencies,

$$\frac{B}{6a^3} = T \quad \text{or} \quad B = 6a^3T \qquad (4\text{-}8)$$

The s^2 term represents a deviation from this desired behavior; therefore, A can be selected to make the coefficient of s^2 zero:

$$A - \frac{11B}{6a} = 0 \quad \text{or} \quad A = 11a^2T \qquad (4\text{-}9)$$

Equation (4-7) becomes

$$D(s) = Ts - \frac{T}{a^2} s^3 + \cdots \tag{4-10}$$

In terms of ω, at low frequencies

$$|D(j\omega)| \longrightarrow T\omega \left(1 + \frac{\omega^2}{a^2}\right) \tag{4-11}$$

The error in differentiation is measured by the ω^2/a^2 term. If the system is to differentiate within 10% over a frequency band from 0 to ω_1,

$$\frac{\omega_1^2}{a^2} = 0.1 \quad \text{or} \quad a = 3.2\omega_1 \tag{4-12}$$

If we return to our earlier example with the signal band extending to 10 rad/sec, the appropriate parameters for a prediction time of 0.1 sec are

$$a = 32 \qquad A = 1,100 \qquad B = 19,700 \tag{4-13}$$

and

$$D(s) = \frac{1,100s^2 + 19,700s}{(s + 32)(s + 64)(s + 96)} \tag{4-14}$$

or

$$D(s) = \frac{1,100s^2 + 19,700s}{s^3 + 192s^2 + 11,000s + 197,000} \tag{4-15}$$

In more general terms

$$D(s) = Ts \frac{\beta s + \gamma}{s^3 + \alpha s^2 + \beta s + \gamma} \tag{4-16}$$

Once the poles are selected (and hence α, β, and γ are known), the numerator polynomial is determined.

A differentiator-filter network

The transfer function of Eq. (4-14) or (4-15) allows us to find a corresponding RC network. This part of the problem is called network synthesis—going from the desired transfer function to an appropriate system or network. In Chap. 6, we discuss several solutions to this problem. Here, we merely show one possible network (Fig. 4.5).*

Other "derivative networks"

The $D(s)$ derived above approximates the desired differentiation property at low frequencies, and the gain falls off with ω at high frequencies. The term "derivative network" is also used in the systems literature for a different type of approximation to the pure differentiator.

Figure 4.6 shows a common derivative network. The transfer function is that of a voltage divider:

$$\frac{E_2}{E_1}(s) = \frac{R_1}{R_1 + R/(RCs + 1)} \tag{4-17}$$

or

$$\frac{E_2}{E_1}(s) = \frac{s + 1/RC}{s + [(R_1 + R)/R_1](1/RC)} \tag{4-18}$$

How does the system behave? It is easiest to consider three separate frequency bands:

(a) low frequencies ($\omega < 1/RC$): then the s terms are negligible and the transfer function is just $R_1/(R_1 + R)$;

(b) high frequencies $\omega > [(R_1 + R)/R_1](1/RC)$; the transfer function is approximately unity (the two constant terms are negligible);

---

*Simply showing the network without indicating how it is found is a letdown after all our work to derive the transfer function. Unfortunately, the author is unable to explain the network synthesis without ten pages of discussion which are really not important. The procedures of Chap. 6 provide a set of different networks for the given $D(s)$—networks which are often preferable to Fig. 4.5.

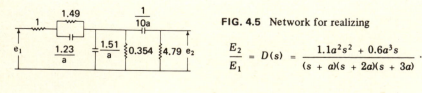

FIG. 4.5 Network for realizing

$$\frac{E_2}{E_1} = D(s) = \frac{1.1a^2 s^2 + 0.6a^3 s}{(s + a)(s + 2a)(s + 3a)} .$$

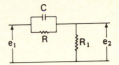

FIG. 4.6 A simple derivative network.

(c) middle frequencies $\omega > 1/RC, \omega < [(R_1 + R)/R_1](1/RC)$; the s term dominates the numerator; the constant term, the denominator; and

$$\frac{E_2}{E_1} \longrightarrow \frac{R_1}{R_1 + R} RC\,s \qquad\qquad (4\text{-}19)$$

The system approximately differentiates.

Figure 4.7 shows the corresponding gain characteristic versus frequency, with a possible set of parameter values.

Devices for differentiation

We have discussed only electric networks for differentiating. In addition, we often use electromechanical devices. For example, if the input signal is the rotary position of a shaft, a device called a *tachometer* can be used; this is an electric generator, with the voltage generated proportional to the rotational velocity of the input.

Indeed, we can use devices based on any physical principle which yields an output proportional to the derivative of the input. If a wire of length l moves with a velocity v through a magnetic field B, the generated voltage is Blv. Hence if the input signal is the position of the wire, the output voltage measures the derivative of the input.

To differentiate, we can also return to the definition of the derivative

$$\frac{dy}{dt} = \lim_{\Delta t \to 0} \frac{y(t) - y(t - \Delta t)}{\Delta t} \qquad\qquad (4\text{-}20)$$

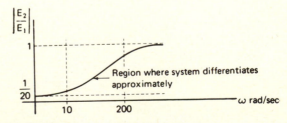

FIG. 4.7 Gain characteristic for the derivative network of Fig. 4.6 with $RC = 1/10$ sec and $R = 19\,R_1$.

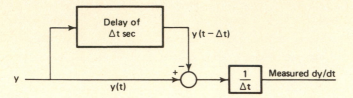

FIG. 4.8   Measurement of derivative.

The system of Fig. 4.8 measures the derivative by using a reasonably small Δt in Eq. (4-20). The system requires a device which gives a delay of Δt seconds.

4.2   INTEGRATORS

Systems which integrate are fully as important and even more common than differentiators. Indeed, Chap. 6 is primarily devoted to using integrators (and adders) to realize any specified transfer function.

Integrators also are essential in an inertial navigation system. Here, the key sensitive element is an accelerometer, with the simplest form shown in Fig. 4.9.* The case and everything outside of it are fixed to the vehicle along its direction of motion. Inside the case, a mass M is restrained by two springs, each with spring constant $K/2$. The case is also filled with oil, so that as the mass moves with respect to the case, there is a friction force opposing the relative velocity $(v_M - v_C)$.

*The same instrument serves as the central element of the seismograph, used to measure earthquakes.

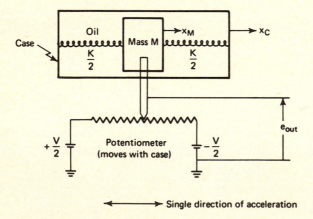

FIG. 4.9   Basic accelerometer.

The system can be analyzed by considering all forces acting on the mass M when the vehicle accelerates. d'Alembert's principle gives

$$M \frac{d^2 x_M}{dt^2} + B \frac{d}{dt}(x_M - x_C) + K(x_M - x_C) = 0 \qquad (4\text{-}21)$$

where B is the damping coefficient. Solution for $(x_M - x_C)$ yields

$$(X_M - X_C) = \frac{-Ms^2 X_C}{Ms^2 + Bs + K}$$

or

$$\frac{X_M - X_C}{s^2 X_C} = \frac{-M}{Ms^2 + Bs + K} \qquad (4\text{-}22)$$

In other words, the input is the acceleration of the case (or vehicle); the output is the relative displacement $x_M - x_C$.

The wiper arm of the potentiometer is attached mechanically to the mass. Hence, as the mass moves a different amount than the case, the pot wiper moves away from the center. The output voltage is proportional to this relative displacement $x_M - x_C$, and

$$\frac{E_{\text{out}}}{s^2 X_C} = \frac{kM}{Ms^2 + Bs + K} \qquad (4\text{-}23)$$

In other words, the voltage output is related by this transfer function to the input acceleration.

In the design of the instrument, we select the M, B, and K to give the desired dynamic characteristics. In an inertial navigation system, the parameters are chosen so that, over the frequency band in which input acceleration has significant energy, Eq. (4-23) reduces to

$$\frac{E_{\text{out}}}{A_C} = k \frac{M}{K} \qquad (4\text{-}24)$$

(the output voltage is directly proportional to case or vehicle acceleration).

Now, we see how an inertial navigation system operates. The output voltage of the accelerometer is integrated twice: the first integration gives vehicle velocity, the second vehicle position (Fig. 4.10). Both integrators have to have inserted the appropriate initial conditions.

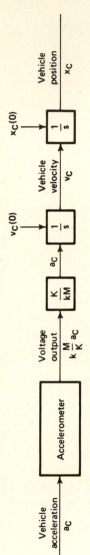

FIG. 4.10 Elements of an inertial navigation system.

The design of accurate inertial navigation systems is one of the monumental engineering achievements of the last two decades. Actually, to measure vehicle position accurately, three accelerometers are used, one in each of three orthogonal directions in space. Each accelerometer is located on a gyro-stabilized platform to maintain its orientation (e.g., north-south) in spite of vehicle motions.

The achievable accuracy has been dramatically demonstrated by an airplane flying with only inertial nagivation from Boston to within a hundred feet of the runway at Los Angeles airport, by the success of the Apollo missions, and by the nuclear submarines navigating for days underwater. In these highly refined systems, the crude accelerometer of Fig. 4.9 is replaced by one in which the springs are provided magnetically, the mass is floated in oil of essentially equal specific gravity, and the output is magnetic rather than the crude potentiometer arrangement.

Systems for integration

The most common electronic integrators are the operational amplifiers described in detail in Chap. 6. Here we mention only three other possibilities.

(1) Since a capacitor voltage is the integral of the current, Fig. 4.11 shows the simplest electrical integrator. A voltage-to-current converter feeds a capacitor, with

$$\frac{E_2}{E_1} = \frac{\alpha}{C} \frac{1}{s} \tag{4-25}$$

(2) We can also integrate by continuous processing of signal values. The integral of $y(t)$ from t_0 to a time t is the area under the curve over this interval. As indicated in Fig. 4.12, if we measure the signal regularly and frequently (so that the change of y in each Δt is small), this area in any interval is very nearly equal to the average of the sample values at the ends of the interval times Δt. For example,

$$\int_{t_0}^{t_1} y(t)\, dt = \frac{y_0 + y_1}{2} \Delta t \tag{4-26}$$

Electronically we can measure $y_0, y_1, y_2, \ldots$ and then calculate the integral every Δt seconds. If Δt is small, the system essentially gives a running value of the integral from the starting time t_0 to t.

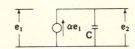

FIG. 4.11 Integrator network.

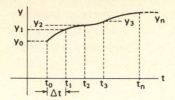

FIG. 4.12 The integral as an area.

(3) Just as in the differentiator case, we often want an "integral network"—a system which essentially integrates over a band of frequencies. The circuit of Fig. 4.13 possesses the transfer function

$$\frac{E_2}{E_1}(s) = \frac{R}{R + R_1} \frac{s + 1/RC}{s + [R/(R + R_1)](1/RC)} \tag{4-27}$$

(essentially the reciprocal of the derivative-network transfer function of the preceding section). Analysis shows that the system approximately integrates in the frequency band

$$\frac{R}{R + R_1} \frac{1}{RC} < \omega < \frac{1}{RC} \tag{4-28}$$

4.3 STEP-FUNCTION RESPONSE

One of the most important test signals in system engineering is the step function: the abrupt change of the input from one value to another. If the systems under study are linear and time-invariant, the size of the step and its exact occurrence in time are not important, and we can focus on the *unit* step function occurring at $t = 0$. (Fig. 4.14). The signal is zero for t negative and unity for t positive, and is denoted by $u(t)$.*

The step function is important in the testing of physical systems because:

(1) The step function is a particularly simple signal to apply. Electrically, a battery or dc source is applied at $t = 0$ (for example, by closing a switch). In mechanical

---

*Some texts use the notation $u_{-1}(t)$, where the -1 subscript is chosen because the Laplace transform is $1/s$ or s^{-1}. Then the unit impulse is $u_0(t)$, the unit ramp function is $u_{-2}(t)$.

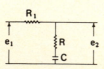

FIG. 4.13 A simple integral network.

FIG. 4.14 The unit step function at $t = 0$: $x(t) = u(t)$.

systems, the input is suddenly displaced; in thermal systems, the temperature is abruptly changed; and in the model for the population of a town, an input (such as the tax rate) is changed in a negligibly short period of time.

(2) The system response to a single step function contains all the input-output dynamics information—in other words, the same information as the transfer function. In actual practice, there is always some unpredictable noise entering with the input or at other points within the system; consequently, we usually apply several step-function inputs and average the separate responses in order to minimize noise effects.

Stability

From our knowledge of transfer functions and system analysis, we can make several general statements about what the step-function response indicates. First, if the system is unstable, the step response eventually grows without bound (the transfer function possesses at least one pole in the right half plane). If those poles are very close to the imaginary axis, a long time may pass before we see the exponentially growing response term (Fig. 4.15). Occasionally the man performing the test may turn off the system before the instability is evident. This situation occasionally occurs in interactive computer simulations where, to economize on computer time, we stop the solution as soon as we believe we have received enough of the response.

If the system is on the borderline of instability (simple poles of the transfer function on the $j\omega$ axis), the step response exhibits a constant-amplitude oscillation.*

Steady state response

A unit-step function drive yields a response with a constant component equal to $T(0)$, the zero-frequency value of the transfer function.

Second-order systems

Many systems either are second order or are higher order with two transfer-function poles dominating the response. Figure 4.16 shows a pole-zero pattern for a system

*If the transfer function possesses a simple pole at $s = 0$, the step response includes a ramp.

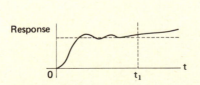

FIG. 4.15 Response with a slowly emerging, unstable mode.

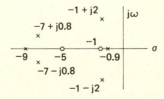

FIG. 4.16 Pole-zero pattern for $T(s)$.

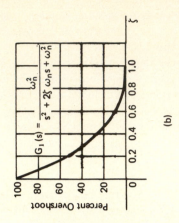

(b)

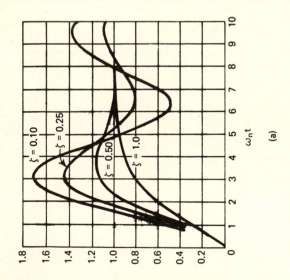

(a)

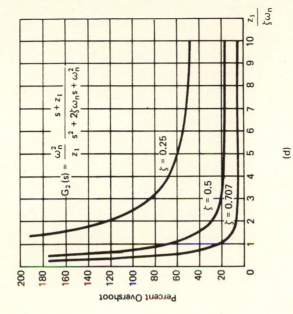

(d)

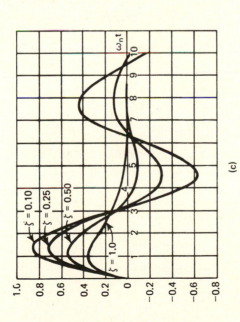

(c)

FIG. 4.17 Step-function response, as a function of ζ and ω_n, for the second order system

$$T(s) = \frac{bs + c}{s^2 + 2\zeta\omega_n s + \omega_n^2}$$

(a) Response with $b = 0$, $c = \omega_n^2$ (transfer function with no finite zero); (b) Overshoot dependence on ζ for system of (a); (c) Response with $b = \omega_n^2$, $c = 0$ [equal to impulse response of system of (a)] ; (d) Overshoot as a function of zero location for various ζ. Zero at $-z_1$ is at $-c/b$.

115

transfer function $T(s)$. Here the two dominant poles are at $-1 \pm j2$, the corresponding response term is a damped sinusoid with an envelope time constant of 1 sec (the reciprocal of the real part of the pole). The other poles are relatively unimportant. The poles at $-7 \pm j0.8$ and -9 have time constants of $1/7$ and $1/9$ sec; these transient terms die out rapidly. The pole at -0.9 is very close to a zero; therefore, the amplitude of the term $e^{-0.9t}$ is small (the residue in the pole at -0.9 is small in the partial fraction expansion of the Laplace transform of the output). Hence, to a first approximation, the system is second-order, with poles at $-1 \pm j2$.

When the system is second order (and stable), the transfer function has the general form

$$T(s) = \frac{bs + c}{s^2 + 2\zeta\omega_n s + \omega_n^2} \qquad (4\text{-}29)^*$$

The denominator is written as a quadratic in terms of the parameters

ω_n: *undamped natural frequency*
ζ: *relative damping ratio* ($\zeta > 1$ means the two poles are real and distinct, $\zeta = 1$ real and equal, $\zeta < 1$ conjugate complex).

Figure 4.17 shows the way the step-function response depends on the parameters in a few important cases. By interpolating between curves, we can often use these figures to sketch the system response with adequate accuracy and thereby avoid the tedium of calculation and plotting.

Several features of these curves are particularly important:

(1) The *percent overshoot* is a convenient measure of the *relative stability* of the system. As the transfer-function poles approach the $j\omega$-axis, the relative stability decreases and the overshoot increases. In certain instrumentation applications, we can tolerate essentially no overshoot; in automatic control applications, it is common to design for less than 20% overshoot to avoid undesirable "ringing" or vibration of the output. For a $T(s)$ with no finite zero, this means a ζ at least 0.5 [Fig. 4.17(b)].

(2) The *settling time* is the total time required for the response to settle within a specified percentage (often 5%) of its final value. For the *underdamped* system with $\zeta < 1$, the settling time is measured by the envelope time constant $1/(\zeta\omega_n)$.

(3) The *rise time* is the time required for the response to move from 10% to 90% of its final value (Fig. 4.18). Sometimes 5% and 95% are used.

(4) The *time delay* measures the lag between the application of the step and the appearance of a significant response. We often use the time to the point where the response is 0.5 of its final value (Fig. 4.18).

---

*We might allow an s^2 term in the numerator. Then $T(s)$ can be written as a constant plus the form of Eq. (4-29). The constant merely means the output includes a step-function component.

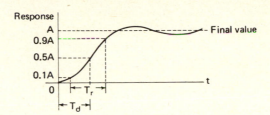

FIG. 4.18 Possible definitions of rise time T_r and time delay T_d.

Higher order systems

For systems of higher order, we can of course still describe the step response in terms of overshoot, settling time, rise time, and time delay. The four measures together constitute a picture of the nature of the step response (and the dynamics of the system). While we might plot curves for third- and fourth-order systems similar to Fig. 4.17, there are so many parameters (numerator and denominator coefficients) that a book of characteristics would be required. Some system analysts have published such characteristics for the most common third- and fourth-order systems. In the author's experience, such compendia have proved of limited value. The published curves never seem to include one for the particular transfer function under study. Consequently, it is usually simpler to simulate the system on an analog computer (Chap. 6) or a digital computer (Chap. 14) and measure the step response.

System identification from step response

Often we can measure the system step response experimentally and we want to determine a transfer function which represents the system at least approximately. We might use a digital computer to determine the Fourier transform of the response, then approximate this transform (as discussed in Chap. 11). Alternatively, we can work directly with the time function, as illustrated below in terms of Fig. 4.19.

We first note that the response includes a damped oscillation approaching the final value of 4. If we assume (optimistically, perhaps) that the curve after $t = 8$ is merely this single, damped sinusoid plus the constant 4,

$$\text{For } t > 8 , \quad y(t) = 4 + Ae^{-\alpha t} \sin(\beta t + \theta) \tag{4-30}$$

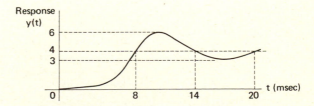

FIG. 4.19 Measured response to unit-step function.

By observation of the times when y crosses the final value 4, we see that the period of the sinusoidal term is 12 msec, or

$$\beta = \frac{2\pi}{\text{Period}} = \frac{2\pi}{0.012} = 520 \text{ rad/sec} \tag{4-31}$$

We next turn to α, which measures the rate at which the envelope is decaying. In 6 msec (the half "period" from maximum to minimum), the envelope decreases by 50%. Therefore,

$$e^{-\alpha(0.006)} = 0.5 \quad \text{or} \quad \alpha = 115 \tag{4-32}$$

Equation (4-30) becomes

$$\text{For } t > 8, \quad y = 4 + Ae^{-115t} \sin(520t + \theta) \tag{4-33}$$

The corresponding poles of the transfer function are at $-115 \pm j520$. In other words,

$$\omega_n = \sqrt{115^2 + 520^2} = 530 \qquad \zeta = \frac{115}{530} = 0.22 \tag{4-34}$$

Now let us compare the actual, measured response with the response of a second order system with this ζ and ω_n and the same final value. We can do this by sketching the two curves on the same graph (Fig. 4.20). Observation of Fig. 4.20 reveals that the primary difference between the two curves is a time delay of 4.7 msec. If we delay the response of the second order system by 4.7 msec, the two curves are reasonably coincident. Hence, an appropriate transfer function is

$$T(s) = \frac{4\omega_n^2}{s^2 + 2\zeta\omega_n s + \omega_n^2} e^{-\beta s} \tag{4-35}$$

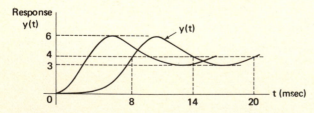

FIG. 4.20 Measured $y(t)$ and analogous response of a second order system with no zeros and $\zeta = 0.22, \omega_n = 530$.

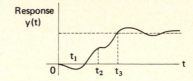

FIG. 4.21 Possible step-function response.

or, in terms of numbers,

$$T(s) = \frac{1.13 \times 10^6}{s^2 + 230s + 2.83 \times 10^5} e^{-0.0047s} \qquad (4\text{-}36)$$

The example warrants several comments. Above all, we clearly were "lucky": the step response of the second order system supplemented by a delay of 4.7 msec matched the measured curve quite well. While very often a higher order system can be approximated by a second order system plus delay, there certainly are systems where this is not possible. Even if the second order approximation is possible, we may have to use a finite zero in the transfer function to achieve a suitable fit.

Figure 4.21 shows a step response where clearly the second order approximation is useless. In the vicinity of t_1, the response starts out negatively (at high frequencies, the transfer function behaves as $-K/s^n$).* Furthermore, around t_2 there is a "plateau" indicating the approximate cancellation of two different response terms.

To find a $T(s)$ or $y(t)$ corresponding to Fig. 4.21, first we hope that the $y(t)$ contains one term with a longtime constant. We then try to approximate the behavior for large t (after t_3) by a single exponential or, in this case, a damped sinusoid. Once this term is found, it is subtracted from $y(t)$ graphically to obtain a sketch of the remaining terms. We continue by working on this remainder in the same way.

As this discussion suggests, there is no simple way to find the transfer function $T(s)$ from the measured step response. The challenge of the problem is emphasized by the fact that many different $T(s)$ functions yield very similar step responses (a problem referred to again in Sec. 11.7).

4.4 LADDER NETWORKS

A particularly important network form is shown in Fig. 4.22. As we move from the source e_1 toward the output e_4, we encounter successively *series* and *shunt* impedances. Z_1 is a series branch, Z_2 a shunt branch, and so on. This structure, with series and shunt branches alternating, is called a *ladder network.*

The particular ladder shown has six branches plus a voltage source. More generally, we can have any number of branches; furthermore, we can start (or end) with either a

---

*This phenomenon is not too unusual in systems. When the elevator surface of an airplane is turned to increased altitude, the initial short-term effect may be a slight decrease in altitude.

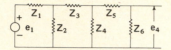

FIG. 4.22 A ladder with six branches.

series or shunt branch, and we may use a current or voltage source at the input (or left port). Finally, each branch may be a complex one-port network; for example, Z_2 in Fig. 4.22 might consist of dozens of circuit elements. In a pure ladder network, each branch is a one-port network (e.g., there is no mutual coupling to other branches and no controlled sources depending on signals elsewhere in the ladder).

Analysis

The ladder network is particularly important because it is easily analyzed. That is, there is a straightforward procedure for finding the Output/Input transfer function. We use the network of Fig. 4.23 as an example.

First we label the different node voltages and the currents. Instead of working from input to output, we assume the output is unity and calculate the corresponding input in the following obvious steps:

$$E_3 = 1 \qquad \text{Basic assumption}$$
$$I_4 = Y_4 \qquad \text{(where } Y_4 \text{ is } 1/Z_4; \text{ this comes from } I_4 = E_3/Z_4\text{)}$$
$$I_3 = Y_4 \qquad \text{(since } I_3 = I_4\text{)}$$
$$E_2 = 1 + Z_3 Y_4 \qquad \text{(from } E_2 = E_3 + I_3 Z_3\text{)}$$
$$I_2 = Y_2 + Y_2 Z_3 Y_4 \qquad \text{(from } I_2 = E_2/Z_2\text{)}$$
$$I_1 = Y_4 + Y_2 + Y_2 Z_3 Y_4 \qquad \text{(node equation at } E_2\text{)}$$
$$E_1 = 1 + Z_3 Y_4 + Z_1 Y_4 + Z_1 Y_2 + Z_1 Y_2 Z_3 Y_4 \qquad \text{(since } E_1 = E_2 + Z_1 I_1\text{)}$$

For $E_3 = 1$, we now know E_1. Since the network is linear, the E_3/E_1 transfer function is independent of the signals. Hence the transfer function is just

$$\frac{E_3}{E_1} = \frac{1}{1 + Z_3 Y_4 + Z_1 Y_4 + Z_1 Y_2 + Z_1 Y_2 Z_3 Y_4} \qquad (4\text{-}37)$$

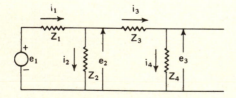

FIG. 4.23 Ladder network.

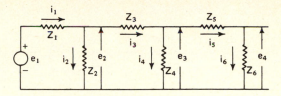

FIG. 4.24 Ladder for problem (1).

Typical problems for ladder networks

(1) For the ladder of Fig. 4.24, we wish to construct a signal flow diagram with the nodes $E_1, I_1, I_2, E_2, I_3, I_4, E_3, I_5, I_6, E_4$, in order from left to right. We do this by writing the equations from right to left as the network of Fig. 4.23 was analyzed.

First, we write the equations by working from the output back toward the input:

$$\begin{cases} I_6 = Y_6 E_4 \\ I_5 = I_6 \\ E_3 = E_4 + Z_5 I_5 \\ I_4 = Y_4 E_3 \\ I_3 = I_4 + I_5 \\ E_2 = E_3 + Z_3 I_3 \\ I_2 = Y_2 E_2 \\ I_1 = I_2 + I_3 \\ E_1 = E_2 + Z_1 I_1 \end{cases} \tag{4-38}$$

We now solve these equations for the successive signals. In other words, these equations are put into the form appropriate for a signal flow graph

$$\begin{cases} E_4 = Z_6 I_6 \\ I_6 = I_5 \\ I_5 = Y_5(E_3 - E_4) \\ E_3 = Z_4 I_4 \\ I_4 = I_3 - I_5 \\ I_3 = Y_3(E_2 - E_3) \\ E_2 = Z_2 I_2 \\ I_2 = I_1 - I_3 \\ I_1 = Y_1(E_1 - E_2) \end{cases} \tag{4-39}$$

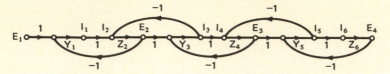

FIG. 4.25 Signal flow graph for six-element ladder.

The corresponding signal flow graph is shown in Fig. 4.25. The solution is interesting because of the regularity of the pattern, and because all branch transmittances are ± 1 except for the six immittances (impedances or admittances).

(2) We wish to determine the transfer function by using Mason's reduction theorem for Problem (1).

For Fig. 4.25, the loop gains are

$$L_1 = -Z_2 Y_1 \qquad L_3 = -Z_4 Y_3 \qquad L_5 = -Z_6 Y_5$$
$$L_2 = -Z_2 Y_3 \qquad L_4 = -Z_4 Y_5$$

and

$$\frac{E_4}{E_1} =$$

$$\frac{Y_1 Z_2 Y_3 Z_4 Y_5 Z_6}{1 + (Z_2 Y_1 + Z_2 Y_3 + Z_4 Y_3 + Z_4 Y_5 + Z_6 Y_5) + (Z_2 Y_1 Z_4 Y_3 + Z_2 Y_1 Z_4 Y_5 + Z_2 Y_1 Z_6 Y_5 + Z_2 Y_3 Z_4 Y_5 + Z_2 Y_3 Z_6 Y_5 + Z_4 Y_3 Z_6 Y_5) + Z_2 Y_1 Z_4 Y_3 Z_6 Y_5}$$

$$(4\text{-}40)$$

(3) For the ladder network shown in Fig. 4.26, we wish to calculate the angular frequencies at which the transfer function E_2/I_1 possesses zeros. Values of $L, C,$ and R are given in millihenrys, picofarads, and kilohms.

In a ladder network, there are two ways that there can be zeros in the transfer function: a shunt branch can have a zero or a series branch a pole. In other words, in the solution of Problem (2), the zeros are the zeros of $Y_1, Y_3,$ and Y_5 or the zeros of $Z_2,$

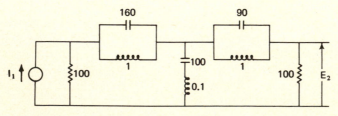

FIG. 4.26 Circuit of Problem (3).

Z_4, Z_6. Hence, in the ladder of Fig. 4.26, zeros of E_2/I_1 occur where

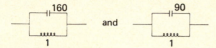

and

have Z poles or parallel resonance, and where

has a Z zero or series resonance. Thus, the network transmission zeros are at

$$\omega_1{}^2 = \frac{1}{(160 \times 10^{-12})(10^{-3})} = \frac{10^{14}}{16} \qquad \omega_1 = \frac{10^7}{4} = 2.5 \text{ megaradians/sec}$$
$$(4\text{-}41)$$

$$\omega_2{}^2 = \frac{1}{(90 \times 10^{-12})(10^{-3})} = \frac{10^{14}}{9} \qquad \omega_2 = \frac{10^7}{3} = 3.33 \text{ megaradians/sec}$$
$$(4\text{-}42)$$

$$\omega_3{}^2 = \frac{1}{(100 \times 10^{-12})(0.1 \times 10^{-3})} = 10^{14} \qquad \omega_3 = 10 \text{ megaradians/sec}$$
$$(4\text{-}43)$$

If we plot network gain, $|E_2/I_1|$, versus ω we can expect a shape such as Fig. 4.27.

4.5 FINAL COMMENT

In this chapter, we have considered a few of those many special-purpose networks which constitute the art of electrical engineering. As in any professional field, electrical engineering mixes science with art: the experienced engineer approaches a problem with the background of a wide variety of past solutions to other problems—in network design, with familiarity with the common networks which possess especially important characteristics.

In a basic text, we can do little more than mention a few of these networks and describe their performances very briefly. The primary sources of information on this art of

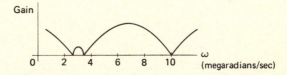

FIG. 4.27 Gain characteristic for ladder of Fig. 4.26.

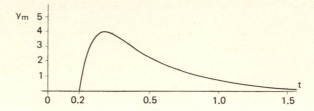

FIG. A4.1   Measured impulse response.

electrical engineering are the various trade journals and manufacturers' publications, with handbooks, textbooks, and occasional publications of the professional societies being the secondary sources.

APPENDIX—IDENTIFICATION FROM IMPULSE RESPONSE

The unit impulse response is measured experimentally, and we obtain the result shown in Fig. A4.1. We wish to find an appropriate transfer function in the simplest possible form.

Actually, the well-behaved curve shown in the figure probably represents a smoothed version of the actual measurement, which is much more likely to have the form in Fig. A4.2. Noise, pickup, and instrument errors all combine to distort the output. Unfortunately, if we only test once with an impulse, we can never be sure how much of the high frequency variation of Fig. A4.2 is actually part of the system impulse response and how much is just noise. In this example, we assume either that we can measure $y_m(t)$ several times (and smooth by using an average), or that we know from our understanding of the system that only slow variations are possible. In either case, we want to work from Fig. A4.1.

An application

This type of response is obtained in dye dilution tests in medical engineering. Blue dye is injected in the arm (the input impulse), and the amount of blue in the ear lobe

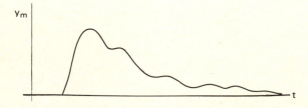

FIG. A4.2   More usual appearance of measured signal.

is measured as the output. The system is the cardiovascular system circulating blood throughout the body, in particular from the arm through the heart to the ear.

Let's be optimistic

We always approach a difficult problem optimistically: we assume the answer is as simple as possible. In this case, since $y_m(t)$ rises and then falls, the simplest possible form is

$$y_m(t) = [Ae^{-\alpha t} - Be^{-\beta t}] \text{ delayed by 0.2 second} \tag{A4-1}$$

We hope to find A, α, B, and β so that the equation for $y_m(t)$ is a reasonable approximation to the measured curve of Fig. A4.1.

On logical grounds, this assumption may be severely criticized. Why not assume $y_m(t)$ is the sum of five exponentials—or ten—or damped sinusoids? Our only answer is: Nature is often kind. It may well happen that we cannot fine a $y_m(t)$ in the form of Eq. (A4-1) and matching Fig. A4.1 reasonably well. If so, it is time to guess more complicated forms for $y_m(t)$.

If we accept Eq. (A4-1), the Laplace transform is

$$Y_m(s) = \left(\frac{A}{s + \alpha} - \frac{B}{s + \beta}\right)e^{-0.2s}$$

$$= \left[\frac{(A - B)s + (\beta A - \alpha B)}{(s + \alpha)(s + \beta)}\right]e^{-0.2s} \tag{A4-2}$$

As we try to find A, B, α, and β, we can work with either Eq. (A4-1) or Eq. (A4-2).

Evaluation of B

We notice first that, at $t = 0.2$ (after the transportation lag), $y_m(t)$ starts at zero. In other words, the part of $y_m(t)$ or $Y_m(s)$ inside the brackets must correspond to a time function with zero initial value. Hence, either equation tells us that

$$A = B \tag{A4-3}$$

and we can now write

$$y_m(t) = [Ae^{-\alpha t} - Ae^{-\beta t}] \text{ delayed by 0.2 sec} \tag{A4-4}$$

$$Y_m(s) = \left[\frac{(\beta - \alpha)A}{(s + \alpha)(s + \beta)}\right]e^{-0.2s} \tag{A4-5}$$

and we have only three parameters left: A, α, and β.

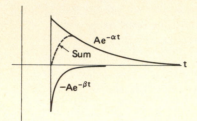

FIG. A4.3 Formation of $y_m(t)$ from two components.

Finding α

Equation (A4-4) shows us that $y_m(t)$ is formed from two equal exponentials—one positive and one negative. Since the total, $y_m(t)$, goes positive, the negative term must die out faster (Fig. A4-3). Consequently, if we go way out in t in Fig. A4.1, the curve must just be

$$Ae^{-\alpha t} \quad \text{(delayed by 0.2 sec)}$$

We should be able to measure α by the way the signal is decaying when t is large.

What do we mean by t large? In other words, in Fig. A4.1 when can we be sure that the curve is a single decaying exponential? To be certain that $-Ae^{-\beta t}$ is negligible, we should look at t as large as possible. Unfortunately, as t exceeds 1.0, however, y_m is so small that the measurement is probably not very accurate. We obtain better accuracy for t smaller. Thus, we need to compromise.*

In one time constant, the single exponential decays to 36.8% (or about 0.37) of its original value. Hence, in Fig. A4.1 we note that

$$y_m = 2 \quad \text{at} \quad t = 0.65 \tag{A4-6}$$

37% of the value is 0.74; this occurs at $t = 1.01$. Therefore, the time constant $(1/\alpha)$ is 1.01–0.65 or

$$\frac{1}{\alpha} = 0.36 \qquad \alpha = 2.8 \tag{A4-7}$$

We can check this by measuring the time for y_m to fall from 1 to 0.37, or from 1.5 to 0.55. If these values agree reasonably well, then we have chosen a region of t within which y_m is a single exponential.

---

*We might plot $\ln y_m$ versus t; when the plot becomes a straight line, we know we have a single exponential. This is more work than really necessary, however.

Thus, we now have

$$y_m(t) = [Ae^{-2.8t} - Ae^{-\beta t}] \text{ delayed by 0.2 sec} \qquad \text{(A4-8)}$$

$$Y_m(s) = \left[\frac{(\beta - 2.8)A}{(s + 2.8)(s + \beta)}\right] e^{-0.2s} \qquad \text{(A4-9)}$$

Finding A
We have discovered above that, at $t = 0.65, y_m$ is 2 and has the form

$$y_m(t) = Ae^{-2.8t} \text{ delayed by 0.2 sec} \qquad \text{(A4-10)}$$

In other words

$$2 = Ae^{-2.8 \times 0.45} \qquad \text{(A4-11)}$$

or

$$A = 2 \times 3.53 = 7.1 \qquad \text{(A4-12)}$$

And we now have

$$y_m(t) = [7.1 e^{-2.8t} - 7.1 e^{-\beta t}] \quad \text{delayed by 0.2 sec} \qquad \text{(A4-13)}$$

$$Y_m(s) = \left[\frac{(\beta - 2.8)7.1}{(s + 2.8)(s + \beta)}\right] e^{-0.2s} \qquad \text{(A4-14)}$$

We need only find β.

Evaluation of β
We can find β in several ways. We might measure the slope of $y_m(t)$ in Fig. A4.1 at the start at $t = 0.2$, for example, and equate this to the value from Eq. (A4-14):

$$(\beta - 2.8)7.1$$

Alternatively, we can simply pick β so that y_m has the correct value at any early time (before $Ae^{-\beta t}$ has died out). If we choose $t = 0.3$, we obtain

$$3.5 = 7.1(e^{-2.8 \times 0.1} - e^{-0.1\beta}) \qquad \text{(A4-15)}$$

or

$$3.5 = 7.1(0.755 - e^{-0.1\beta})$$

$$e^{-0.1\beta} = 0.755 - 0.493 = 0.262$$

$$0.1\beta = 1.34 \tag{A4-16}$$

$$\beta = 13.4$$

Final expressions for $y_m(t)$ and $Y_m(s)$

$$y_m(t) = [7.1\,e^{-2.8t} - 7.1\,e^{-13.4t}] \quad \text{delayed by 0.2 sec} \tag{A4-17}$$

$$Y_m(s) = \frac{75.2}{(s + 2.8)(s + 13.4)}\,e^{-0.2s} \tag{A4-18}$$

Imprecision of the numbers

Since we know that transfer-function poles away from the $j\omega$ axis can be moved appreciably without changing the time function very much, it is not surprising that the numbers in Eqs. (A4-17) and (A4-18) can be quite different while $y_m(t)$ is changed very little. Indeed, the author developed this problem in the following steps:

(1) $y(t)$ was chosen arbitrarily as

$$y(t) = [8e^{-3t} - 8e^{-12t}] \quad \text{delayed by 0.2 sec} \tag{A4-19}$$

The corresponding transform is

$$y(s) = \frac{72}{(s + 3)(s + 12)}\,e^{-0.2s} \tag{A4-20}$$

(2) $y(t)$ was plotted (Fig. A4.1) just by substituting in values of t;

(3) then the analysis of these notes was used to derive Eqs. (A4-17) and (A4-18)—which seem quite different from the original (A4-19) and (A4-20).

If we plot Eq. (A4-17), however, we find the curve is essentially that of Fig. A4.1.

PROBLEMS

4.1 The circuit shown has been proposed as an integrator; i.e., the output e_2 is to be proportional to the integral of the input e_1.

(a) Determine the condition s must satisfy for proper operation as an integrator. Assume that $x + y$ is approximately y if $y = 5x$.

(b) If the largest available resistance is 10M and the largest available C is 200 μF, what is the minimum sinusoidal frequency for which the circuit yields reasonable integration?

(c) If we apply a step function, sketch the output.

4.2 The circuit shown has been proposed as a differentiator; the output e_2 is to be proportional to the derivative of the input e_1.

(a) Determine the condition s must satisfy for operation as a differentiator. Assume that $x + y$ is essentially y if $y = 5x$.

(b) If $C = 0.1$ μF and $R = 100K$, what is the maximum sinusoidal frequency for which the circuit yields reasonable differentiation?

(c) If a ramp function is applied, sketch the output.

(d) Sketch the response to a step function.

4.3 The circuit shown has been proposed as an improved integrator, compared to the system of Prob. 4.1. The quantity A, the gain of the ideal voltage amplifier, is always positive.

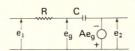

(a) Determine the condition s must satisfy for proper operation as an integrator. Assume that $x + y$ is essentially y if $y = 5x$.

(b) If the largest available R is 10M, the largest C is 1 μF, and A is 500,000, what is the minimum sinusoidal frequency for which the circuit yields reasonable integration. (These values are reasonably typical for an analog computer.)

4.4 In the analog computers built during the early years (the 1930s), integration was accomplished with a mechanical ball-and-disk arrangement. As late as 1960, such integrators were sometimes still used because of the accuracy achievable, and they still appear in applications where electrical signals are apt to cause dangerous sparking. The sketch shows such a device. Describe how integration is achieved: what are the input and output signals; what determines the gain constant? (Reference books on analog computers often include descriptions of various components.)

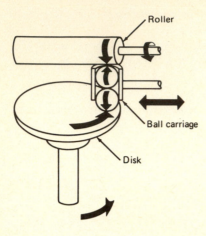

4.5   Write the equations for the network shown, using the ladder-network approach.

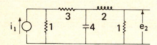

4.6   The graph represents the response of a d'Arsonval recording galvanometer when an excitation of 2 a dc is suddenly applied.

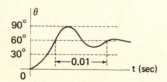

(a) Write the response function $\theta(t)$ in algebraic form.
(b) Write $\theta(t)$ with all constants evaluated numerically.
(c) Sketch the pole-zero plot for $\theta(s)/I(s)$.
(d) Determine the transfer function with all constants evaluated numerically.

4.7   The block diagram of a shaft-positioning servomechanism is given. The transfer functions of the elements are known, with the exception of the amplifier gain K and the motor time constant τ, which are to be determined. The system must meet the following performance requirements:
(a) Static accuracy: With a constant input $e(t)$, the error in the output shaft position due to coulomb friction, stiction, and other spurious load torques must not exceed 0.1°. The magnitude of these coulomb forces is such that they can be

represented by a voltage of 0.2 v applied to the motor. In other words, the servo output shaft will not rotate unless the voltage applied to the motor exceeds 0.2 v. Determine the smallest value of K which permits meeting this requirement.

(b) Velocity error: With a unit ramp input, the steady state positional lag error of the output shaft must not exceed $0.1°$. (Coulomb forces can be neglected at constant velocity.) If the amplifier gain is adjusted as in (a), determine the smallest value of τ which permits the servo to meet this velocity error requirement.

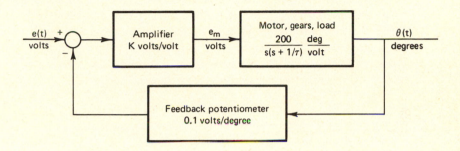

(c) With the constants K and τ selected, what is the numerical value of the steady state output for a unit-step input? What is the form of the transient term? (By "form," we mean you should give the numerical value of all terms except those determined by initial energy storage.) Neglect coulomb forces for this part (c).

4.8 An important electric circuit is shown, one which is widely used in prediction networks for tracking problems.

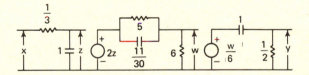

(a) Construct a signal flow diagram.

(b) Determine the overall transfer function.

(c) Write the transfer function with each polynomial in ascending powers of s; then divide the denominator into the numerator to obtain a Maclaurin-series expansion. Estimate the frequency range over which the system will yield approximate differentiation.

(d) Why do we choose a denominator of higher degree than the numerator in the transfer function?

(e) What modifications would be required in the circuit if we wished the output to be the predicted value of the input 0.4 sec into the future?

(f) In a practical prediction system, why do we rarely use the second derivative of the signal in order to measure the rate at which the first derivative is changing?

4.9 A seismograph is used to measure ground displacements during earthquakes. The essential parts of the instrument are shown in the figure. The output signal z is $x - y$, the relative displacement of the ground and the seismographic mass M.

(a) Find the transfer function Z/X.

(b) With $K/M = 1$ and $B/M = 0.1$, sketch the plots of gain and phase versus frequency—in other words, the magnitude and angle of Z/X versus ω.

(c) If the instrument is to measure ground displacements, two conditions must be satisfied over the frequency band of interest: (1) Gain is constant with frequency; (2) Phase shift is proportional to frequency. Over what range of frequencies is this seismograph an accurate instrument?

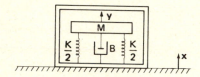

4.10 In automobile safety studies, we are interested in how a car should be rapidly stopped to minimize the danger to human life from the deceleration. Automotive and biomedical engineers often use the Gadd severity index I defined as:

$$I = \int \left(\frac{a}{g}\right)^{2.5} dt$$

where a is the deceleration and g is the acceleration of gravity. Values of I greater than 1000 sec are dangerous for humans.

If the initial velocity V_0 and the allowable stopping distance L are specified, we can find the $a(t)$ function (the acceleration profile) which minimizes I. The answer is

$$a_{\text{opt}}(x) = \frac{5V_0^2}{8L}\left(1 - \frac{x}{L}\right)^{1/4}$$

where x is the distance through which the vehicle travels as it stops (i.e., x varies from 0 to L). Compare the severity index achieved with this deceleration profile with that realized by a mechanical spring in which the deceleration is proportional to distance.

4.11 There exists a class of circuit design problems in which two conflicting conditions must be met: (1) Once the transient has died out, the desired relation between input and output is given by a particular transfer function, say $T(s)$; (2) We cannot tolerate the long duration of the transient which is implied by the poles of $T(s)$. As a specific example, the network shown realizes the desired transfer function. When i_1 is applied at $t = 0$, the consequent transient lasts ten times as long as allowable.

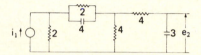

In order to decrease the transient duration by a factor of 10, we can insert additional network elements by means of switches which close or open at $t = 0$ and remain in this state for a predetermined length of time. In order to reduce every RC product by 10, we can multiply either every R or each C by 0.1. If the switching at the end of the interval is not to cause a transient itself, the stored energy distribution must be the same in the highspeed and the normal networks.

(a) Explain the last sentence above on the basis of the reason for transients. Illustrate by explaining what happens in a simple RC network with a step applied and with the network speeded up by reduction of either R or C. Show the networks required to reduce R and C.

(b) For our circuit of this problem, show the circuit diagram for the final circuit which includes the highspeed transient response feature.

(c) Determine the approximate duration of the transient for the highspeed network.

5 ROOT-LOCUS DESIGN

The system engineer is often confronted with the problem depicted in Fig. 5.1. A system is described by the transfer function $T(s)$, which depends on a parameter K (possibly a capacitance, the gain of an amplifier, or a potentiometer setting). The engineer might face either of two questions:

(1) The system is being designed. What value should be chosen for K to give the best system performance?

(2) The system has been designed. During operation, K is known to vary (because of environmental conditions, changes in the power supply, aging of components, or other reasons). What is the effect on system performance?

In either case, the engineer is interested in a study of how system characteristics depend on K. This term "system characteristics" may take many different forms. If

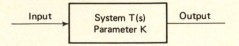

FIG. 5.1 A problem in analysis or design.

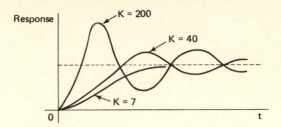

FIG. 5.2 Possible dependence of system step response on a parameter K.

the system is a servomechanism, the important performance characteristic might be the step response, and the dependence on K can be depicted by a family of response curves (Fig. 5.2). If the system is a low-pass filter, the cutoff frequency may be the important performance measure (Fig. 5.3).* If the system is the control for landing planes on an aircraft carrier flight deck, the only important performance measure may be the percentage of planes which fail to land safely.

To a considerable degree, system engineering is the study of this problem: how performance characteristics depend on particular parameters. In this chapter, we want to consider the root-locus method of analysis and design, a method which focuses on one form of the problem:

> How do the poles (and possibly also the zeros) of the transfer function $T(s)$ depend on the parameter K?

5.1 ROOT LOCI

Figure 5.4 shows a single-loop feedback system in which the error is sent through two networks, each with a transfer function $1/(s + a)$, and also amplified by K. In

*This sort of a filter with controllable cutoff frequency can be used in a record player. When there is no high-frequency signal, the cutoff frequency is lowered to decrease the high-frequency noise output.

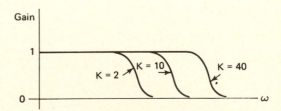

FIG. 5.3 Filter characteristics as K varies.

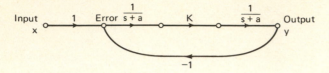

FIG. 5.4 Single-loop feedback system.

the design of the total system, we are free to choose K. In the next paragraphs, we study how the choice of K affects the dynamic characteristics of the overall system, Y/X.

The overall transfer function, $T(s)$, is

$$T(s) = \frac{Y}{X} = \frac{K/(s + a)^2}{1 + K/(s + a)^2} \tag{5-1}$$

or

$$T(s) = \frac{K}{(s + a)^2 + K} \tag{5-2}$$

The first thing that Eq. (5-2) shows is that the choice of K affects the poles of $T(s)$. These poles are the roots of the system characteristic equation

$$1 + \frac{K}{(s + a)^2} = 0 \quad \text{or} \quad (s + a)^2 + K = 0 \tag{5-3}$$

As K is varied from 0 to a very large positive value, the poles of T *move* in the s plane.

The second equation in (5-3) allows a more definitive statement about how the poles of T vary with K. This equation can be rewritten

$$(s + a)^2 = -K \tag{5-4}$$

Taking the square root gives

$$s + a = \pm j\sqrt{K} \tag{5-5}$$

or

$$s = -a \pm j\sqrt{K} \tag{5-6}$$

The pole motion as K varies from 0 to ∞ is shown graphically in Fig. 5.5.

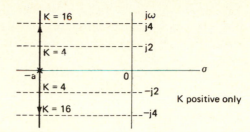

FIG. 5.5 Root-locus plot for $1 + \dfrac{K}{(s + a)^2}$.

Such a picture of pole motion as it depends on a particular system parameter is called a *root-locus plot*. The lines representing motion of the two poles are *root loci*. The word "root" comes from the fact we are looking at the roots of the system characteristic equation

$$1 + \frac{K}{(s + a)^2} = 0 \qquad\qquad (5\text{-}7)$$

The term "locus" refers to the fact we show the location of the roots as K is varied.

Solution of the design problem

In our example, a typical design problem would be: K is to be as large as possible, but with the constraint that the system should have a relative damping ratio of at least 0.5 (i.e., the step-function response should have less than 18% overshoot?).* Once we have the root-locus plot, we can answer this question almost by inspection.

Figure 5.6 shows the root loci redrawn. The relative damping ratio of 0.5 means that the angle θ should be $60°$. In other words, the poles of T are

$$-a \pm ja\sqrt{3}$$

*When we study the uses of feedback, we will see that the larger K, the more effective the feedback in achieving its goals.

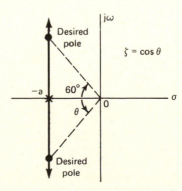

FIG. 5.6 Root loci redrawn for first example.

The corresponding K is $3a^2$. Smaller values of K bring the poles toward the $-\sigma$ axis (i.e., give larger damping ratios); larger values of K reduce the ζ to less than 0.5. Thus, we can meet our design specifications with

$$\sqrt{K} = \sqrt{3}\,a$$

$$T(s) = \frac{3a^2}{(s+a)^2 + 3a^2} \qquad (5\text{-}8)$$

$$K = 3a^2$$

Even this very simple example illustrates the essential ideas of the root-locus method of system design. The kind of design problem we are considering is the following. We are given the system; we have freedom merely to choose one parameter; we must select the best value for this parameter.*

In the root-locus method, we determine first how the poles of the overall transfer function move as the parameter is changed. By observation of these plots, we select the best value for the parameter.

5.2 A SECOND EXAMPLE OF A DESIGN PROBLEM

In the network of Fig. 5.7, C is the only element value not specified. We wish to select C so that the circuit is resonant with the maximum possible Q.

Straightforward circuit analysis reveals that

$$\frac{E_2}{E_1} = \frac{4Cs^2}{16Cs^2 + (16C+1)s + 4} \qquad (5\text{-}9)$$

The zeros of E_2/E_1 are both at $s = 0$; the poles, dependent on the value of C, are the roots of the characteristic equation

$$16Cs^2 + (16C+1)s + 4 = 0 \qquad (5\text{-}10)$$

As we shall see in later sections, the root loci can be determined simply (and are as shown in Fig. 5.8).

*Historically, the root-locus method was first presented in 1948 by Walter Evans of North American Aviation. In aircraft control, we very often encounter this type of design problem. The airplane, control surfaces, and type of control system are determined from aeronautical engineering considerations. The control engineer has relatively little freedom to modify the basic system.

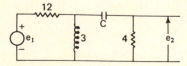

FIG. 5.7 Network for second example.

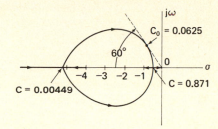

FIG. 5.8 Root loci for example 2.

The network is resonant for that range of C values within which the poles are conjugate complex $(0.00449 < C < 0.871)$. The maximum Q is realized when C is chosen so that the poles are at a maximum angle from the negative real axis. In this example, the optimum is

$$C_0 = 0.0625 \tag{5-11}$$

For this value of capacitance, the network poles are at $-1 \pm j\sqrt{3}$, and the maximum Q is unity (hardly a sharply resonant system).

This example is interesting for several reasons. First, it points out that root loci can be constructed to show the effects of the variation of any one circuit parameter (any L, C, R, or gain of a controlled source). Second, we see that, if we are to use the root-locus method, we must have some simple techniques for constructing loci such as Fig. 5.8. In the first example, the loci were obvious from the characteristic equation. In this second example, Eq. (5-10) does not readily show how the system poles move as C varies from 0 to ∞. We could simply substitute various values of C into the quadratic equation and solve for s; this would be a tedious procedure, particularly for more complex systems.

Thus, we now turn to the problem: given the characteristic equation with one undetermined parameter, how do we find the root loci?

5.3 TWO FORMS FOR THE CHARACTERISTIC EQUATION

The poles of the system function are the roots of the characteristic equation. When one parameter is unspecified, the form of this equation is illustrated by Eq. (5-10)

$$16Cs^2 + (16C + 1)s + 4 = 0 \tag{5-12}$$

There are two equivalent forms which prove convenient. The first is obtained if we separate all terms involving C:

$$(s + 4) + C(16s^2 + 16s) = 0 \tag{5-13}$$

More generally, this form has the appearance

$$q(s) + Cp(s) = 0 \tag{5-14}$$

where C is the parameter, $q(s)$ and $p(s)$ are polynomials in s.

The second form is obtained if we divide throughout by the polynomial *not* multiplying the parameter C. Then Eqs. (5-13) and (5-14) become

$$1 + C \frac{16s(s + 1)}{s + 4} = 0 \tag{5-15}$$

$$1 + C \frac{p(s)}{q(s)} = 0 \tag{5-16}$$

A slight modification of the last equation is also useful:

$$C \frac{p(s)}{q(s)} = -1 \tag{5-17}$$

The ratio $p(s)/q(s)$ is a transfer function itself. Let us call this $G(s)$. Thus, the characteristic equation as a function of a parameter C can be written in the various forms

$$
\begin{array}{ll}
1 + CG(s) = 0 & \text{(a)} \\
CG(s) = -1 & \text{(b)} \\
q(s) + Cp(s) = 0 \qquad \text{where } G = \dfrac{p}{q} & \text{(c)}
\end{array}
\tag{5-18}
$$

The poles of the system function are the roots of any of these three forms of the characteristic equation. The root loci show how these roots move as the parameter C varies from zero to infinity.

5.4 RULES FOR DRAWING ROOT LOCI

If the root loci are to be useful in design (that is, in selecting a value for the parameter called C above), we must be able to sketch these loci easily. Naturally, we could substitute every possible value of C into Eq. (5-18) and solve for the roots (the poles of the system function). The root-locus method of design is attractive because in many cases we can avoid such tedious calculations. Indeed, we can often sketch the root loci approximately with very little work.

This possibility of drawing the root loci rapidly is based on a set of rules which we now describe in terms of the notation of Eq. (5-18).

(1) There are n loci, where n is the degree of $q(s)$—or the degree of $p(s)$ if, as occasionally happens, p is of higher degree than q.

This rule follows from Eq. (5-18c) and the fact that a polynomial of degree n has n zeros.

(2) As C increases from 0 to ∞, the roots move from the poles of $G(s)$ to the zeros of G in the s plane.

Equation (5-18b) reveals this rule. When C is very small, $|G|$ must be very large (near a pole) if the $|CG|$ is to equal unity. When C tends to ∞, $|G|$ must tend to zero. If we include zeros (or poles) at ∞, $G(s)$ has the same number of zeros as poles. Hence, each root locus goes from a pole to a zero of G as C varies from zero to ∞.

(3) The loci, when complex, always occur in conjugate pairs and are continuous.

If $q(s)$ and $p(s)$ in Eq. (5-18c) have real coefficients and C is real, the total polynomial on the left side of the equation has real coefficients. Hence, the complex zeros occur in conjugate pairs. Furthermore, the zeros of a polynomial are continuous functions of the coefficients. Hence, as C is varied from 0 to ∞, the roots of Eq. (5-18) move continuously through the s plane.

(4) No value of s corresponds to more than one value of C.

Equation (5-18c) is linear in C. When any s is substituted, a single C is determined. Thus, there is no looping of the loci as shown in Fig. 5.9 (where s_1 would correspond to two different values of the parameter C).

(5) The portions of the σ axis to the left of an *odd* number of critical frequencies (zeros or poles) of G are parts of the loci.

Equation (5-18b) shows that, for C positive, G must be negative and real. If $G(s)$ is in the form

$$G(s) \;=\; K \frac{(s - z_1)(s - z_2)}{(s - p_1)(s - p_2)(s - p_3)} \tag{5-19}$$

where K is positive, each term of G is positive to the right of all poles and zeros on the σ axis. As we move leftward on the σ axis, the sign of G changes when we traverse a pole or a zero. Hence, on the σ axis G is negative when we see an odd number of real zeros and poles as we look to the right.

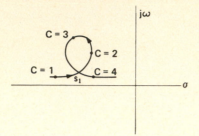

FIG. 5.9 An impossible root locus.

⑥ If $G(s)$ has a zero of order m at ∞, m loci approach infinity as C becomes very large. The asymptotes for these loci are straight lines at angles $(\pm 180° \pm k\,360°)/m$ where $k = 0, 1, 2, \ldots$. When extended to the σ axis, these asymptotes meet at

$$\left(\sum p_j - \sum z_j\right)/m\,.$$

The proof of this rule follows from Eq. (5-19). (We assume K is positive.) For very large $|s|$, G behaves as

$$K\,\frac{s^q - \left(\sum z_j\right)s^{q-1} + \cdots}{s^{q+m} - \left(\sum p_j\right)s^{q+m-1} + \cdots}$$

Dividing the numerator into the denominator gives

$$\frac{K}{s^m - \left(\sum p_j - \sum z_j\right)s^{m-1} + \cdots}$$

Now if we use Eq. (5-18b), we find

$$\frac{CK}{s^m - \left(\sum p_j - \sum z_j\right)s^{m-1} + \cdots} = -1 \qquad (5\text{-}20)$$

or

$$s^m - \left(\sum p_j - \sum z_j\right)s^{m-1} + \cdots + CK = 0 \qquad (5\text{-}21)$$

The left side is a polynomial with m zeros, the sum of which is $\sum p_j - \sum z_j$.

Hence, the asymptotes meet at the point

$$\frac{\sum \text{poles of } G - \sum \text{zeros of } G}{(\text{number of finite poles}) - (\text{number of finite zeros})}$$

This point is called the *centroid* of the pole-zero constellation of $G(s)$.

There are additional rules which we develop later, but these six are enough to work many simple problems. We now illustrate the use of these in four examples.

5.5 EXAMPLE 1 (ROOT LOCUS)

Figure 5.10 shows a single-loop feedback system in which the forward transfer function $H(s)$ includes the parameter K. We wish to determine how the closed-loop poles (the poles of Y/X) vary as K changes from 0 to ∞. We proceed in the following steps.

(1) We first write Y/X as a function of K:

$$\frac{Y}{X} = \frac{K(s + 4)/[s(s + 1)(s + 2)]}{1 + K(s + 4)/[s(s + 1)(s + 2)]} \tag{5-22}$$

We are lucky: the denominator of Y/X is already in the form of Eq. (5-18)

$$1 + KG(s)$$

and we are ready to draw the root loci. In this example,

$$G(s) = \frac{s + 4}{s(s + 1)(s + 2)} \tag{5-23}$$

(2) We first mark on the s plane the finite poles and zeros of $G(s)$—the function multiplying the parameter K. There are three poles and one finite zero (Fig. 5.11). We expect three loci starting from 0, -1, and -2. One ends at -4; hence, the other two must both go to ∞ [indeed, $G(s)$ has a double zero at ∞].

(3) Rule 5 tells us that the parts of the real axis from 0 to -1 and from -2 to -4 are on the loci (Fig. 5.12). $G(s)$ is negative along only these segments of the σ axis.

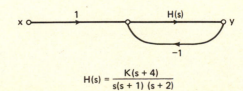

FIG. 5.10 First example of the rules.

$$H(s) = \frac{K(s + 4)}{s(s + 1)(s + 2)}$$

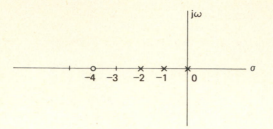

FIG. 5.11 Poles and zeros of $G(s)$. (Pole indicated by x, zero by $\bigcirc$.)

(4)   Next we find the asymptotes (i.e., how do the two loci go to ∞?). The asymptotes go out at angles of $\pm 180/2$ or $\pm 90°$ since there are two. The centroid is

$$\frac{\sum \text{poles of } G - \sum \text{zeros of } G}{3-1} = \frac{[0 + (-1) + (-2)] - [-4]}{2} = \frac{1}{2} \tag{5-24}$$

Thus, the two asymptotes meet at $s = 0.5$ and are at angles of $\pm 90°$ (Fig. 5.13).

(5)   Remembering that the loci go from poles to zeros, we can now sketch these loci approximately (Fig. 5.14). As K increases from 0 to ∞, one pole of Y/X moves from -2 to -4; one leaves from 0, another from -1; these meet and then become complex.* As K increases further, they move into the right half plane.

Figure 5.14 shows how the system poles depend on K. The figure reveals, for example, that the system is unstable for all K beyond a critical value where the loci cross the $j\omega$ axis.

We can easily find the K when the two loci are on the $j\omega$ axis. The characteristic equation [i.e., Eq. (5-18) for this sytem] is

$$1 + K \frac{s+4}{s(s+1)(s+2)} = 0 \quad \text{or} \quad s^3 + 3s^2 + (2+K)s + 4K = 0 \tag{5-25}$$

*Loci can only leave the negative real axis in conjugate pairs, one up and the other down, since complex loci must appear in conjugate pairs. Hence, any departure from the real axis must be preceded by two loci coming together—in this case from 0 and –1.

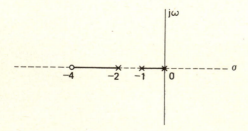

FIG. 5.12 Loci on the real axis.

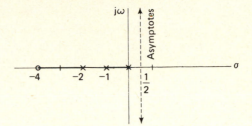

FIG. 5.13 Asymptotes of loci added.

The second form shows that the sum of the roots is always -3 (the negative of the coefficient of s^2). Hence, when two loci are on the $j\omega$ axis, the other must be at -3. If we substitute $s = -3$ into either Eq. (5-25), we obtain

$$1 + K\frac{(1)}{(-3)(-2)(-1)} = 0 \quad \text{or} \quad K = 6 \tag{5-26}$$

Hence $K = 6$ is the borderline between stability and instability. When $K > 6$, the system is unstable.*

The interesting feature of this example is that we can sketch the approximate root loci without doing any real work. Even if the loci are not accurately constructed, we can begin to see how the dynamic characteristics of the system depend on the parameter K. Finally, when we have this rough sketch, we can easily find the loci with greater accuracy in any region of particular interest, as illustrated by the next example.

The example does emphasize the usefulness of one additional rule (to be added to our six) for root-locus construction.

*In Chap. 15 we discuss the Routh test. This is an algebraic method for determining simply when there are roots of the characteristic equation on the $j\omega$ axis—a method which is useful in systems much more complicated than our present example.

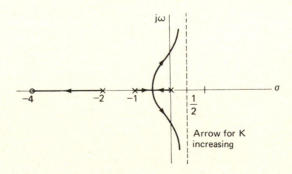

FIG. 5.14 Sketch of the root loci for first example.

⑦ If $G(s)$ has at least a double zero at ∞, the sum of the loci is constant.

In other words, if the degree of the denominator of G exceeds the degree of the numerator by 2 or more, the sum of the roots of the characteristic equation is constant. Hence, if one locus moves leftward, the others overall must move rightward.

5.6   EXAMPLE 2 (ROOT LOCUS)

The system for the second example is again single-loop, but here there are both forward and feedback transfer functions. The parameter K appears in the feedback path (Fig. 5.15). We wish to choose K so that the complex poles of Y/X have a relative damping ratio ζ of 1/2.

(1)   Our first step is always to write the system function Y/X as it depends on the parameter, here K:

$$\frac{Y}{X} = \frac{(s + 5)/s^2}{1 + K[(s + 5)(s + 8)]/[s^2(s + 1)]} \tag{5-27}$$

and the poles of Y/X are determined by the characteristic equation

$$1 + K\frac{(s + 5)(s + 8)}{s^2(s + 1)} = 0 \tag{5-28}$$

Thus,

$$G(s) = \frac{(s + 5)(s + 8)}{s^2(s + 1)} \tag{5-29}$$

and we are ready to construct the root loci: the variation of the poles of Y/X as K changes from 0 to ∞.

(2)   The poles and zeros of $G(s)$ are shown on the s plane. There is a double pole at $s = 0$, a pole at -1, and finite zeros at -5 and -8 (Fig. 5.16).

(3)   The portions of the real axis on the loci are indicated (Fig. 5.17).

(4)   We find the asymptotes (the loci for large K and s). Since there is only one zero of G at ∞, the only asymptote is at $180°$—or the negative real axis. We can now add the

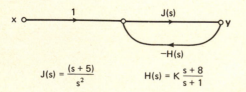

$$J(s) = \frac{(s + 5)}{s^2} \qquad H(s) = K\frac{s + 8}{s + 1}$$

FIG. 5.15 System for second example.

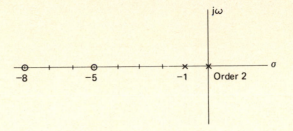

FIG. 5.16 Critical frequencies of $G(s)$.

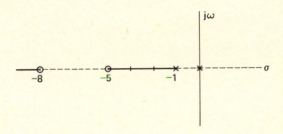

FIG. 5.17 Real axis investigated.

arrows indicating the loci for large K: each locus (of the three) approaches one of the zeros at -5, -8, and $-\infty$ (Fig. 5.18).

(5) Now we can sketch in the loci very approximately. Two loci leave the origin as K increases from zero. Since the real axis in this vicinity is *not* part of the locus, they must leave as a conjugate complex pair. Indeed, for very small $|s|$, $G(s)$ looks like $40/s^2$, so the loci leave at $\pm 90°$ (Fig. 5.19).

As s increases, the $(s + 1)$ term becomes significant, and G looks like $40/[s^2(s + 1)]$. Hence, the complex loci bend to the right, and the real locus leaves -1 to the left (Fig. 5.20).

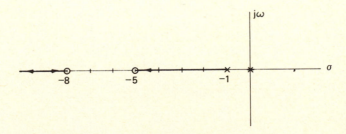

FIG. 5.18 Behavior for K large indicated.

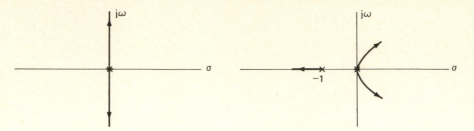

FIG. 5.19 Loci for

$$G = \frac{40}{s^2}.$$

FIG. 5.20 Loci for

$$G = \frac{40}{s^2(s + 1)}.$$

As K is increased further, the complex loci must bend around to the left so that they can eventually end up at -8 and $-\infty$. Therefore, the complete (though very approximate) form is shown in Fig. 5.21.

(6) The design specification is a relative damping ratio of $1/2$ for the complex poles of Y/X. We now need to find the value of K_1 and s_1—at the point where the locus meets the $\zeta = 1/2$ line in Fig. 5.21. In other words, we want to find the locus more accurately in the region shown in Fig. 5.22. Somewhere along the darkened line segment, there is a point on the locus.

To find this point, we notice that the locus really is defined by the equation

$$KG(s) = -1 \tag{5-30}$$

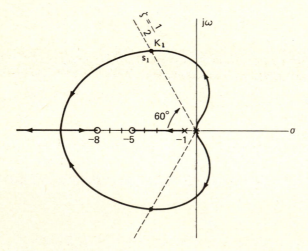

FIG. 5.21 Sketch of root loci for second example.

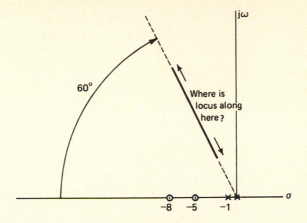

FIG. 5.22 Line segment where we are looking for the locus.

Since K is positive, $G(s)$ must be negative and real; the angle of $G(s)$ must be $\pm 180°$. But in our example

$$G(s) = \frac{(s + 5)(s + 8)}{s^2(s + 1)} \tag{5-31}$$

Along the line segment of Fig. 5.22, the angle of G is

$$\angle G = \angle(s + 5) + \angle(s + 8) - 2\angle s - \angle(s + 1)$$
$$= \angle(s + 5) + \angle(s + 8) - 240° - \angle(s + 1) \tag{5-32}$$

The angle of $(s + 1)$ is just the angle of the vector from -1 (the pole) to the s point of interest. Figure 5.22 shows this angle is slightly less than $120°$. Then approximately

$$\angle G = \angle(s + 5) + \angle(s + 8) - 360° \tag{5-33}$$

To satisfy $\angle G = \pm 180°$, we must have $\angle(s + 5) + \angle(s + 8) = 180°$. In other words, the point on the locus must be about midway between -5 and -8, or with a real part of -6.5 (Fig. 5.23).

Actually, the point shown in Fig. 5.23 is not quite correct. At this point,

$$\angle G = 180° - 240° - \angle(s + 1) \tag{5-34}$$

and the angle of $(s + 1)$ is slightly less than $120°$. We could try another point along the $\zeta = 1/2$ line slightly toward the origin in an attempt to find the locus precisely.

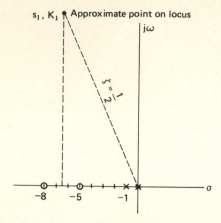

FIG. 5.23 Location of intersection of locus and $\zeta = 1/2$ line.

Such accuracy is, however, ordinarily ridiculous; we simply do not know the system parameters or the poles and zeros that precisely.

We still need to find K_1 (Fig. 5.23). The loci (Fig. 5.24) are only very rough guesses, except in the vicinity of the $\zeta = 1/2$ line, where we have made a more careful intrepretation. The value of the closed-loop pole there can be measured directly. If we measure the distance from the origin to s_1 on the same scale used in the plot, we find the distance is 13. Hence,

$$s_1 = 13 \angle 120° \tag{5-35}$$

Now we can find K_1 since

$$K_1 |G(s_1)| = 1 \tag{5-36}$$

$|G(s_1)|$ is just the product of the distances from the two zeros to s_1 divided by the product of the distances from the three poles $(0, 0, -1)$. Direct measurements on the

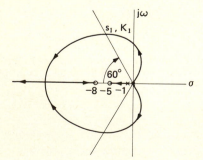

FIG. 5.24 Complete root locus for second example.

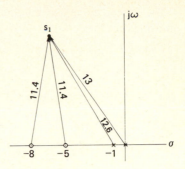

FIG. 5.25 Terms in the evaluation of

$$|G(s_1)| = \frac{|s_1 + 8| |s_1 + 5|}{|s_1|^2 |s_1 + 1|}$$

the magnitude of $G(s)$ at s_1.

graph (which we would normally draw much larger) give the data shown in Fig. 5.25, or the relation

$$|G(s_1)| = \frac{11.4 \times 11.4}{13 \times 13 \times 12.6} = \frac{1}{16.4} \tag{5-37}$$

Thus,

$$K_1 = 16.4 \tag{5-38}$$

The design example is now completed. When we adjust K to the value 16.4, we obtain a Y/X with complex poles with a ζ of 1/2—poles at $13 \underline{/\pm 120}°$. Equation (5-27) from the beginning of the example shows that

$$\frac{Y}{X} = \frac{(s + 5)(s + 1)}{s^3 + (1 + K)s^2 + 13Ks + 40K} \tag{5-39}$$

When $K = 16.4$, we know that there are complex poles at $13 \underline{/\pm 120}°$, corresponding to a complex factor

$$s^2 + 13s + 169$$

Hence, the real factor has a constant term $40K/169$ or $40(16.4)/169$ or 3.9. With $K = 16.4$,

$$\frac{Y}{X} = \frac{(s + 5)(s + 1)}{(s^2 + 13s + 169)(s + 3.9)} \tag{5-40}$$

We close this example with two comments. First, the system is interesting because it has this unusual property: if the gain K is reduced enough, the system becomes unstable. (Normally, increasing gain makes a feedback system unstable.) Such a system is called "conditionally stable," and is discussed again in Chap. 16.

The second comment on this example really amounts to two additional rules for drawing root loci.

⑧ If we want to determine the locus more accurately in a particular region, we can use the fact that $\angle G(s)$ must be $\pm 180°$ along the locus. For the calculation of the angle of $G(s)$, we can use $\angle G(s) = \sum_{\text{zeros}}$ angle of vector from the zero to s $- \sum_{\text{poles}}$ angle of vector from the pole to s $\qquad$ (5-41)

⑨ Once the loci are known, the value of K at any locus point s_1 can be determined by

$$K\,|G(s_1)| = 1 \qquad\qquad\qquad (5\text{-}42)$$

Here $|G(s_1)|$ is calculated from

$$|G(s_1)| = \frac{|s_1 - z_1||s_1 - z_2|\cdots}{|s_1 - p_1||s_1 - p_2|\cdots} \qquad\qquad (5\text{-}43)$$

for zeros at z_1, z_2, ... and poles at p_1, p_2, ... The $|s_1 - p_1|$ term is the magnitude of the vector from the pole to s_1.

5.7 EXAMPLE 3 (ROOT LOCUS)

There is one more rule which is occasionally useful in drawing root loci—a rule which comes naturally from a discussion of the example shown in Fig. 5.26. When we proceed to construct the root loci as a function of K, we find:

(a)   Real axis from 0 to -4 is on the loci.

(b)   Asymptotes go to infinity at angles of $\pm 45°$ and $\pm 135°$ from a centroid at $(-4-2)/4$ or $-3/2$.

We can then drawn Fig. 5.27.

The next question is: which of the two patterns of Fig. 5.28 is correct? Do the two loci emanating from the poles at $-1 \pm j2$ go to the right asymptotes or to those on

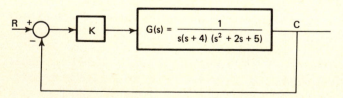

FIG. 5.26 Feedback system.

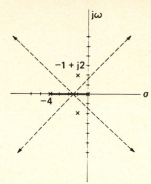

FIG. 5.27 Information readily deduced from our rules.

the left? A hint of the answer is available if we find the angle at which the locus leaves the pole at $-1 \pm j2$. Since the locus cannot loop back on itself, if it leaves to the right, it probably heads for the $+45°$ asymptote.

In order to find the angle of departure of a locus from a pole, we look for the locus for K very small, but not zero. The locus is *very* near the pole. All vectors from the *other* poles and the zeros to the s on the locus are essentially vectors to the pole. Then the equation

$$\angle G(s) = 180° \qquad\qquad (5\text{-}44)$$

satisfied everywhere on the locus, allows us to find the angle from the pole to the locus.

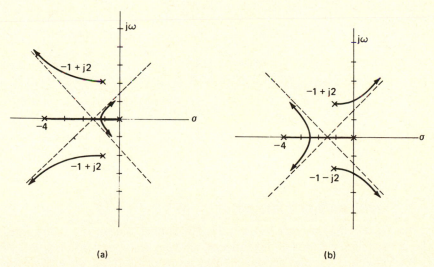

(a)   (b)

FIG. 5.28 Two possibilities for the root loci.

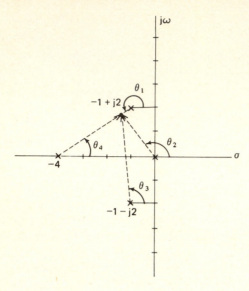

FIG. 5.29 Angles involved very near $-1 \pm j2$.

For example, in Fig. 5.29, we want the angle of departure of the locus from $-1 \pm j2$. For s_1 very near $-1 \pm j2$, Eq. (5-44) gives

$$\theta_1 + \theta_2 + \theta_3 + \theta_4 = 180° \tag{5-45}$$

But θ_2 is essentially the angle from the origin to $-1 \pm j2$, or 116.5°; θ_3 is 90°; and θ_4 is 33.7° (arc tan 2/3). Equation (5-45) becomes

$$\theta_1 + 240.2° = 180° \tag{5-46}$$

or θ_1 is $-60.2°$. The locus leaves the pole at $-1 \pm j2$ at an angle of $-60.2°$. Consequently, possibility (b) of Fig. 5.28 is likely, and we can now sketch the root loci (Fig. 5.30).

The example shows the importance of the final rule:

⑩ The departure angle for a locus leaving a pole can be found by considering a locus point s_1 *very* close to that pole p_1. We then apply Rule ⑧. The only angle undetermined is that from p_1 to s_1.

5.8 EXAMPLE 4 (ROOT LOCUS)

The last example is shown in Fig. 5.31. The normal value of L is 0.203, but L varies. We are to find the effects of this variation on the overall transfer function E_2/E_1.

We write first the transfer function in terms of L. If we work from the output backward, we obtain

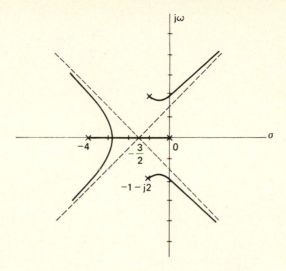

FIG. 5.30 Root loci for the system of Fig. 5.26.

$$i_1 = \frac{1}{9} e_2$$

$$e_C = e_2 + \frac{Ls}{9} e_2 = \frac{9 + Ls}{9} e_2$$

$$i_{in} = \frac{1}{9} e_2 + 0.0004111s \frac{9 + Ls}{9} e_2 = \frac{1 + 0.00370s + 0.0004111Ls^2}{9} e_2$$

$$e_1 = \frac{(9 + Ls) + 55 + 0.2035s + 0.0226Ls^2}{9} e_2$$

$$= \frac{64 + (L + 0.2035)s + 0.0226Ls^2}{9} e_2$$

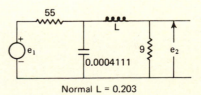

FIG. 5.31 Circuit for fourth example (values in ohms, henrys, farads).

Normal L = 0.203

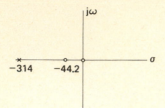

FIG. 5.32 Poles and zeros of $G(s)$.

Hence,

$$\frac{E_2}{E_1} = \frac{398}{Ls^2 + (44.2L + 9)s + 2832} \tag{5-47}$$

We see that the zeros of E_2/E_1 are independent of L (both are at ∞). Only the poles vary as L changes. Therefore, we want the denominator in the form $1 + LG(s)$. To obtain this, divide both numerator and denominator in Eq. (5-47) by $(9s + 2832)$— the part of the denominator not multiplying L:

$$\frac{E_2}{E_1} = \frac{398/(9s + 2832)}{1 + L[s(s + 44.2)/9(s + 314)]} \tag{5-48}$$

The poles and zeros of $G(s)$ are shown in Fig. 5.32. As L varies from 0 to ∞, the E_2/E_1 poles (i.e., the loci) travel from ∞ and -314 to 0 and -44.2. The real axis lies on the loci from 0 to -44.2 and from -314 to $-\infty$. The shape of the loci is indicated in Fig. 5.33.

Where is the normal value of L ($L = 0.203$) along the loci? We know that

$$L|G(s)| = 1 \tag{5-49}$$

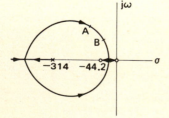

FIG. 5.33 Root loci for

$$G(s) = \frac{1}{9} \frac{s(s + 44.2)}{s + 314}.$$

Hence, the $L = 0.203$ value occurs when $|G| = 4.93$, or

$$\left| \frac{s(s + 44.2)}{s + 314} \right| = 44.4 \qquad (5\text{-}50)$$

To find such a value of s along the loci, we have to hunt. If a point between 0 and -44.2 is tried we find that the maximum value of the left side (the minimum L) above is about

$$\underset{0 > \sigma > -44.2}{\text{Max}} \left| \frac{s(s + 44.2)}{s + 314} \right| = \frac{22 \times 22}{292} = 1.66 \quad \text{Min} L = \frac{9}{1.66} = 5.4 \qquad (5\text{-}51)$$

Thus, $L = 0.203$ occurs before the loci rejoin the real axis.

Next we might try a value around A in Fig. 5.33. The vectors involved in finding $|G|$ are shown in Fig. 5.34. Here roughly (we don't know the locus accurately, so there is no point in being precise),

$$|G(s)| = \frac{1}{9} \frac{200 \times 200}{250} = 17.8 \quad \text{or} \quad L = 0.056 \qquad (5\text{-}52)$$

Hence, we would estimate the normal value of L occurs at about B in Fig. 5.33. The correct value [which we can find by factoring the denominator quadratic of Eq. (5-47) with the normal L value] is $-44.2 \pm j109.5$ (the pole lies directly above the G zero at -44.2).

The example is somewhat artificial since the characteristic polynomial

$$Ls^2 + (44.2 + 9)s + 2832$$

is only a quadratic and can easily be factored for each value of L. The example is included here for two reasons:

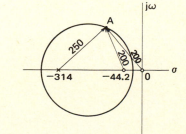

FIG. 5.34 Evaluation at A of

$$G(s) = \frac{1}{9} \frac{s(s + 44.2)}{s + 314}$$

(Lengths of vectors are very approximate.)

(a) It shows that the root-locus method can be used when the parameter is a circuit element (here L). We write the denominator of the system transfer function in the form $1 + LG(s)$, then determine how the roots of

$$1 + LG(s) = 0 \qquad (5\text{-}53)$$

move as L varies from 0 to ∞.

If the system transfer function has zeros dependent on the parameter L, we can also write the numerator as

$$1 + LH(s)$$

and then plot the corresponding "root loci" to determine how the zeros of the system function move as the parameter varies.

(b) The network happens to be of practical importance as an impedance–matching system to maximize power delivered to the load (the 9-ohm resistor). Actually, the author chose the C and the normal value of L to accomplish the matching at 100 radians/sec.

Dependence of system function on a parameter

As indicated just above, this fourth example illustrates that the root-locus method can be used to determine how the poles (or zeros) of a system function depend on a particular parameter. The parameter may be the "gain" of an amplifier, a circuit element value (resistance, inductance, or capacitance), the gain of a controlled source, or another system constant (as the moment of inertia of a mechanical load).

The basis of the root-locus method is a theorem in circuit analysis which states: If the circuit diagram can be drawn to display a parameter as the descriptive constant for only one element, the system transfer function is a bilinear transformation of that parameter. In other words, the system function $T(s)$ can be written in the form

$$T(s) = A(s) \frac{1 + KB(s)}{1 + KC(s)} \qquad (5\text{-}54)$$

where K is the parameter and A, B, and C are transfer functions independent of K.*

---

*Equation (5-54) is actually not quite general if one is a stickler for details. If $T(s)$, for example, is proportional to K, we can have the form

$$\frac{KB(s)}{1 + KC(s)}$$

or, if T is inversely proportional to K, the form would be

$$\frac{1 + KB(s)}{KC(s)}$$

These are both bilinear transformations. In both cases, the root-locus method can be used to find how the zeros and poles of T depend on K.

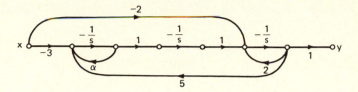

FIG. 5.35 Feedback system with zeros and poles depending on the parameter α.

In all cases where $T(s)$ takes this form, the root-locus method permits determination of how the zeros and poles of T vary with K. In general, we need two root-locus plots: one for $B(s)$ for the T zeros and the other for $C(s)$ for the poles in Eq. (5-54).

The system of Fig. 5.35 shows how we first develop $T(s)$ in the form of Eq. (5-54). (Here we start from a signal flow diagram rather than a circuit model, but the approach is the same.) Direct application of Mason's reduction theorem gives

$$T(s) = \frac{Y}{X} = \frac{3/s^3 + (2/s)(1 + \alpha/s)}{1 + (\alpha/s + 2/s + 5/s^3) + 2\alpha/s^2} \tag{5-55}$$

Clearing of fractions leads to

$$T(s) = \frac{2s^2 + 2\alpha s + 3}{s^3 + (2 + \alpha)s^2 + 2\alpha s + 5} \tag{5-56}$$

We want the numerator in the form $[1 + \alpha B(s)]$. To obtain this, we divide out of the numerator the factor representing all terms *not* involving α—i.e., the factor $(2s^2 + 3)$:

$$T(s) = (2s^2 + 3)\frac{1 + \alpha 2s/(2s^2 + 3)}{s^3 + (2 + \alpha)s^2 + 2\alpha s + 5} \tag{5-57}$$

Repeating this procedure with the denominator gives the desired form:

$$T(s) = \frac{2s^2 + 3}{s^3 + 2s^2 + 5}\frac{1 + \alpha 2s/(2s^2 + 3)}{1 + \alpha(s^2 + 2s)/(s^3 + 2s^2 + 5)} \tag{5-58}*$$

5.9 PARAMETER NEGATIVE

Throughout this chapter, we have developed the root-locus method to study the roots of the characteristic equation in the form

$$1 + KG(s) = 0 \tag{5-59}$$

---

*There is a partial check. The function in front should be $T(s)$ with $\alpha = 0$, which can also be determined from Fig. 5.35.

with K varying over positive values from zero to infinity, and with $G(s)$ in the form

$$G(s) = k \frac{(s - z_1)(s - z_2)}{(s - p_1)(s - p_2)(s - p_3)} \qquad (5\text{-}60)$$

where k is positive. In other words, the root loci define the values of s where

$$\angle G(s) = 180° \qquad (5\text{-}61)$$

Whenever Eq. (5-61) is satisfied, we are on the loci and K can be determined.

There are occasional situations in which K varies over *negative* values from zero to $-\infty$. This is equivalent to k in Eq. (5-60) being negative. (We can then pull the negative sign out of k and assign it to K.) In such a case, the loci are all values of s at which

$$\angle G(s) = 0° \qquad (5\text{-}62)$$

The rules for constructing the loci must then be changed. The parts of the real axis on the loci are those to the left of an *even* number of poles and zeros. The asymptotes go to infinity at angles of $0°/m$, $\pm 360°/m$, and so on (where m is the order of the open-loop zero at infinity). Except for this change from Eq. (5-61) to (5-62), the root loci are constructed by the same techniques described for positive values of the parameter.

5.10 FINAL COMMENT

When is the root-locus method useful in design? When the system is first or second order, the characteristic polynomial is easily factored for different values of the parameter—then we do not need the root-locus approach. When the system order is higher than six, the root loci are often so difficult to construct that we have to use a computer for help.

Thus, the root-locus method is primarily useful for systems of orders three through six. Actually, many of the systems the control engineer is asked to design fall into this category. Even if the actual system is more complex, we often approximate it by a system in this range, design the approximate system, then analyze the corresponding performance of the more detailed model of the system.

In addition to its usefulness in design, the root-locus approach is important conceptually. The system engineer needs to understand not only the importance of the poles and zeros of the system transfer function, but also how these move in the s plane as a parameter changes.

There are standard programs available for computer calculation of the root loci for systems of order greater than six, so that the engineer can work with systems over a very wide range of difficulty. When *two* or more parameters can be adjusted in the design or when they vary during operation, the engineer faces the difficult task of studying system performance with many different combinations of parameter values. Root loci as a function of K_1 may have to be constructed when parameter K_2 takes on a succession of values covering its operating range.

The specter of such a problem emphasizes one of the fundamental problems in teaching engineering. In a text, there is restriction to problems or examples which can be completed in a reasonable time, measured in minutes, or at most in hours. In actual engineering practice, it is not unusual to spend weeks and months on the preliminary system design before any equipment is actually built. Under such circumstances, we can work with systems of great complexity and develop an enormous variety of graphical analyses of system operation.

Finally, we should conclude this chapter with a comment that is perhaps self-evident from the examples considered. We have used a variety of rules in constructing the root loci in these specific cases—indeed, some ten rules in all. In spite of these formal rules, a primary asset in constructing root loci is experience, and an awareness of the normal flow of the loci. Hopefully, the examples and the problems at the end of the chapter begin to develop this experience. In general, here, as in so many other facets of engineering, the correct solution is one which is natural-looking and aesthetically pleasing— qualities which are very difficult to describe in analytical terms.

PROBLEMS

5.1 For the system shown, draw the root loci for each of the following $G(s)$ functions. In each case, indicate a few values of K along the loci, and consider the variation of K over positive values.

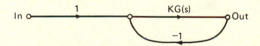

(a) $G(s) = \dfrac{1}{s + 2}$ (b) $G(s) = \dfrac{1}{(s + 2)^2}$ (c) $G(s) = \dfrac{1}{(s + 2)^3}$

(d) $G(s) = \dfrac{1}{(s + 2)^4}$ (e) $G(s) = \dfrac{1}{(s + 2)^5}$

5.2 Repeat Prob. 5.1 for the two transfer functions

$$G(s) = \dfrac{s}{s^2 + 4} \qquad G(s) = \dfrac{s}{s^2 + s + 4}$$

Compare the answers to the two parts. Notice that the two are nearly identical, except for the values of K. Why?

5.3 Draw the root loci for the system shown, with K as the positive parameter.

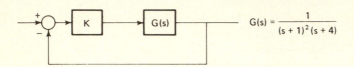

$$G(s) = \frac{1}{(s+1)^2(s+4)}$$

5.4 Construct the root loci for the system shown, with R the parameter. Consider separately the cases when R is positive and when R is negative.

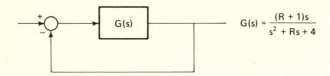

$$G(s) = \frac{(R+1)s}{s^2 + Rs + 4}$$

5.5 Draw the root loci for the same configuration as Prob. 5.3 with

$$G(s) = \frac{(s+4)^2}{s(s+1)(s+2)(s+8)}$$

A primary objective in working a series of root-locus problems is to obtain a feeling for the various forms possible for the loci. As these problems try to indicate, root loci are similar to flow lines, in that they are symmetrical, natural-flowing, and aesthetically pleasing (in addition to being mathematically correct).

5.6 For the system shown in Prob. 5.3, construct the root loci when

$$G(s) = \frac{1}{s(s^2 + 2s + 2)}$$

Give particular attention to the direction along which the loci leave the G pole at $-1 + j$. To find this, assume that we are looking for a point on the locus very close to $-1 + j$. Then the angles to the other two, G poles can be found, and, only unknown is the angle to $-1 + j$.

5.7 For the system of Prob. 5.3, construct the root loci when $G(s)$ has each of the two forms:

$$G_a(s) = \frac{1}{s(s+1)(s+2)} \qquad G_b(s) = \frac{100(s+20)^2}{s(s+1)(s+2)(s+200)^2}$$

Notice that $G_b(s)$ behaves essentially as $G_a(s)$ when $|s|$ is less than 4. As $|s|$ increases beyond this, G_b can be approximated by assuming $(s + 200)^2$ is essentially 200^2 until $|s|$ exceeds 40.

5.8 A feedback control system is described by the diagram shown.

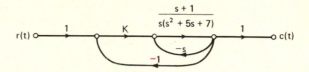

(a) Find $T(s)$.

(b) Construct the root loci showing the poles of $T(s)$ as a function of K for K varying from zero to infinity through positive values.

(c) Estimate K for a relative damping ratio of $1/2$ for the dominant poles. (Very often in higher order systems, we find that the main characteristics of the response are determined by a pair of conjugate complex poles, which we call the *dominant poles*. All other poles are of secondary importance, either because the terms have short time constants, or because the poles are so near zeros that the residues are negligible.)

5.9 We consider a feedback control system as shown, where K is an adjustable, positive gain constant.

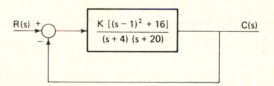

(a) Determine the closed-loop transfer function $C(s)/R(s)$ in terms of K.

(b) Find the value of K for which the closed-loop system is on the threshold of stability; compute the corresponding pole location.

(c) Determine the value of K for which C/R has a double pole, and determine that pole.

(d) Based on the results of (b) and (c), construct a sketch of the root loci for K varying from zero to infinity.

5.10 The feedback system shown is characterized by the overall transfer function $T(s)$ defined as C/R. Show how the poles of T move as K is varied from zero to very large positive values.

For very large gain, the two poles of T near the imaginary axis zeros may be assumed to cancel these zeros, and the system is then second order. What are the

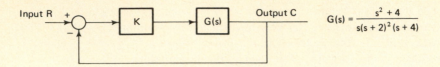

poles then for a relative damping ratio of 0.5? What is the corresponding value of K? (Notice that, with regard to the discussion of Prob. 5.8 on dominant poles, the poles near the imaginary axis in this case are negligible because of near-cancellation.)

5.11 For the configuration shown, sketch the root loci as a function of K when the G_1 compensation block has zeros at $-1 \pm j10$, and poles at -10 and -20, and when the G_2 block has poles at $0, -10$, and either of the two locations:

$$-1 \pm j12 \qquad\qquad -1 \pm j8$$

(In other words, two root-locus diagrams are desired.)

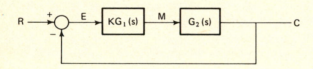

The system represents what happens when we attempt to insert a compensation network (G_1) to cancel the nearly unstable poles of the process being controlled, G_2. When the process poles move (e.g., from $j12$ to $j8$ as in this example), the cancellation is not exact, and we can anticipate stability problems. This situation arises in attempts to control an airplane, where the system poles move over a wide range as the speed of the plane changes from near-landing conditions to cruising operation.

6 OPERATIONAL AMPLIFIERS (OP-AMPS) AND SYNTHESIS

Given a transfer function, for example

$$T(s) = \frac{3s^2 + s + 2}{s^4 + 4s^3 + 12s^2 + 16s + 8} \qquad (6\text{-}1)$$

can we find a network? Can we determine an interconnection of circuit elements which is described by this $T(s)$?

This is the problem of *network synthesis* or the problem of synthesizing (putting together) a circuit with specified characteristics. Only a decade ago, the subject of network synthesis was largely restricted to graduate work in electrical engineering. The task of finding a combination of R, L, and C elements to realize the $T(s)$ of Eq. (6-1) is exceedingly difficult and tedious.

In the last few years, an additional "circuit element," the operational amplifier (or op-amp), has become available at low cost. As we shall see in this chapter, synthesis of a network of Rs, Cs, and op-amps to realize the $T(s)$ of Eq. (6-1) is a simple task. Thus, with the availability of op-amps, the engineer is able to design readily a wide variety of electronic systems.

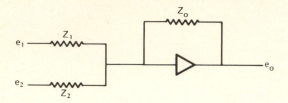

FIG. 6.1 Linear op-amp circuit.

Why is synthesis important? Why do we ever want to find a network with a specified transfer function? There are two primary answers:

(1) In signal processing, we often want to modify a signal according to a selected transfer function. For example, we might want to separate the desired signal from noise. In many such cases, we can find a transfer function which performs this separation. Then we need to build a system with this transfer function.

(2) In analog simulation studies, we want to build a network with the transfer functions of the system for the purposes of design. For instance, in the design of an aircraft, the aeronautical engineer selects tentatively the dimensions and characteristics of the plane. Rather than build test airplanes immediately on which the dimensions can be changed slightly, he simulates his system with an electronic network in which potentiometer variations can represent the changing parameters of the aircraft, while "performance" is observed on an oscilloscope.

In both these areas, network or system synthesis is a key problem.

Thus, in this chapter, we will develop a solution of the synthesis problem: how to go from a given transfer function (or a set of transfer functions if there are multiple inputs or outputs) to an appropriate electronic network.

6.1 OP-AMPS

The basic operational-amplifier (op-amp) circuit is shown in Fig. 6.1. There are two inputs, e_1 and e_2, and the output e_o; all voltages are measured with respect to ground which is not shown. The impedances (Z_1, Z_2, and Z_o) are normally networks of R, L, and C components if we are interested in linear systems.

The triangular symbol (Fig. 6.2) represents an "ideal voltage amplifier" with the characteristics:

(a) large voltage gain: $|e_o/e_i| \gg 1$; typically, the gain is at least 20,000 and often more than 10^5;

FIG. 6.2 Ideal amplifier.

(b) zero input current (or infinite input impedance);
(c) zero output impedance.

As a result of these characteristics, analysis of the circuit of Fig. 6.1 to find how e_o depends on e_1 and e_2 is trivial—indeed, so trivial that *design* of a circuit for many desired transfer characteristics is simple.

Circuit analysis

Figure 6.3 shows the circuit with node A labeled. The output e_o is just the amplifier gain times e_A (the voltage from A to ground). If e_o is of reasonable size, e_A must be almost zero (since the gain is very large). *The key to simple analysis of the circuit is to assume e_A is zero.*

Then we write a single node equation at A. Three currents enter A, one through each Z. Through Z_1 the current entering A is E_1/Z_1 (again since e_A is zero). Summing the three currents gives

$$\frac{E_1}{Z_1} + \frac{E_2}{Z_2} + \frac{E_o}{Z_o} = 0 \qquad (6\text{-}2)$$

(No current flows into the amplifier). In other words,

$$\boxed{E_o = -\frac{Z_o}{Z_1}E_1 - \frac{Z_o}{Z_2}E_2} \qquad (6\text{-}3)$$

Equation (6-3) is the description of the circuit.

There is an alternate way to look at the circuit. The current I (Fig. 6.3) is $E_1/Z_1 + E_2/Z_2$, since A is essentially at ground potential. This current can only flow through Z_o. The voltage across Z_o is then $-IZ_o$ or

$$E_o = -Z_o\left(\frac{E_1}{Z_1} + \frac{E_2}{Z_2}\right) \qquad (6\text{-}4)$$

which is just Eq. (6-3) again. The two approaches are equivalent.

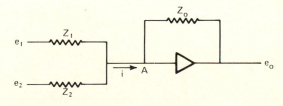

FIG. 6.3 Circuit with node A labeled.

Uses of the op-amp circuit

Equation (6-3) reveals that, by choosing various Z_o and Z_1 and Z_2 impedances, we can use the op-amp to realize different system characteristics. Two cases are particularly important.

(1) If Z_o, Z_1, and Z_2 are all resistances, the circuit is an *adder*. The output is a linear combination of the various inputs. We can have any number of inputs merely by adding e_3 and Z_3, e_4 and Z_4, and so on.

(2) If Z_o represents a capacitor

$$Z_o = \frac{1}{C_o s} \tag{6-5}$$

and Z_1 and Z_2 are resistors, the output is the sum of the integrals of the input; the circuit is an *integrator*.

In Sec. 6.2, we will discover this astonishing fact: any transfer function can be realized by an interconnection of adders and integrators. In other words, any transfer function can be realized by a combination of op-amps of the two basic types.

Of course, we are not restricted to use only single elements for Z_o, Z_1, and Z_2 in the op-amps, Fig. 6.3. We can use much more complex networks. In such a case, Eq. (6-3) shows that we can realize much more complex relations between e_o and the inputs.

Adders

Figure 6.4 is a circuit diagram of the basic adder. The single node equation at the amplifier input yields

$$\frac{e_1}{R_1} + \frac{e_2}{R_2} + \frac{e_3}{R_3} + \frac{e_o}{R_o} = 0 \tag{6-6}$$

or

$$e_o = -\frac{R_o}{R_1} e_1 - \frac{R_o}{R_2} e_2 - \frac{R_o}{R_3} e_3 \tag{6-7}$$

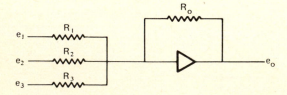

FIG. 6.4 Adder with three inputs.

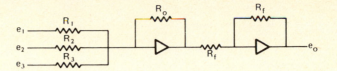

FIG. 6.5 Realization of

$$e_o = \frac{R_o}{R_1} e_1 + \frac{R_o}{R_2} e_2 + \frac{R_o}{R_3} e_3$$

(all plus signs).

The output is the negative of a linear algebraic combination of the inputs. If we dislike the minus sign, we can follow this op-amp with another circuit with a gain of -1 (Fig. 6.5).

Two examples illustrate the design procedure.

(1) In the first example, we want to realize a simple amplification

$$e_o = 16e_1 \tag{6-8}$$

The circuit of Fig. 6.6(a) gives the desired gain, but with a minus sign. (Often the minus is immaterial—for example, if the output is to be observed on a scope or used to drive a loudspeaker.) If we must have the correct sign, the circuits of Fig. 6.6(b) and (c) are appropriate. Obviously, other combinations of breaking up the total of 16 into two factors can be used.

In Fig. 6.6, the resistances R, R_1, and R_2 can be chosen to have *any* convenient value. We can use resistors which we have on hand. The only important value is the ratio of the feedback and input resistances.

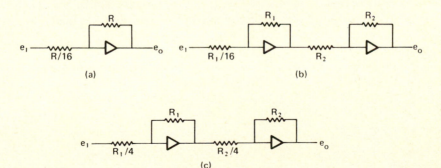

FIG. 6.6 Three systems for achieving a gain of 16: (a) $e_o = -16e_1$; (b) $e_o = 16e_1$; (c) $e_o = 16e_1$.

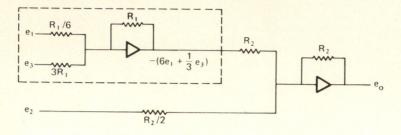

FIG. 6.7 $e_o = -2e_2 + 6e_1 + \frac{1}{3}e_3$.

(2) In the second example, we wish to build a circuit with

$$e_o = 6e_1 - 2e_2 + \frac{1}{3}e_3 \tag{6-9}$$

We recognize that each op-amp yields a negative sign. Hence e_o can be rewritten

$$e_o = -\left[2e_2 - \left(6e_1 + \frac{1}{3}e_3\right)\right] \tag{6-10}$$

First, we realize $-\left(6e_1 + \frac{1}{3}e_3\right)$ by the circuit within the dashed lines in Fig. 6.7, then realize the overall system as shown. Again, R_1 and R_2 can be chosen arbitrarily.*

The two examples show that the realization of

$$e_o = \alpha e_1 + \beta e_2 + \gamma e_3 \tag{6-11}$$

where α, β, and γ are constants (positive or negative) can be accomplished with at most two op-amps, regardless of the number of different inputs. The resistive op-amps can be used as adders, with the design requiring essentially no computation.

Integrators

In a manner exactly parallel, we can design integrators if the feedback element is a capacitor rather than a resistor. Figure 6.8 shows the basic integrator circuit with three inputs. The node equation at the amplifier input is

$$\frac{E_1}{R_1} + \frac{E_2}{R_2} + \frac{E_3}{R_3} + CsE_o = 0 \tag{6-12}$$

*This arbitrariness is not quite complete, as indicated in Sec. 6.6, where some practical limitations are discussed.

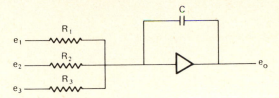

FIG. 6.8 Basic integrator.

or

$$E_o = -\frac{1}{R_1 Cs} E_1 - \frac{1}{R_2 Cs} - \frac{1}{R_3 Cs} E_3 \qquad (6\text{-}13)$$

The output is the sum of the integrals of the inputs, each multiplied by a different constant. Again, a minus sign is associated with each term. In the design of a circuit for a specified output, we can choose C arbitrarily, then find the resistance values required. Figure 6.9 shows the realization of

$$E_o = -\frac{2}{s} E_1 + \frac{1/3}{s} E_2 \qquad (6\text{-}14)$$

Modification of design

We usually do not have resistances available of precisely 500K, and so forth. As a result, it is desirable to have fine adjustments available, so that values of R can be set accurately in the laboratory. Figure 6.10 shows one common circuit in the case of an adder originally designed as in part (a) of the figure.

To understand the circuit operation, we can replace the circuit to the left of A (as an example of one of the three inputs) by Thevenin's theorem. There is a voltage source of βx_1 (where β varies from 0 to 1 as the potentiometer is changed) in series with a Thevenin resistance (the two parts of the potentiometer in parallel.) Usually the potentiometer R is small compared to 100K, so the major effect is to reduce the effective input by β. Thus, the gain of the adder proper must be increased (and the

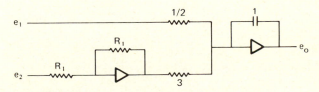

FIG. 6.9 Realization of Eq. (6-14). If C is in μF, R is in $M\Omega$.

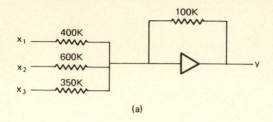

(a)

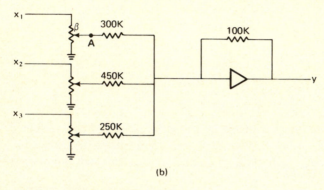

(b)

FIG. 6.10 Adder circuit: (a) Realization of

$$y = -\frac{1}{4}x_1 - \frac{1}{6}x_2 - \frac{2}{7}x_3$$

(b) An alternate realization for (a) to permit fine adjustment of coefficients in

$$y = -\frac{1}{4}x_1 - \frac{1}{6}x_2 - \frac{2}{7}x_3.$$

400K resistance is changed to 300K). Then the three potentiometers are adjusted experimentally to realize exactly

$$y = -\frac{1}{4}x_1 - \frac{1}{6}x_2 - \frac{2}{7}x_3 \qquad (6\text{-}15)$$

Clearly, this modification is a practical detail which does not change the basic procedure of adder design.

6.2 ANALYSIS OF AN OP-AMP CIRCUIT

For an example of the analysis of an op-amp circuit, we use a circuit involving operational amplifiers as shown in Fig. 6.11. Three stages of amplification are included,

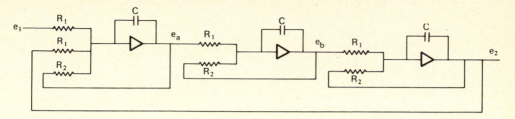

FIG. 6.11 Complete circuit diagram for the example of op-amp circuit analysis.

in this case, identical. It is desired (1) to formulate the signal flow diagram; then (2) to find the transfer function E_2/E_1; and finally (3) to interpret this transfer function in terms of the characteristics of the circuit.

Formulation of signal flow diagram

Analysis of this system begins with the formulation of a signal flow diagram. As we observe the flow of signals through the circuit, it is noted that e_1 and e_2 cause e_a, e_a leads to e_b, and e_b causes e_2. Hence, we might guess that the signal flow diagram includes the corresponding nodes as shown in Fig. 6.12.

It is now necessary to find the equations by which e_a, e_b, and e_2 are determined. If we arbitrarily start with e_b, Fig. 6.13 shows this portion redrawn. Here, by inspection, we can write

$$e_b = -\frac{1}{Cs} i_1 \qquad\qquad (6\text{-}16)$$

since the current i_1 flows through C (zero input current to the amplifier) and point A is essentially at ground potential. Likewise,

$$i_1 = \frac{1}{R_1} e_a + \frac{1}{R_2} e_b = G_1 e_a + G_2 e_b \qquad\qquad (6\text{-}17)$$

Hence the diagram of Fig. 6.12 is now partially filled in, as shown in Fig. 6.14, which represents Eqs. (6-16) and (6-17). We have introduced a new node (i_1) so that the equations can be written by inspection.

$$
\begin{array}{cccc}
e_1 & e_a & e_b & e_2 \\
\circ & \circ & \circ & \circ
\end{array}
$$

FIG. 6.12 Original set of nodes.

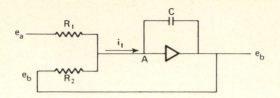

FIG. 6.13 Portion of the circuit determining e_b.

The portion of the circuit determining e_2 is exactly the same, and we can move directly to Fig. 6.15.

Finally, we must represent the equations permitting solution for the other dependent variable e_a in terms of e_1, e_b, and e_2. Inspection of Fig. 6.11 reveals that the only novel feature of the input op-amp circuit is that there are three components of the current $[(1/R_1)e_1, (1/R_1)e_2,$ and $(1/R_2)e_a]$. Thus, the total signal flow diagram is shown in Fig. 6.16.

We observe that the signal flow diagram involves branch transmittances (or gains) which have dimensions. Any branch from a voltage to a current (e.g., e_1 to i_3) must be a conductance (G_1) or 1/resistance. Any branch from i to e must be an impedance (here $-1/Cs$).

Overall transfer function

The overall transfer function

$$T(s) = \frac{E_2}{E_1} \tag{6-18}$$

can be written directly from the reduction theorem (in this case there is a non-touching triplet):

$$T(s) = \frac{-G_1{}^3 (1/C^3 s^3)}{1 - (-3G_2/Cs - G_1{}^3/C^3 s^3) + 3G_2{}^2/C^2 s^2 - (-G_2{}^3/C^3 s^3)} \tag{6-19}$$

Multiplication by $C^3 s^3$ gives

$$T(s) = \frac{-G_1{}^3}{C^3 s^3 + 3G_2 C^2 s^2 + 3G_2{}^2 Cs + G_1{}^3 + G_2{}^3} \tag{6-20}$$

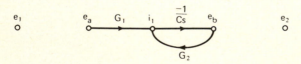

FIG. 6.14 Nodes with e_b described.

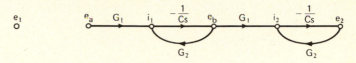

FIG. 6.15 Partial diagram with e_b and e_2 described.

Interpretation of transfer function

Our particular interest in the circuit and $T(s)$ results from the fact that the system can behave as an oscillator if we select the parameters (G_1, G_2, C) appropriately. If the circuit is to be an oscillator, $T(s)$ must have poles on the $j\omega$-axis; hence the denominator has a factor of the form

$$s^2 + \omega_o^2$$

where ω_o is the angular frequency of oscillation. In order to find out if this is possible, we divide the denominator of Eq. (6-20) by $(s^2 + \omega_o^2)$ and determine when the remainder is zero:

$$
\begin{array}{l}
\quad C^3 s \;\; + 3G_2 C^2 \\[4pt]
\hline
s^2 + \omega_o^2\,\big|\; C^3 s^3 \;+\; 3G_2 C^2 s^2 \;+\; 3G_2{}^2 C s \qquad\qquad +\, G_1{}^3 + G_2{}^3 \\[4pt]
\quad C^3 s^3 \qquad\qquad\qquad +\, \omega_o{}^2 C^3 s \\[4pt]
\hline
\qquad 3G_2 C^2 s^2 \;+\; (3G_2{}^2 C - \omega_o{}^2 C^3) + G_1{}^3 + G_2{}^3 \\[4pt]
\qquad 3G_2 C^2 s^2 \qquad\qquad\qquad +\, 3\omega_o{}^2 G_2 C^2 \\[4pt]
\hline
\qquad (3G_2{}^2 C - \omega_o{}^2 C^3)\, s + (G_1{}^3 + G_2{}^3 - 3\omega_o{}^2 G_2 C^2)
\end{array}
$$

For the remainder to be zero,

$$
\left.
\begin{aligned}
3G_2{}^2 C - \omega_o{}^2 C^3 &= 0 \\
G_1{}^3 + G_2{}^3 - 3\omega_o{}^2 G_2 C^2 &= 0
\end{aligned}
\right\}
\tag{6-21}
$$

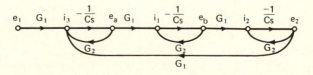

FIG. 6.16 Complete signal flow diagram for Fig. 6.11.

The first equation gives

$$\omega_o^2 = 3 \frac{G_2^2}{C^2} \tag{6-22}$$

Substitution of this in the second relation gives

$$G_1^3 + G_2^3 = 9G_2^3 \tag{6-23}$$

or

$$G_1^3 = 8G_2^3 \qquad \boxed{G_1 = 2G_2} \quad \text{or} \quad \boxed{R_2 = 2R_1} \tag{6-24}$$

Thus, if $G_1 = 2G_2$, the circuit is an oscillator, with the oscillation frequency given by the equation:

$$\boxed{\omega_o = \frac{\sqrt{3}}{R_2 C}} \tag{6-25}$$

Comments

The above analysis reveals certain important features:

(a) The circuit is one example of an oscillator constructed with resistors, capacitors, and active elements (in this case, ideal voltage amplifiers of high gain). No inductors are used. The ability to build an oscillator without Ls is important in applications where inductors are expensive, overly large, nonlinear, and so forth—e.g., particularly for low-frequency applications, as below one Hz.

(b) The oscillation frequency

$$\omega_o = \frac{3}{R_2 C} \tag{6-26}$$

is proportional to $1/C$, in contrast to the LC or tuned-circuit oscillator where

$$\omega_o = \frac{1}{\sqrt{LC}} \tag{6-27}$$

Consequently, if C can be varied over a 10:1 range, the LC oscillation frequency can only be controlled over about a 3:1 range*; if we require a wider range, we have to switch bands (switch to a different L or C). In the RC oscillator, however, a 10:1 variation of C permits a 10:1 change in frequency.

(c) The circuit of Fig. 6.11 and the above analysis assume that all capacitances are equal (C), and all resistances are R_1 or $2R_1$. To change the frequency, it is then necessary to vary the three capacitors together. It is possible to design active RC circuits in which the oscillation frequency is controlled by only a single capacitance; however a 10:1 variation in capacitance then does not permit as much variation in ω_0. In Fig. 6.11, for example, when only one capacitor is variable, the maximum frequency range is $\sqrt{10}:1$, realized when that capacitance is significantly less than the other two. The circuit of Fig. 6.11 is only one of many possible oscillators using Rs, Cs, and active elements.

In this section and Sec. 6.1, we have introduced op-amps, with particular emphasis on the adder and the integrator. Essentially, we have changed the working tools of the circuit designer. Instead of R, L, and C, we are proposing to use R, C, and op-amps.

In the next three sections, we see how any desired transfer function can be realized with op-amps. In Sec. 6.6, we resume the discussion of why op-amps provide a more attractive solution to the synthesis problem than the classical R-L-C networks.

6.3 TRANSFER-FUNCTION REALIZATION

In this section, we wish to find a network to realize a given transfer function; in other words, in Fig. 6.17 we know $T(s)$ or Y/X, and we want to determine a system which has this transfer function.

Any given $T(s)$ can be realized by a tremendous variety of different networks or systems. In order to limit the number of possible different solutions, we specify at the outset the use of only resistors, capacitors, and operational amplifiers. This is not really a great restriction, however; there are still many possible solutions. For simplicity, we consider only three different forms—which we call types I, II, and III in this section and the next two.

The solutions we find are really *analog computer programs* or *analog simulations* for the system. In control engineering, they are termed *state models*. In electronics, they are called op-amp realizations for the specified $T(s)$ and are often used today as

*This is the reason the AM broadcast band was originally chosen from about 550 KHz to 1600 KHz. With this range of about 3:1, a *single*, variable air capacitor could be used in the receiver to tune to the desired station.

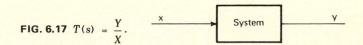

FIG. 6.17 $T(s) = \dfrac{Y}{X}$.

particularly attractive and economical filters or equalizers for communication systems. Even in digital circuitry, we often design these circuits first and then convert to the digital form.

Thus, we are solving a fundamental problem of electrical engineering: given a transfer function $T(s)$ which has the system characteristics we want, how do we find a network with this $T(s)$? How do we build such a system in the laboratory?

Further restrictions on the system

The limitation to RC op-amp circuits still leaves us with a multitude of possible solutions, and we wish to be even more restrictive. Designing will be done by going from the transfer function $T(s)$ to a signal flow diagram (or block diagram) and then to the network. Hence, it is convenient to impose additional restrictions in terms of the form of the signal flow diagram.

For convenience in building the system in the laboratory, we impose the conditions:

(1) s can only appear in a branch of transmittance $-1/s$ (the minus sign is used because the simple integrator always involves sign reversal, as shown in Sec. 6.1).

(2) if the degree of the denominator of T is n, there are n different branches, each of transmittance $-1/s$.*

(3) No other signal enters the node at the end of each $-1/s$ branch.

Constraint (1) means that the signal flow diagram consists of integrators and gain constants. Condition (2) means that a third-order system (for example) is realized by three integrators. Constraint (3) means that, if we call the output of the $-1/s$ branch the variable z, the input node is $-z'$, since

$$z = -\frac{1}{s}(-z') \tag{6-28}$$

(There is no other branch entering z; hence z is just $-1/s$ times the other node variable shown in Fig. 6.18.)

These three restrictions may seem irrational; we shall see below, however, that they still permit many different solutions and that they lead to systems particularly easy to build and analyze.

---

*We assume here the degree of the denominator of T is at least as great as the degree of the numerator. In most cases, the denominator degree exceeds the numerator (the gain tends toward zero at very high frequencies).

FIG. 6.18 The $-1/s$ branch.

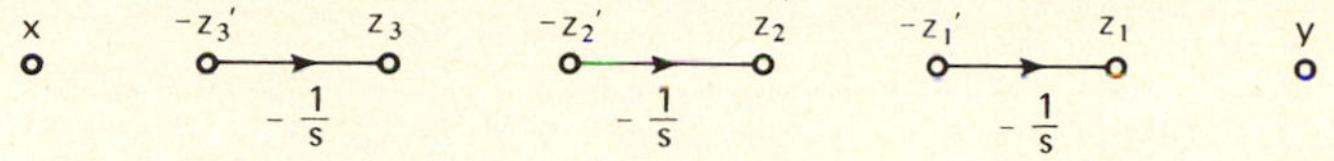

FIG. 6.19 Realization of the system order.

Type I realization

In order to illustrate the realization, we use the particular example

$$\frac{Y}{X} = T(s) = \frac{2s^2 + 6s + 8}{s^3 + 7s^2 + 14s + 8} \tag{6-29}$$

We proceed to find the signal flow diagram in the following steps:

(a) This is a third-order system, we need three integrators. (Fig. 6.19.) We arbitrarily call the three integrator outputs z_1, z_2, and z_3, as indicated in the figure. Then the integrator inputs are $-z'_1$, $-z'_2$, and $-z'_3$, respectively.

(b) We now connect z_3 to $-z'_2$, z_2 to $-z'_1$, and z_1 to y by unity branches (Fig. 6.20).

(c) Now we divide by numerator and denominator of $T(s)$ by s^3.

$$T(s) = \frac{2/s + 6/s^2 + 8/s^3}{1 + 7/s + 14/s^2 + 8/s^3} \tag{6-30}$$

(d) We wish to add feedback loops to Fig. 6.20 in such a way that the denominator of $T(s)$ is realized. Mason's reduction theorem indicates that the simplest situation is when all loops are touching; then the denominator is one minus the sum of all loop gains:

$$\Delta = 1 - \sum T_j \tag{6-31}$$

To be sure all loops touch, we start them all from z_1 in Fig. 6.20. Then we desire the loop gains

$$-\frac{7}{s} \qquad -\frac{14}{s^2} \qquad -\frac{8}{s^3}$$

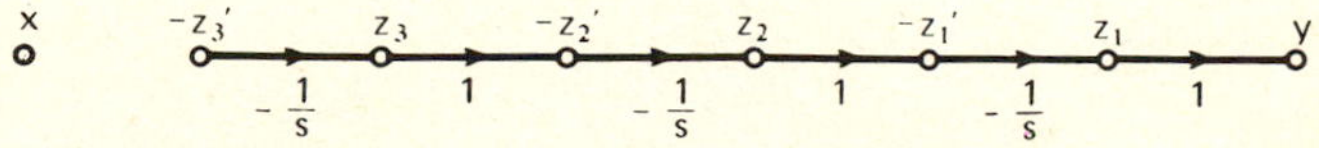

FIG. 6.20 Insertion of unity branches.

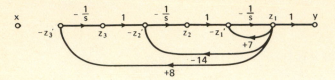

FIG. 6.21 Denominator of $T(s)$ is realized.

The first can be realized by a branch from z_1 to $-z_1'$ of gain +7; the second, z_1 to $-z_2'$ of gain -14; and the last, z_1 to $-z_3'$ of gain +8. Hence, Fig. 6.21 realizes the desired denominator of $T(s)$ in Eq. (6-30).

(e) Now we turn to the numerator. Again we take the simplest approach and make sure all direct paths (x to y) pass through z_1 so that the numerator is just

$$\sum P_i$$

(i.e., all $\Delta_i = 1$). Equation (6-30) reveals we want path gains

$$\frac{2}{s} \qquad \frac{6}{s^2} \qquad \frac{8}{s^3}$$

The last is realized by a branch of gain -8 from x to $-z_3'$; the second, gain 6 from x to $-z_2'$; and the first, gain -2 from x to z_1'. The final system is shown in Fig. 6.22.

Comments

Two comments should be made on this realization. First, we have made a number of totally arbitrary decisions—starting with the use of three unity branches in Fig. 6.20, and then forcing all loops and later all forward paths to pass through z_1. Changing these decisions would result in different diagrams; indeed, an infinity of correct solutions is possible for the specified $T(s)$. Figure 6.22 happens to be a particularly simple .

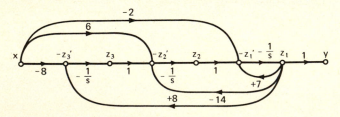

FIG. 6.22 Realization of

$$T(s) = \frac{2s^2 + 6s + 8}{s^3 + 7s^2 + 14s + 8}.$$

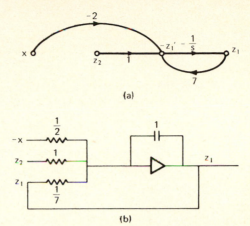

FIG. 6.23 Circuit for the rightmost integrator in Fig. 6.22; (a) Portion of signal flow diagram realized by the integrator; (b) Integrator circuit (values in megohms and microfarads).

form. Second, the signal flow diagram is drawn *without* factoring the polynomials of $T(s)$. Regardless of the degree of these polynomials, the signal flow diagram can be written by inspection.

The circuit

Once we have the signal flow diagram (Fig. 6.22), we can draw the circuit diagram by inspection. The realization of the last (rightmost) integrator is shown in Fig. 6.23. The circuit realizies the $-1/s$ branch and the three branches coming into $-z_1'$. The circuit requires the inputs $-x$ and z_2, and gives z_1 as the output.

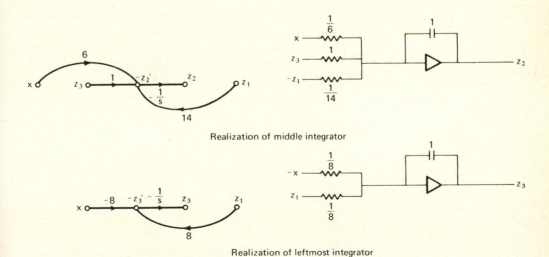

Realization of middle integrator

Realization of leftmost integrator

FIG. 6.24 Other integrators of Fig. 6.22.

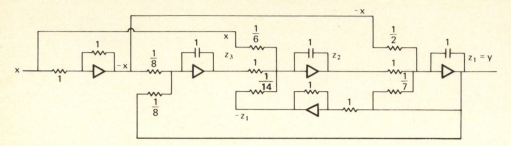

FIG. 6.25 Op-amp realization of

$$T(s) = \frac{2s^2 + 6s + 8}{s^3 + 7s^2 + 14s + 8}.$$

In the same way, the middle and leftmost integrators are designed (Fig. 6.24). Now the three circuits of Figs. 6.23 and 6.24 require the inputs $-x$, z_2, z_1, x, z_3, and $-z_1$. We have x as the input signal, and z_1, z_2, and z_3 are available as integrator output signals. To generate $-x$ and $-z_1$, two sign changers are necessary. The complete, interconnected circuit is shown in Fig. 6.25.

Comments on the circuit

(1) We require three integrators because the system is third order. In addition, two sign changers are usually needed (in special cases, fewer may be required). Thus, the circuit demands $(n + 2)$ ideal amplifiers, where n is the system order.

(2) In the circuit diagram of Fig. 6.25, R and C values are shown in consistent units. If the Rs are in megohms, the Cs must be microfarads. The RC products (or the R ratios in sign changers) are the critical quantities.

(3) In the description above, we have included every step. In practice, one would draw the signal flow diagram from the $T(s)$, then immediately construct the circuit diagram. With Rs, Cs, and ideal amplifiers, we can now synthesize any transfer function in a straightforward procedure (e.g., no factoring of the characteristic polynomial is needed). Furthermore, the circuit is practical.*

6.4 A SECOND REALIZATION

While Sec. 6.3 is a complete solution to the problem of transfer function realization, we always enjoy alternative networks. If there are several different solutions available for a design problem, we can select the best on the basis of criteria which are difficult to phrase mathematically—for example, lowest cost.

---

*This sounds like a ridiculous advantage. Obviously, we want a circuit which is practical, that can be easily built with available components. Unfortunately, one of the problems plaguing network synthesis before the availability of op-amps was that synthesized circuits often involved impractical element values.

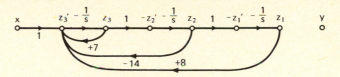

FIG. 6.26 Denominator of $T(s)$ realized.

A synthesis procedure similar to the above leads to a different realization (which we call Type II) if we arbitrarily decide that all feedback loops and all parallel paths should go through $-z_3'$ rather than z_1. Again we write

$$T(s) = \frac{2/s + 6/s^2 + 8/s^3}{1 + 7/s + 14/s^2 + 8/s^3} \tag{6-32}$$

Now the denominator is realized as shown in Fig. 6.26. (We note that in this case the unity branches include x to $-z_3'$, rather than z_1 to y, so that all paths from x to y go through $-z_3'$ and hence touch all loops.)

To realize the numerator *simply*, we want all x-to-y paths to go through $-z_3'$. Hence, the parallel paths are realized by branches from $z_1, z_2,$ and z_3 to y, and the final signal flow diagram is shown in Fig. 6.27.

In terms of the difficulty of the synthesis, there is really no difference between Type I and Type II forms. Both realizations involve three integrators. Type II is perhaps found more often in textbooks, Type I is somewhat more common in the laboratory. (There is a difference when we consider initial conditions in Sec. 9.2.)

The network

The signal flow diagram of Fig. 6.27 leads directly to the circuit diagram of Fig. 6.28. The three integrators are drawn first, then the adder to form the output is inserted. Again, in general we require two sign changers: one to realize the feedback paths with minus signs (only the -14 branch in Fig. 6.27), the other to realize the forward paths

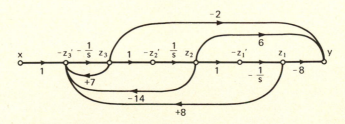

FIG. 6.27 Type II realization of

$$T(s) = \frac{2s^2 + 6s + 8}{s^3 + 7s^2 + 14s + 8}.$$

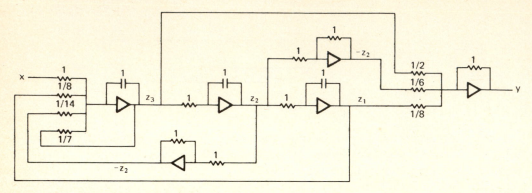

FIG. 6.28 Circuit for Fig. 6.27 or

$$T(s) = \frac{2s^2 + 6s + 8}{s^3 + 7s^2 + 14s + 8}.$$

to y with plus gains (only the +6 branch in this example). The entire circuit requires $n + 2$ ideal amplifiers, where n is the order of the system. (We could eliminate one sign changer in Fig. 6.28, since $-z_2$ is needed for both the feedback and feed-forward branches.)

Again, the fascinating feature of this op-amp realization of a specified $T(s)$ is the simplicity of the synthesis.

6.5 A THIRD SOLUTION

The third common form is based upon the partial fraction expansion. In our particular example,

$$T(s) = \frac{2s^2 + 6s + 8}{s^3 + 7s^2 + 14s + 8} \tag{6-33}$$

can be written with the denominator factored

$$T(s) = \frac{2s^2 + 6s + 8}{(s + 1)(s + 2)(s + 4)} \tag{6-34}$$

A partial fraction expansion gives

$$T(s) = \frac{4/3}{s + 1} + \frac{-2}{s + 2} + \frac{8/3}{s + 4} \tag{6-35}$$

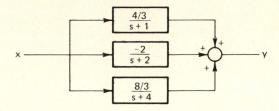

FIG. 6.29 Three parallel elements for $T(s)$.

In other words, $T(s)$ can be realized as the parallel connection of three, first-order systems (Fig. 6.29). If we realize each of these using feedback around a single integrator, we recognize that $k/(s + a)$ is realized by the system of Fig. 6.30. Hence, the total system of the $T(s)$ of Eq. (6-35) is shown in Fig. 6.31, and the circuit diagram in terms of ideal amplifiers, R and C elements is shown in Fig. 6.32.

We should note that the Type III realization requires that we factor the system characteristic polynomial (the denominator of T)—the previous two forms do not require this step, which is often tedious.

Furthermore, the partial-fraction-expansion approach has to be modified if there are conjugate complex or multiple poles of $T(s)$. The approach can be illustrated by the example

$$G(s) = \frac{26s}{(s^2 + 4s + 13)(s + 2)} \tag{6-36}$$

The corresponding partial fraction expansion is

$$G(s) = \frac{(13\sqrt{13}/9)\angle\theta}{s + 2 - j3} + \frac{(13\sqrt{13}/9)\angle-\theta}{s + 2 + j3} + \frac{-52/9}{s + 2} \tag{6-37}$$

$$\theta = -\tan^{-1}\frac{3}{2}$$

A network for the first term requires complex values of gain (or resistance or capacitance).

FIG. 6.30 Realization of

$$\frac{k}{s + a}.$$

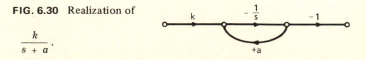

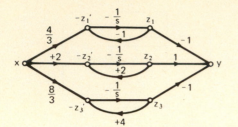

FIG. 6.31 Type III (normal form) realization of

$$T(s) = \frac{2s^2 + 6s + 8}{s^3 + 7s^2 + 14s + 8}.$$

This difficulty can be avoided if the first two terms are combined in the partial fraction expansion. In other words, we want to write $G(s)$ of Eq. (6-37) in the form

$$\frac{26s}{(s^2 + 4s + 13)(s + 2)} = \frac{As + B}{s^2 + 4s + 13} + \frac{k_{-2}}{s + 2} \tag{6-38}$$

k_{-2} is just the residue of G in the pole at -2, or $-52/9$. A and B can be evaluated in any convenient manner [e.g., by selecting two values of s and substituting in Eq. (6-38)]:

$$
\left.
\begin{array}{l}
\text{At } s = 0 \qquad 0 = \dfrac{B}{13} - \dfrac{52/9}{2} \quad \text{ or } \quad B = \dfrac{338}{9} \\[3mm]
\text{As } s \to \infty, \text{ the left side behaves as } \dfrac{26}{s^2}; \text{ hence } \dfrac{A}{s} - \dfrac{52/9}{s} = 0 \qquad A = \dfrac{52}{9}
\end{array}
\right\}
$$
$$\tag{6-39}$$

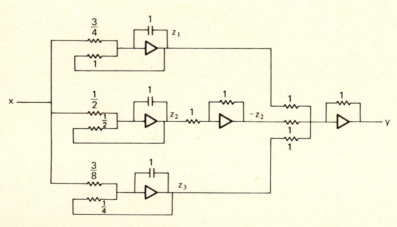

FIG. 6.32 Circuit for realization of normal form for

$$T(s) = \frac{2s^2 + 6s + 8}{s^3 + 7s^2 + 14s + 8}.$$

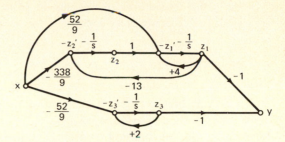

FIG. 6.33 Quasi-partial fraction expansion form for

$$T(s) = \frac{26s}{(s^2 + 4s + 13)(s + 2)}.$$

and

$$G(s) = \frac{(52/9)s + 338/9}{s^2 + 4s + 13} + \frac{-52/9}{s + 2} \tag{6-40}$$

Figure 6.33 shows the corresponding system, with the first term of Eq. (6-40) realized by a Type I form and thus with all real coefficients in the complete system.*

Final comment

In this section and the two preceding, we have developed procedures for going from a given transfer function, $T(s)$, to a corresponding electrical network (using resistors, capacitors, and ideal, high-gain, voltage amplifiers). We have solved a *network synthesis* problem; we can now synthesize a network to realize a desired transfer function.

Furthermore, there is an infinity of different solutions. We arbitrarily chose three as examples. Others can be derived by changing our unity branches to any other non-zero gains, or by using different configurations.†

Finally, we should mention that the solutions developed are also called *analog computer programs*. An electronic analog computer is merely an aggregate of amplifiers, Rs, and Cs, which can be wired up to realize a specified transfer function.

---

*A directly analogous procedure is applicable if the transfer function contains poles of order greater than unity. The partial fraction expansion then includes terms of the form

$$\frac{A}{(s + a)^n}$$

and we need the tandem connection of n integrators.

†A delightful feature of the synthesis problem is this multiplicity of correct solutions. We can choose one solution which is attractive in terms of cost, simplicity, availability of components, and so forth. In contrast, there is only one correct solution to the analysis problem (Given the network what is the response?).

In the actual laboratory construction of such circuits, there are several additional (practical) problems which we consider in the remainder of this chapter.

6.6 PRACTICAL PROBLEMS IN OP-AMPS

In the laboratory, building op-amp circuits to realize a given transfer function, the engineer finds that several practical problems arise. In this section, we discuss the effects of real (non-ideal) components. In following sections, we consider problems imposed by the desirability of a reasonable time scale, and by the need for signal levels which are neither too small nor too large.

If we had ideal amplifiers and Rs and Cs which could be chosen at exactly the desired values, we could build the circuits described in the preceding sections. Unfortunately, if we use a 300 K resistance, the actual R value is likely to be anywhere within 5% (±15 K). The same problem exists with capacitances. Also, no real amplifier is truly *ideal* within our definition (zero input admittance, zero output impedance, and effectively infinite voltage gain).

Incorrect R and C values

Figure 6.34 shows the circuit for realization of a first-order transfer function. If the capacitance is in microfarads, the resistances must be in megohms. In other words, if C is 1 μF, the two resistances needed are 250 K and 500 K.

If values of R cannot be selected precisely, the circuit can be modified to permit experimental adjustment of the two integrator gain constants. Instead of using x and y as the two integrator inputs, we use the outputs of potentiometers fed by x and y (Fig. 6.35).

The analysis of Fig. 6.35 can be carried through by applying Thevenin's theorem to the left of A and B. To the left of A, the open circuit voltage is αx and the Thevenin resistance is the parallel combination of the two parts of the potentiometer (Fig. 6.36). If the total potentiometer resistance is much less than R_1', the $R_a R_b/(R_a + R_b)$ resistance in Fig. 6.36(b) is negligible and the input current i is just

$$\frac{\alpha}{R_1'} x$$

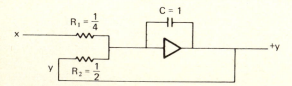

FIG. 6.34 Op-amp with

$$\frac{Y}{X}(s) = \frac{-4}{s + 2}.$$

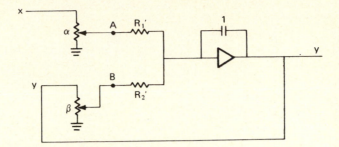

FIG. 6.35 Integrator with adjustable gains.

In other words, the potentiometers in Fig. 6.35 effectively multiply R_1' and R_2' by $1/\alpha$ and $1/\beta$, respectively.

We can now see how to redesign the circuit of Fig. 6.34 to incorporate the potentiometers. If the α and β settings are to be about midway along the pots, we choose R_1' and R_2' equal to $R_1/2$ and $R_2/2$, respectively. (The two gain constants of the integrator are doubled—then to be halved by α and β.) Figure 6.37 shows the resulting circuit, with α and β adjusted experimentally to realize precisely the constants 4 and 2 in the transfer function.

In the actual construction of any of the circuits described in preceding sections, each parameter in the transfer function depends on a single resistance in the network. If that parameter must be set precisely, the potentiometer arrangement can be used.

Furthermore, the potentiometer circuit allows variation of the parameter value in order to observe the effect on system characteristics. This capability is really the primary goal of analog simulation in system design. Before we actually go to construction of a prototype system (airplane, production process, and so on), we can use a simulator to test the effects of different parameter values.

Non-ideal amplifiers

No amplifier is ideal in the sense that we have been using the term. We have assumed the amplifier has the three properties shown in Fig. 6.38(a). Under these constraints, the transfer function of the representative op-amp of (b) is

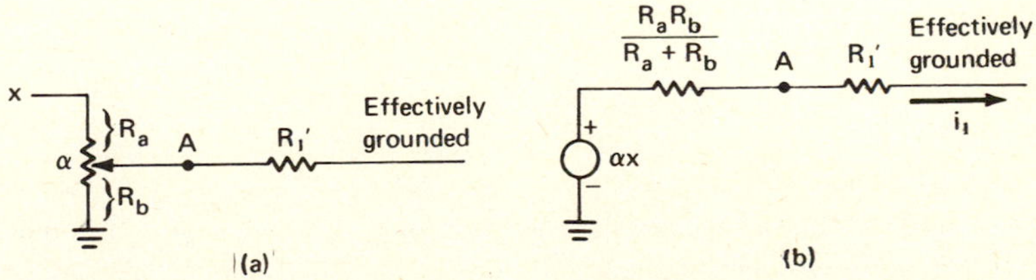

FIG. 6.36 Analysis of part of Fig. 6.35: (a) Original circuit for x input; (b) Thevenin equivalent of (a).

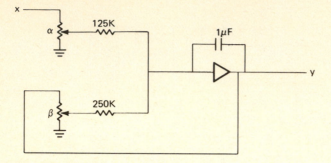

FIG. 6.37 Adjustable circuit for realization of

$$\frac{Y}{X} = \frac{-4}{s + 2}.$$

$$\frac{Y}{X} = -\frac{Z_f}{Z_i} \tag{6-41}$$

derived from a single node equation at the u node. If all three conditions of (a) are violated, a circuit model is that of Fig. 6.38(c). From this, we can calculate the actual transfer function Y/X.

Now two node equations (at u and y) are needed:

$$\left.\begin{array}{c} \dfrac{U - X}{Z_i} + \dfrac{U}{Z_{\text{in}}} + \dfrac{U - Y}{Z_f} = 0 \\[2ex] \dfrac{Y - AU}{Z_{\text{out}}} + \dfrac{Y - U}{Z_f} = 0 \end{array}\right\} \tag{6-42}$$

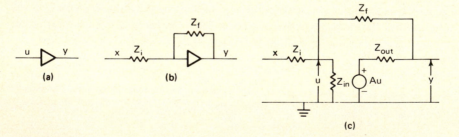

FIG. 6.38 Op-amps: (a) Ideal amplifier, satisfying: (1) Input $Y = 0$, (2) Output $Z = 0$, (3) |Gain| = ∞; (b) Ideal op-amp

$$\frac{Y}{X} = -\frac{Z_f}{Z_i}$$

(c) Actual op-amp.

Algebraic elimination of U gives

$$\frac{Y}{X} = \frac{AZ_f + Z_{\text{out}}}{(1 - A)Z_i + Z_f + Z_{\text{out}} + (Z_i Z_f + Z_i Z_{\text{out}})/Z_{\text{in}}} \qquad (6\text{-}43)$$

We can use this expression to investigate the effects of any one or more deviations from the ideality assumptions of Fig. 6.38(a). Actually, substitution of typical values shows that the first terms of both denominator and numerator really dominate—particularly if $|A|$ is greater than 1000 and if Z_i and Z_f are each much larger than Z_{out}, much smaller than Z_{in} (in other words, in a typical voltage amplifier if Z_i and Z_f are around 100 K).

Frequency dependence of A

A much more difficult deviation from ideal behavior occurs if we try to use the op-amps at high frequencies where the gain A has an associated phase shift. This is particularly important when integrated circuits are used,* since it is then common for the amplifier to involve as many as 18 resistors and 18 transistors and diodes, in order to realize the required gain and stability. Under these circumstances, the gain falls off at high frequencies, often as low as 50 KHz, and the phase shift may be significant at even lower frequencies.†

Two inputs for op-amps

Finally, many commercially available op-amps provide two possible inputs—one involving a sign reversal, the other not. Under such circumstances, we can modify our synthesis procedures of the preceding three sections. There is now no need to realize the integrators by $-1/s$ blocks; instead, $1/s$ blocks can be used.

Furthermore, since the total system is constructed of adders and integrators, it is useful to simplify the diagram. Instead of working in terms of Rs, Cs, and ideal amplifiers, we use the adders and integrators as building blocks, with the symbols shown in Fig. 6.39(a).

6.7 TIME SCALING

In using analog computers or simulators in engineering work, we often wish to change the time scale in which the simulation responds—without altering the form of the dynamic characteristics. For example, an airplane may oscillate at a low frequency (less than one Hz) in what is called the phugoid mode. To study this oscillation, we derive the differential equation or transfer function describing the phenomenon, then

---

*And a major advantage of op-amps is in integrated circuits, where we can avoid inductors. This possibility is also advantageous in networks for very low frequencies, where inductors become large, heavy, and expensive.

†The manufacturer provides gain (and occasionally phase) information versus frequency.

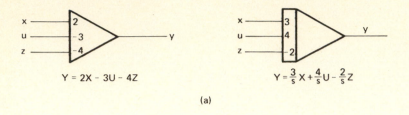

$$Y = 2X - 3U - 4Z \qquad\qquad Y = \frac{3}{s}X + \frac{4}{s}U - \frac{2}{s}Z$$

(a)

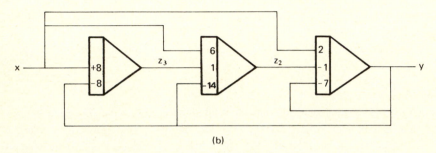

(b)

FIG. 6.39 Simplified representation of op-amp realization of a transfer function: (a) Symbols for basic op-amps; (b) Realization of Fig. 6.25,

$$T(s) = \frac{2s^2 + 6s + 8}{s^3 + 7s^2 + 14s + 8}.$$

build an analog simulation in the laboratory. In the simulator, we can easily vary the system parameters or constants (by varying potentiometers) and observe the corresponding changes in the oscillation. The analog simulation avoids the expensive and awkward step of changing airplane dimensions and then flying the modified plane. We "fly" the analog simulator instead.

Once the simulator is designed, we often find the response time inconvenient electronically. The system may respond so slowly that we have to wait an inconveniently long time to observe the behavior; the response may be too slow or too fast for the recording equipment or signal generators available. In such a case, we wish to *time scale* the simulation—to change the duration of the system response without changing the form. Figure 6.40 is an example of the type of change desired.

Modification of T(s)

Time scaling of the system simulation can be accomplished by a simple modification of the transfer function $T(s)$. To speed up the response by a factor a, we simply replace each s by s/a.

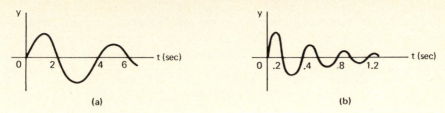

FIG. 6.40 Example of time scaling: (a) Response of original system; (b) Response of system speeded up by factor of 10.

In other words, if the system is described by

$$T(s) = \frac{2s + 3}{s^2 + 8s + 20} \tag{6-44}$$

a system which responds in the same way but 20 times as fast (or in $1/20$ of the time) has

$$T_1(s) = \frac{2s/20 + 3}{(s/20)^2 + 8s/20 + 20} \tag{6-45}$$

or, if we clear of fractions,

$$T_1(s) = \frac{40s + 1200}{s^2 + 160s + 8000} \tag{6-46}$$

The meaning of this time scaling is illustrated in Fig. 6.41. The original system, $T(s)$, gives a steady state response indicated in (a) when driven by a signal $4\cos(2t + 10°)$, The time-scaled system, $T_1(s)$, is shown in part (b): when we drive the system 20 times

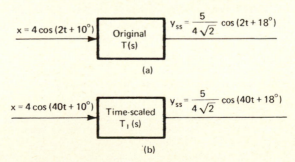

FIG. 6.41 Meaning of time scaling: (a) Original system; (b) Time-scaled system.

as fast, the response is the original response speeded up by the factor of 20. We can measure the characteristics of $T_1(s)$ in 1/20 the time required to measure $T(s)$.

Thus, in terms of the system, time is scaled simply by changing all frequencies proportionally.

Modification of the circuit

The preceding paragraphs show that we can scale time by modifying the transfer function. If the circuit is already designed, we can also modify the circuit elements $(R, C, \text{ and } L)$ to achieve the same result.

To time scale, we replaced s by s/a in order to speed up the characteristics by a. But every s in the transfer function comes from a term of the form Cs or Ls when we analyze the network by node equations or any other technique.* Therefore, replacing s by s/a in the transfer function is equivalent to replacing each C by C/a and each L by L/a:

$$Cs \longrightarrow C\frac{s}{a} = \left(\frac{C}{a}\right)s$$

$$\tag{6-47}$$

$$Ls \longrightarrow L\frac{s}{a} = \left(\frac{L}{a}\right)s$$

Thus, to speed up the response of a network by a, we simply replace each C and L by C/a and L/a, respectively.

The process in terms of an RLC network is illustrated by Fig. 6.42, and in terms of an analog simulation by Fig. 6.43. In both cases, the response is speeded up by a factor of 20. (Once the new simulator network is designed, we can change the Cs by any desired factor, as long as we change the resistances at the integrator inputs by the reciprocal factor to keep the RC products unchanged.)

---

*We assume here that the network is electrical and the elements are R, L, C, and ideal amplifiers. If we have other elements, we may need to expand the origins of s terms; for example, simple mechanical circuits lead to terms of the form Ms and $(1/K)s$ where M and K represent the mass and the spring constant.

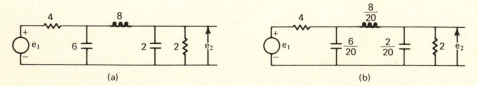

(a)   (b)

FIG. 6.42 Network response speeded up by 20: (a) Original network; (b) Time-scaled network.

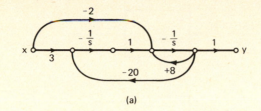

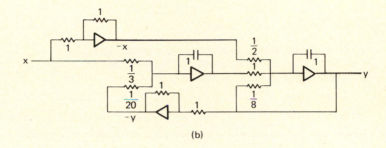

(a)

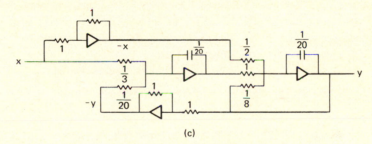

(b)

(c)

FIG. 6.43 Analog computer time-scaled: (a) Signal flow diagram for

$$T(s) = \frac{2s + 3}{s^2 + 8s + 20}$$

(b) Simulation of $T(s)$ with R's, C's, ideal amplifiers; (c) Simulation of $T_1(s)$—system speeded up by factor of 20.

Signal applied to time-scaled network

When the system (the network or the analog simulation) is time-scaled as described above, we can proceed with our laboratory tests in order to observe the characteristics as particular parameters are changed, and so forth. We note that testing the time-scaled system requires, of course, that we time-scale the drive signals by the same factor as the network. For example, Fig. 6.44 shows a possible test signal, and the way that signal has to be speeded up when the network response is speeded up. If $T(s)$ is converted to $T_1(s)$ by substituting s/a for each s in $T(s)$, the drive function is changed by replacing each t by (at)—so that the time signal occurs correspondingly faster.

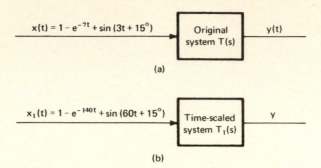

(a)

(b)

FIG. 6.44 Change in test signal when system speeded up by factor of 20: (a) Original system with possible test signal; (b) Time-scaled system with new test signal.

There is one situation in which we encounter trouble. If the drive signal contains an impulse, we have to consider the way this impulse is actually applied in practice (e.g., as a large pulse of short duration). To speed up the signal by a, we merely cut the duration of this pulse by the factor a (Fig. 6.45). Then, the area under the pulse is also reduced by the factor a; hence, to the time-scaled network we apply a test signal which includes an impulse of area $(1/a)$ times the impulse applied to the original network. The response is then unchanged in form.

In other words, if the input signal for the original system is

$$x(t) = 4\delta(t) + 6 - 6e^{-2t} + 7\sin(3t + 125°) \tag{6-48}$$

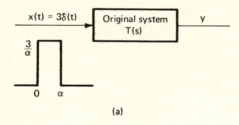

(a)

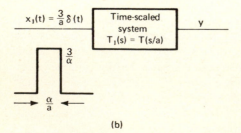

(b)

FIG. 6.45 The problem when we wish to apply an impulse to a time-scaled system: (a) Original system with a "real impulse" shown as the applied signal (α is small); (b) Time-scaled system speeded up by a factor a. Now the applied pulse is of duration α/a; hence $x_1(t)$ is an "impulse" of area $3/a$ rather than 3.

and if we decide in the laboratory to speed up the response of the system by the factor 8, we proceed as follows:

(a) Each C and L is replaced by $C/8$ and $L/8$, respectively.

(b) The drive signal applied is

$$x_1(t) = \frac{4}{8}\delta(t) + 6 - 6e^{-16t} + 7 \sin(24t + 125°) \tag{6-49}$$

Equation (6-49) is derived from Eq. (6-48) by:

(1) replacing t by $8t$

(2) dividing the impulse amplitude by 8

Step (2) must not be omitted; if it is, the response of the time-scaled system may bear little resemblance to the response of the original system.

6.8 AMPLITUDE SCALING

Amplitude scaling is the process of changing the levels of signals at various points within a system without altering the overall dynamics. Figure 6.46 is a very simple example. In each case, the overall input-output transfer function is

$$\frac{Y}{X} = \frac{100s}{(s + 1)(s + 100)} \tag{6-50}$$

In the original system, the signal u is between the two blocks. If noise is added to the system at this point, the signal u may be so small that it is swamped by the noise; the output y then is primarily the noise modified by the system $100s/(s + 100)$. Figure 6.46(b) shows one way to decrease this difficulty if it is possible to realize the signal $10u$, rather than u. Now the signal at the input of B' (i.e., $10u$) is ten times larger and a given noise signal has less effect.

Thus, amplitude scaling refers to a redistribution of the constant multipliers among the various parts of the system.* The objective is to change the levels of various internal signals, without changing the overall dynamics.

There are three common purposes for amplitude scaling:

(1) to ensure that the signal at each point is much larger than the noise;

(2) to avoid saturation and nonlinear operation;

(3) to improve the required element values (Rs and Cs).

*Occasionally we wish to do something like amplitude scaling, but with the process frequency-dependent (e.g., to increase the high-frequency components of the signal early in the system and then restore them to normal level later). FM radio transmission uses this approach, called pre-emphasis and deemphasis. The term *amplitude scaling* refers only to frequency-independent changes effected by modifying the constant multipliers of component transfer functions.

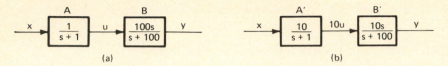

FIG. 6.46 Trivial example of amplitude scaling: (a) Original system; (b) System after scaling.

The second objective is simply illustrated by Fig. 6.47, three tandem amplifiers. When the system is built in the laboratory, we find that amplifier B saturates (the signal u is too large). By shifting the constant multipliers as indicated in part (b) of the system, the two internal signals are each divided by 10, while the overall Y/X is unchanged.*

The third possible goal above of amplitude scaling is illustrated in the example given in the appendix of this chapter.

In open-loop structures involving the tandem or parallel connection of components, amplitude scaling is straightforward: we merely distribute the constant multipliers to ensure desired signal levels at each point.

Feedback systems

Amplitude scaling in a feedback system is a somewhat more difficult task because the various multipliers (gain constants) must be adjusted to leave unchanged not only the direct transmission from input to output, but also all loop gains. For example, in the system of Fig. 6.48,

$$\frac{Y}{X} = \frac{G_0 G_1 G_2}{1 + G_1 G_2 H} \tag{6-51}$$

Hence, we can introduce scaling constants only if we leave both products

$$G_0 G_1 G_2 \quad \text{and} \quad G_1 G_2 H$$

unchanged.

---

*Amplitude scaling to avoid nonlinearity may also require raising signal levels. Many mechanical components exhibit a dead zone (i.e., very small signals cause no output at all). The input signals for such devices must be kept large enough to ensure proper operation.

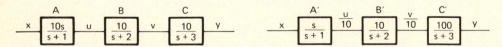

FIG. 6.47 Amplitude scaling to avoid saturation: (a) Original system; (b) System scaled to avoid saturation in B.

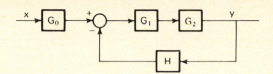

FIG. 6.48   Single-loop feedback system.

When we go to much more complex feedback systems, the approach of listing all the loop gains, which must be kept constant, is tedious and uninviting. Instead, we accomplish amplitude scaling in a step-by-step procedure, which is illustrated by the example in Fig. 6.49. Part (a) of the figure shows the original system in the form of a signal flow diagram, with z_1 and z_2 as the two integrator outputs.* We wish to realize the same overall transfer function, but with the two integrator outputs as az_1 and bz_2 (a and b are any positive constants).

The first step is to realize az_1 instead of z_1. *We can do this if every branch entering the z_1 node is multiplied by a and each outgoing branch is divided by a*—part (b) of the figure.

Usually it is desirable that the integrating branches have unity gain; therefore, we change the $\dot{z}_1$ node to $a\dot{z}_1$ as indicated in part (c). Again, each incoming branch to $\dot{z}_1$ is multiplied by a, each outgoing branch is divided by a.

Next, we realize bz_2 rather than z_2 as portrayed in (d), and finally return to $1/s$ integrating branches in the final realization displayed in Fig. 6.49(e).

The example of Fig. 6.49 is shown in terms of signal flow diagrams only because of the simplicity of this form; exactly the same procedure can, of course, be carried out in block diagrams as shown in Fig. 6.50.

Miscellaneous comments

Inspection of either Fig. 6.49 or Fig. 6.50 reveals that the second integrator (from $\dot{z}_1$ to z_1) and the associated feedback of -2 are unchanged when we scale both the input ($a\dot{z}_1$) and the output (az_1) of the integrator. In other words, as shown in Fig. 6.51, we can scale leaving everything within the dashed lines unchanged. This is a refinement, however, which need not be used. If we just proceed with the amplitude scaling on a step-by-step basis as shown in Fig. 6.49 or 6.50, we obtain the desired modification of the feedback system.

In analog computers or simulators, the operational amplifiers are often provided with overload lights to indicate when the signal exceeds the allowable value. The system under study is then represented on the computer; the specified input signals and initial conditions are applied, and the simulator is allowed to run. If any internal signals are too large, the overload indicators operate and we can experimentally determine the required factors of amplitude scaling needed to bring the signals down to the

---

*z_1 and z_2 are the two state variables: the output y for $t > 0$ depends on $x(t)$ for $t > 0$ and the initial conditions $z_1(0)$ and $z_2(0)$.

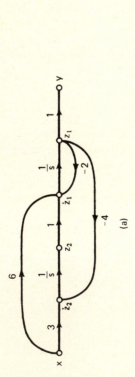

(d)

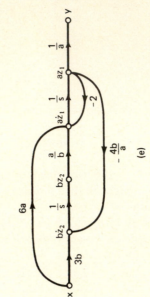

(a)

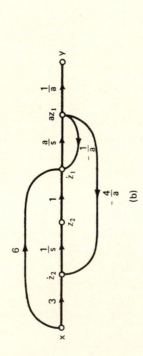

(e)

FIG. 6.49 Amplitude scaling in a second order two-loop feedback system: (a) Feedback system: analog-computer simulation for realizing

$$\frac{Y}{X} = \frac{6s + 3}{s^2 + 2s + 4}$$

(b) az_1 realized instead of z_1; (c) System of (b) with two unity-gain integrators again; (d) bz_2 realized instead of z_2; (e) Final system with amplitude scaling completed.

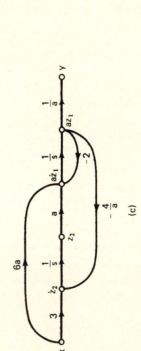

(b)

(c)

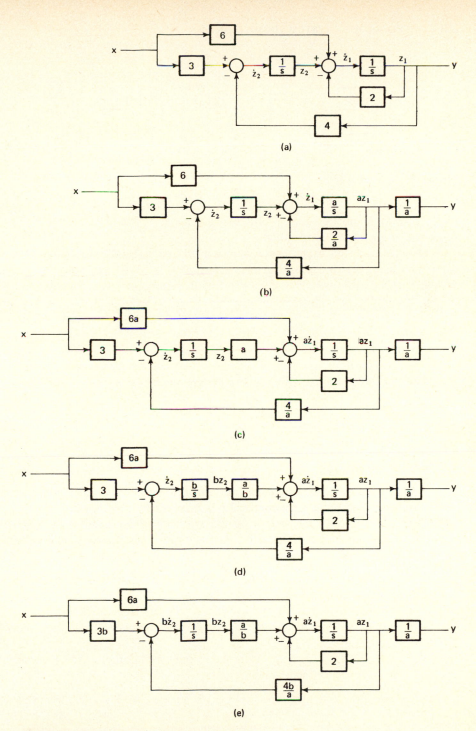

FIG. 6.50 Block-diagram equivalent of Fig. 6.49.

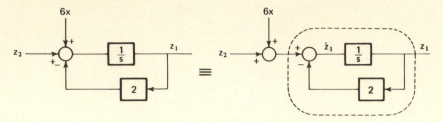

FIG. 6.51 One portion of Fig. 6.50.

required levels. An oscilloscope or voltmeter can be used to locate signals which remain too small during simulator operation.

In an actual case, this experimental approach to finding amplitude-scaling factors is often necessary. To find these factors analytically, we need to know the internal system signals during operation: if we can find these signals analytically, there is little reason to build the experimental simulator.*

Final example

The system shown in Fig. 6.52 serves as a final example of amplitude scaling in a feedback configuration. Here, either analysis or experiment indicates that z_2 and $\dot{z}_2$ should be replaced by $6z_2$ and $6\dot{z}_2$ respectively (i.e., the actual signals should be larger), while the simulation should realize $\frac{1}{8}z_1$ and $\frac{1}{8}\dot{z}_1$. The required changes in the realization are shown.

6.9 FINAL COMMENTS

This chapter represents a complete solution to the transfer function synthesis problem: Given $T(s)$, determine a network realizing that transfer function.

We have placed certain restrictions on the synthesis problem and its solution (in addition to the constraints of linearity, time invariance, and lumped network which are inherent in a transfer function which is a ratio of polynomials in s):

(1) The numerator of $T(s)$ cannot be of higher degree than the denominator (otherwise, differentiators are required).

(2) The system is inert—no energy storage at $t = 0$ or initial conditions.

(3) The system has one input and one output; it is described by a single transfer function.

(4) The realization uses op-amp adders and integrators; in other words, the available components are Rs, Cs, and ideal amplifiers.

*There are exceptions to this statement—cases in which internal signal levels are known à priori. We often build an analog simulator or op-amp feedback circuit to drive a piece of equipment to be tested. In such cases, all signals in the simulator may be known in detail, and the scaling factors can be selected before the equipment is built.

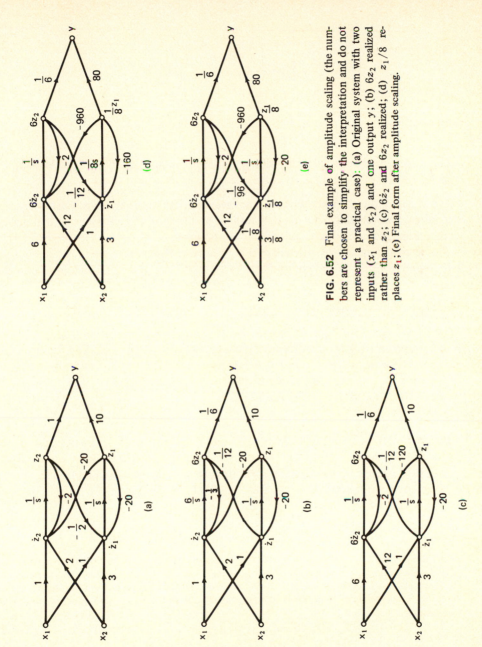

FIG. 6.52 Final example of amplitude scaling (the numbers are chosen to simplify the interpretation and do not represent a practical case): (a) Original system with two inputs (x_1 and x_2) and one output y; (b) $6z_2$ realized rather than z_2; (c) $6\dot{z}_2$ and $6z_2$ realized; (d) $z_1/8$ replaces z_1; (e) Final form after amplitude scaling.

203

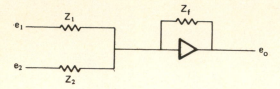

FIG. 6.53 Basic op-amp.

(5) Indeed, we have used a very simple form of op-amp (Fig. 6.53): the elements Z_1 and Z_2 are always resistors, Z_f is either a single resistor or capacitor.

In the next three chapters, we find that these restrictions are really not necessary. We can often find valuable and novel solutions to the synthesis problem or the analog-computer programming by relaxing these restrictions.

The awesome aspect of this chapter is, however, that within these restrictions we have found an infinite variety of solutions to the problem of transfer function realization.

APPENDIX—BUTTERWORTH FILTERS

There is a class of filter networks* which are particularly important and useful in both analog and digital circuits for communications and control. These are named *Butterworth filters* since the class was first described in the 1930s by Butterworth of Great Britain. A specific Butterworth filter will be described here, then some of the properties of the entire class discussed briefly.

A simple filter

Figure A6.1 shows a circuit which acts as a low-pass filter (i.e., the network passes low-frequency signals, blocks or stops high-frequency signals). The input is the current i_1 and the response is the voltage e_2. For this system, the transfer function E_2/I_1 can be found by node equations (or any other method of circuit analysis):

$$T(s) = \frac{E_2}{I_1} = \frac{64}{s^3 + 8s^2 + 32s + 64} \tag{A6-1}$$

*A filter is a network which allows selection of a desired signal from the sum of desired and undesired signals (signals plus noise, for example); the selection is accomplished by the differences in the frequency components of the desired and undesired signals. In other words, a filter has a high gain over certain bands of frequencies (the *pass bands*), a low gain over other frequency bands (the *stop bands*).

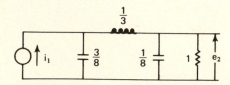

FIG. A6.1 Low-pass Butterworth filter.

Factoring the denominator polynomial yields

$$T(s) = \frac{E_2}{I_1} = \frac{64}{(s + 4)(s^2 + 4s + 16)} \tag{A6-2}$$

The poles are located at -4, $-2 \pm j2\sqrt{3}$, as shown in Fig. A6.2—on a circle of radius 4 and spaced $60°$ from the negative real axis.

In order to see the filtering characteristic of the circuit, it is necessary to calculate the way in which the gain $|T(j\omega)|$ varies with frequency ω. Equation (A6-2) gives $T(j\omega)$ as:

$$T(j\omega) = \frac{64}{(j\omega + 4)[(j\omega)^2 + 4j\omega + 16]} \tag{A6-3}$$

We find $|T(j\omega)|$ by determining the magnitude of the denominator by ordinary algebra. Alternatively, we can recognize that

$$|T(j\omega)|^2 = T(j\omega)\,T(-j\omega) \tag{A6-4}$$

(the magnitude squared of a complex number is just the number times its conjugate). Either calculation gives

$$|T(j\omega)| = \frac{64}{\sqrt{4^6 + \omega^6}} \tag{A6-5}$$

or, in somewhat more convenient form,

$$|T(j\omega)| = \frac{1}{\sqrt{1 + (\omega/4)^6}} \tag{A6-6}$$

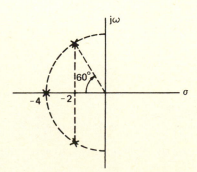

FIG. A6.2 Poles for $T(s)$ of Eq. (A6-2).

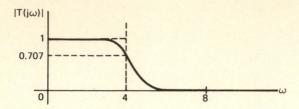

FIG. A6.3 Gain versus frequency for circuit of Fig. A6.1.

Figure A6.3 shows a sketch of $|T(j\omega)|$ versus ω.

Either Eq. (A6-6) or Fig. A6.3 reveals that the circuit is a *low-pass filter* with a cutoff frequency of 4 rad/sec. In other words, at frequencies less than 4 rad/sec, the gain is nearly unity; at frequencies above 4 rad/sec, the gain is close to zero. If the input signal is the sum of two sinusoids, one at 2 rad/sec and the other at 8 rad/sec, the output approximates just the low-frequency component of the input. The filter removes all components above the *cutoff frequency* of 4 rad/sec.

Gain at cutoff frequency

In the vicinity of the cutoff frequency, the gain is in transition from the low-frequency value of unity to the high-frequency value of zero [Fig. A6.3 or Eq. (A6-6)]. At the cutoff frequency

$$|T| = \frac{1}{\sqrt{1+1}} = 0.707 \tag{A6-7}$$

or the gain is down 3dB from the $\omega = 0$ value.

General Butterworth characteristic

The above example is a third order Butterworth filter: the system is of order three [the characteristic polynomial of $T(s)$ is a cubic] and

$$|T(j\omega)| = \frac{1}{\sqrt{1 + (\omega/\omega_c)^{2 \times 3}}} \tag{A6-8}$$

where ω_c is the cutoff angular frequency. The general Butterworth characteristic of order n is described by

$$|T(j\omega)| = \frac{1}{\sqrt{1 + (\omega/\omega_c)^{2n}}} \tag{A6-9}$$

n can be any positive integer.

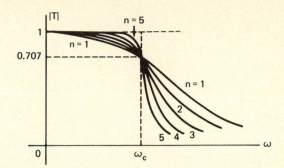

FIG. A6.4 Gain characteristics for various Butterworth filters.

The gain characteristics for $n = 1, 2, 3, 4,$ and 5 are shown in Fig. A6.4, while Fig. A6.5 shows the poles of the corresponding transfer functions. Clearly, the higher the order, the sharper the filter cutoff and the more complex the transfer function and the network. Each of the gain characteristics starts at 1 at $\omega = 0$, and each passes through 0.707 at the cutoff frequency. All n poles are located in the s plane on a circle of radius ω_c and separated by $180/n$ degrees around this circle. When n is odd, there is a pole on the negative real axis; when n is even, all poles are in conjugate complex pairs.

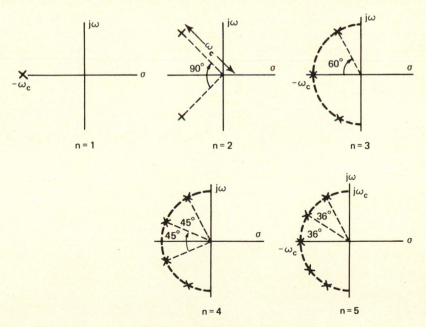

FIG. A6.5 Pole locations of Butterworth transfer functions.

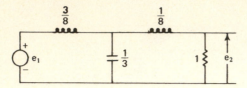

FIG. A6.6 Circuit to realize

$$\frac{E_2}{E_1} = \frac{64}{s^3 + 8s^2 + 32s + 64} .$$

Other transfer functions

The circuit of Fig. A6.1 realizes the Butterworth filter characteristic by the transfer function

$$T(s) = \frac{E_2}{I_1} \tag{A6-10}$$

(the output voltage divided by the input or drive current). If the input signal source is a voltage (rather than a current), a different network is required to realize the same transfer function, as shown in Fig. A6.6. If the input signal comes from a voltage source with an internal resistance of one ohm, the circuit of Fig. A6.7 realizes the Butterworth characteristic.

The derivation of the circuits in Figs. A6.1, A6.6, and A6.7 from the given transfer function is discussed in detail in courses on passive network synthesis. We emphasize only that the particular network depends on what is given as the source or input signal and what the output is. In all three cases above, the output is the voltage across a 1-ohm resistor. We might also consider cases where the output is a current in a specified circuit element; in general, then, the network would be different.

Change of impedance level

The element values of any of the above circuits are not practical; in Fig. A6.6, for example, we have a resistance of 1 ohm, inductances of 3/8 and 1/8 henry, and a capacitance of 1/3 farad. We can multiply the impedance of each element by any desired quantity and not change the Butterworth characteristic.

For example, Fig. A6.8 shows both Fig. A6.1 and Fig. A6.6 modified by raising the impedance level by a factor of 1000. Since impedances are Ls, R, and $1/Cs$, each L and R is multiplied by 1000, each C is divided by 1000. The voltage ratio E_2/E_1 is unchanged by this impedance transformation; the transfer impedance E_2/I_1 is multiplied by the same factor of 1000, but the frequency variation is unchanged.

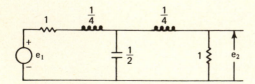

FIG. A6.7 Circuit to realize

$$\frac{E_2}{E_1} = \frac{32}{s^3 + 8s^2 + 32s + 64} .$$

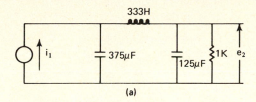
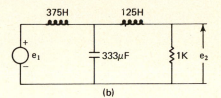

(a)   (b)

FIG. A6.8 Impedance level raised by 1000:
(a) Circuit of Fig. A6.1 modified

$$\frac{E_2}{I_1} = \frac{64000}{s^3 + 8s^2 + 32s + 64} ;$$

(b) Circuit of Fig. A6.6 modified

$$\frac{E_2}{E_1} = \frac{64}{s^3 + 8s^2 + 32s + 64} .$$

Change of cutoff frequency

All circuits in the above paragraphs possess a cutoff frequency of 4 rad/sec; this value was arbitrarily chosen in the original example of Fig. A6.1 simply so that we had a convenient number with which to work. In an actual filter, the cutoff frequency is determined by the specifications. For example, if we are building an amplifier for speech, we might wish to choose

$$f_c = 4000 \text{ Hz} \quad \text{or} \quad \omega_c = 8000\pi \text{ rad/sec} \qquad \text{(A6-11)}$$

since normal speech possesses very few significant frequency components above 4000 Hz.

To convert any of the above circuits from $\omega_c = 4$ to $\omega_c = 8000\pi$, we recognize that ω appears only in the forms

$$L\omega \quad \text{or} \quad C\omega$$

in the evaluation of the transfer function. Hence, the desired change is effected by dividing each L and each C by 2000π. Figure A6.9 shows the circuit of Fig. A6.6 modified to have a cutoff frequency of 8000π rad/sec, with a low-pass, $n = 3$ Butterworth characteristic. This frequency modification is an example of the time scaling discussed earlier.

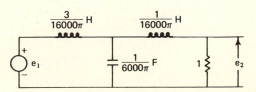

FIG. A6.9 Circuit of Fig. A6.6 with cutoff frequency multiplied by 2000π.

Filter design

In designing a low-pass Butterworth filter for a specified signal source and response variable and for a given cutoff frequency, we proceed in the steps:

(1)   the order n required for the desired rate of cutoff is determined;
(2)   the filter is synthesized for any convenient ω_c and impedance level;
(3)   the cutoff frequency is adjusted to the desired value (by time or frequency scaling);
(4)   the impedance level is adjusted to give convenient values for the Rs, Ls, and Cs.

Op-amp realization

For an example, we wish to realize a low-pass Butterworth filter with op-amps according to the following specifications:

$n = 3$ to assure sufficiently rapid cutoff
$\omega_c = 2000$ rad/sec (cutoff frequency of 318 Hz)
Gain at zero frequency is unity
Input and output both voltage signals.

The design proceeds in the following steps.

(1)   First we find the transfer function with any convenient cutoff frequency. If we select an ω_c of unity, the three poles are on the unit circle every $60°$, or at

$$1\underline{/180°}, \quad 1\underline{/120°}, \quad 1\underline{/240°}$$

Then the desired transfer function is

$$T(s) = \frac{1}{(s + 1)(s^2 + s + 1)} \tag{A6-12}$$

or

$$T(s) = \frac{Y}{X} = \frac{1}{s^3 + 2s^2 + 2s + 1} \tag{A6-13}$$

(2)   One corresponding op-amp realization is shown in Fig. A6.10.
(3)   The cutoff frequency can be raised to 2000 rad/sec by dividing each C by 2000 (Fig. A6.11).
(4)   The gain at zero frequency is -1 in Fig. A6.11; we must drive with $-x$ to obtain the y given by Eq. (A6-13). Alternatively, we can drive with x and simply take the output after the sign changer (Fig. A6.12).
(5)   Finally, we can adjust the impedance level of the system (indeed, of each op-amp separately, if we wish). If all Rs above are measured in megohms, then the Cs are in microfarads, and Fig. A6.13 shows the final system.

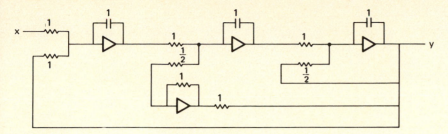

FIG. A6.10 System for cutoff frequency of 1 rad/sec [actual T realized is negative of Eq. (A6-13)].

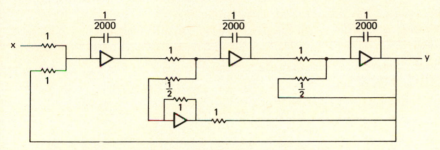

FIG. A6.11 Butterworth filter: $\omega_c = 2000$, gain at zero frequency is –1.

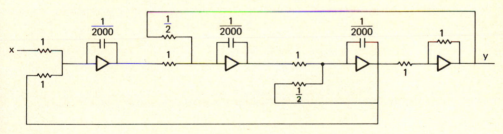

FIG. A6.12 System modified to give $T(0) = 1$.

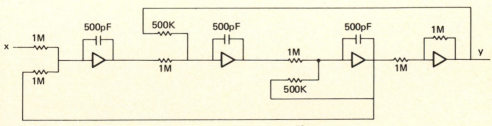

FIG. A6.13 Op-amp Butterworth filter for $n = 3$, $\omega_c = 2000$, $\dfrac{Y}{X}(0) = 1$.

PROBLEMS

6.1 Using capacitors of $2\ \mu F$ only, find a circuit realization (also using resistors and high-gain voltage amplifiers) for the transfer function

$$\frac{E_2}{E_1} = \frac{2s + 3}{s^2 + 8s + 24}$$

Redraw the circuit with the response ten times as fast (but still with $2\ \mu F$ capacitors only).

Redraw the *original* circuit with the output of the first (leftmost) integrator four times greater (but still with $2\ \mu F$ capacitors).

6.2 The figure shows the analog computer diagram for the simulation of a dynamical system. Determine the transfer function and the differential equation describing the relation between $f(t)$ and $y(t)$. Note that the symbols are defined to the right of the diagram.

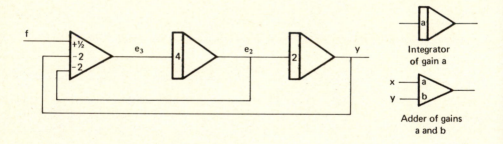

6.3 How many op-amps will be required for integrators in the realization of

$$T(s) = \frac{s^4 + s^3 + s^2 + s + 1}{s^8 + 2s^7 + 6s^6 + 20s^5 + 24s^4 + 16s^3 + 6s^2 + 2s + 1}$$

How many sign changers? (It is not necessary to find the flow diagram or circuit.)

6.4 (a) Using operational amplifiers, realize with all feedback from the output

$$T(s) = \frac{2s^2 + 3s + 4}{s^3 + 6s^2 + 11s + 6}$$

(b) Repeat (a) with all feedback to the input of the first integrator.

(c) Repeat (a) with a partial fraction expansion of $T(s)$ and a parallel realization of each term (notice that the denominator has a zero at -1).

6.5 (a) The network shown has a response $y(t)$ when the input is $x(t)$. We want the network to operate 80 times as fast: i.e., when the input is $x(80t)$, the output should be $y(80t)$. Find the new network.

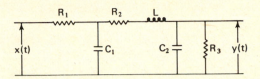

(b) If $x(t) = 3\delta(t) + 4 - 6e^{-2t}\sin 7t$, what signal should be applied to the time-scaled network?

6.6 In the solution of Prob. 6.4(a), rework with the response speeded up by 8. Then call the three state variables $z_1, z_2,$ and z_3, from left to right. Modify the realization to obtain $4z_1, z_2/2,$ and $5z_3$ as integrator outputs.

6.7 For the system shown, determine an equivalent system (same Y/X) which realizes

$8z_2$ and $8\dot{z}_2$ rather than z_2 and $\dot{z}_2$

$4z_3$ and $4\dot{z}_3$ rather than z_3 and $\dot{z}_3$

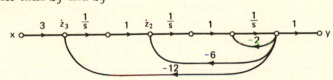

6.8 The circuit shown is taken from the pamphlet, "Fairchild Second Generation Linear Integrated Circuits," and utilizes a high-gain, monolithic operational amplifier. Input 3 is the noninverting input; input 2 is the inverting input.

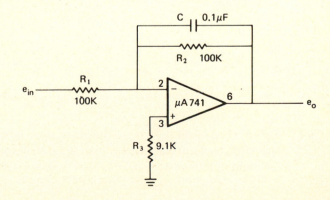

(a) Why is the resistor R_3 included?

(b) If the input is a square wave at 1 KHz and of amplitude 5 volts (peak-to-peak), sketch the output, indicating the peak-to-peak amplitude (the manufacturer promises that this will be linear within 1%).

6.9 Design a low-pass Butterworth filter to realize the following characteristics (the filter is to separate tracking signals from noise picked up in the radar and amplifiers).

(a) Input is the voltage from an amplifier with zero output impedance.

(b) Output is voltage.

(c) Circuit to contain only resistors, capacitors, and high-gain, ideal voltage amplifiers.

(d) Number of amplifiers is to be minimized if possible.

(e) Potentiometers are to be available to permit laboratory adjustment of R values.

(f) Cutoff frequency is 20 rad/sec.

(g) System is third order.

6.10 It is possible to construct systems with oscillatory behavior from modules consisting of resistors, capacitors, and op-amplifiers. An example is the system shown, in which the blocks T_1 and T_2 represent the two modules indicated. (They are identical except for the values of the resistances.)

(a) What is the transfer function $T(s)$ of the system?

(b) Can you suggest a set of values a, b, and c for which the system is oscillatory? Can you suggest one for which it is not? Can you suggest a condition on a, b, and c which ensures oscillatory behavior?

(c) Suggest values of a, b, and c which will produce a damping ratio of 0.5 and a resonant frequency of 2 in $T(s)$.

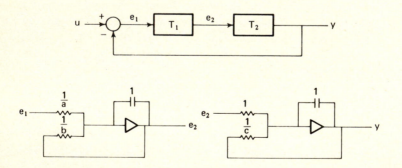

6.11 The block diagram portrays a control system in which the human operator is measuring the error and thereby controlling an output position. The two transfer functions of interest are

$$H(s) = K \frac{T_1 s + 1}{(T_2 s + 1)(T_3 s + 1)} e^{-Ts} \qquad P(s) = \frac{10}{s(s^2 + s + 10)}$$

(a) The $H(s)$ given is a commonly used representation of the human controller. In terms of the intuitive explanation of the way a human being reacts to a visually observed error, explain the source of each of these terms.

(b) Construct an analog simulation for this system in which each parameter $(K, T_1, T_2,$ and $T_3)$ appears only once as a multiplicative constant. Realize the transportation lag as a distinct block.

(c) Discuss various ways in which we might realize the transportation lag in the laboratory.

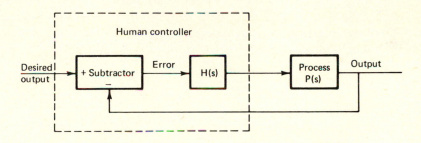

6.12 The flow diagram represents the simulation of a given dynamic system.

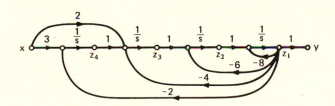

(a) What is the transfer function of the system?

(b) Because of saturation problems and noise considerations, we wish to reprogram the simulation to realize $z_1, 10z_2, 4z_3,$ and $z_4/2$. Determine the appropriate signal flow diagram.

(c) The response of the simulation is too slow. Return to the original signal flow diagram and redraw this, so that the simulator operates 100 times as fast.

6.13 Using operational amplifiers, resistors, capacitors, and potentiometers, determine two analog simulations for the transfer function

$$\frac{Y}{X} = \frac{2s^2 + 4s + 4}{s^4 + 4s^3 + 8s^2 + 12s + 6}$$

In each case, each integration should be represented by $-1/s$ and each adder must include a minus sign.

How many operational amplifiers are required in your simulation? Is this a minimum for realization of the given transfer function subject to the constraints on minus signs?

6.14 In the realization of a high order transfer function, it is desirable to use a tandem connection of simpler subsystems, each realizing one pair of poles. This form permits the poles to be adjusted (the network "tuned") in steps; otherwise, tuning tends to be exceedingly difficult, particularly for filter networks where often poles are very close to the imaginary axis.

The systems shown have been suggested (in an article by J. Tow, in the *IEEE Spectrum,* Dec. 1969) as basic building blocks. What are the two transfer functions realized?

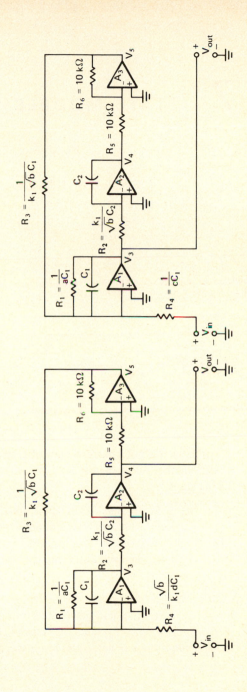

7 STATE MODELS

The input-output characteristics of a system are portrayed in the transfer function model. In the study of a system, we often arrive at the transfer function in the following steps:

(1) the scientific laws which describe each component are identified;
(2) the components and interconnections are described by differential equations;
(3) these equations are transformed;
(4) algebraic manipulation yields the transfer function.*

The transfer function yields simple and tremendously powerful procedures for calculating system response or for designing a system with desirable response characteristics. The simplicity and power stem in part from our success in disregarding details of the interior of the system. The transfer function is an input-output model; the same transfer function may correspond to many different systems.

---

*Obviously, there are many times when we cannot follow this procedure—for example, because the system is too complex or because the scientific laws are not known. Then we may have to apply a test signal and try to measure the transfer function directly.

There are important problems which require that we retain these interior details throughout the analysis. In other words, we need to work directly from the set of differential equations which describes the detailed components and their interconnections.

In this chapter, we consider such differential equation models, although we restrict the form very severely so that we obtain a set of equations called the state model. State models are important in system engineering for three primary reasons:

(1) They show important properties of the system which may not be evident from the transfer function.

(2) They are particularly appropriate for computer analysis.

(3) Many of the analysis techniques can be extended to the study of time-varying and nonlinear systems.

In the following sections, we define state models first, then show how they are used, with the overall objective of justifying these three motivations for the topic.

In the recent history of systems engineering, state models play a very minor role in the published literature before 1960. During most of the 1960s, control engineers were particularly intrigued by the state model concept, and this system description tended to dominate (actually state models had been widely used in mechanics for centuries, but represented a new tool in control system studies). In the 1970s, we find a more sane approach in which the engineer uses transfer functions and state models interchangeably, depending on which is preferable for the particular task.

7.1 THE STATE MODEL

Figure 7.1 shows a two-port network: the input port (or terminal pair) at which e_1 is applied, and the output port where the signal e_2 is available. To write the differential equations describing this system, there are several possible approaches.

We might write two node equations:

$$\left. \begin{aligned} \frac{1}{R_1}(e_C - e_1) + C\frac{de_C}{dt} + \frac{1}{L}\int_0^t (e_C - e_2)\, dt + i_L(0) &= 0 \\[2ex] \frac{1}{R_2}e_2 + \frac{1}{L}\int_0^t (e_2 - e_C)\, dt - i_L(0) &= 0 \end{aligned} \right\} \qquad (7\text{-}1)$$

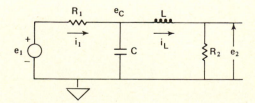

FIG. 7.1 A two-port electrical network.

Alternatively, we can write two equations from the Kirchhoff voltage law:

$$\left.\begin{aligned}
-e_1 + R_1 i_1 + \frac{1}{C} \int_0^t (i_1 - i_L)\,dt + e_C(0) &= 0 \\[2em]
R_2 i_L + L \frac{di_L}{dt} + \frac{1}{C} \int_0^t (i_L - i_1)\,dt - e_C(0) &= 0 \\[2em]
e_2 &= R_2 i_L
\end{aligned}\right\} \qquad (7\text{-}2)$$

Both pairs of simultaneous equations have both differentiation and integration. If we differentiate once, we obtain both second and first derivatives.

There is another possibility: we can try to write the equations in terms of the variables for which the initial values must be known, if we are to define the initial energy storage within the system. Since energy stored in the capacitor is measured by e_C, in the inductor by i_L, can we write the circuit equations in terms of the input e_1, the output e_2, and the two energy storage measures e_C and i_L?

At the e_C node, we have

$$\frac{1}{R_1}(e_C - e_1) + C\frac{de_C}{dt} + i_L = 0 \qquad (7\text{-}3)$$

Around the right loop through L we have

$$-e_C + L\frac{di_L}{dt} + R_2 i_L = 0 \qquad (7\text{-}4)$$

Finally, the output is defined by

$$e_2 = R_2 i_L \qquad (7\text{-}5)$$

If we rewrite these three equations so that the first derivative of an energy-storage variable appears alone on the left side of the first two, we obtain

$$\frac{de_C}{dt} = -\frac{1}{R_1 C}e_C - \frac{1}{C}i_L + \frac{1}{R_1 C}e_1 \qquad (7\text{-}6)$$

$$\frac{di_L}{dt} = \frac{1}{L}e_C - \frac{R_2}{L}i_L \qquad (7\text{-}7)$$

$$e_2 = R_2 i_L \qquad (7\text{-}8)$$

This is the form called a *state model*.

State model for a linear system

What is the general form of the state model for a linear, time-invariant system? To answer this question, we consider first a second order system (two energy-storage or *state variables*, x_1 and x_2), with one input u and a single output y. Then the state model has the form*

$$
\begin{cases}
\dot{x}_1 = a_{11} x_1 + a_{12} x_2 + b_1 u & (7\text{-}9) \\[2mm]
\dot{x}_2 = a_{21} x_1 + a_{22} x_2 + b_2 u & (7\text{-}10) \\[2mm]
y = c_1 x_1 + c_2 x_2 + du & (7\text{-}11)
\end{cases}
$$

In other words, the first derivative of each state variable is a linear sum of the state variables and input; the output is also a similar, linear combination.

Writing the equations in the form of (7-9)–(7-11) is obviously tedious, particularly as the number of state variables increases. To avoid this annoyance, we can use matrix notation. The equations above are then represented by

$$\dot{\mathbf{x}} = \mathbf{A}\mathbf{x} + \mathbf{B}u \qquad (7\text{-}12)$$

$$y = \mathbf{C}\mathbf{x} + Du \qquad (7\text{-}13)$$

For the second order system described by Eqs. (7-6)–(7-8), the network of Fig. 7.1, we have

$$
\mathbf{x} = \begin{bmatrix} e_C \\ i_L \end{bmatrix} \qquad u = e_1 \qquad y = e_2
$$

$$(7\text{-}14)$$

$$
\mathbf{A} = \begin{bmatrix} -\dfrac{1}{R_1 C} & -\dfrac{1}{C} \\[4mm] \dfrac{1}{L} & -\dfrac{R_2}{L} \end{bmatrix} \qquad \mathbf{B} = \begin{bmatrix} \dfrac{1}{R_1 C} \\[4mm] 0 \end{bmatrix} \qquad \mathbf{C} = \begin{bmatrix} 0 & R_2 \end{bmatrix} \qquad D = 0
$$

---

*Hereafter we will often use the notation $\dot{x}_1$ in place of dx_1/dt.

In more general situations, the system may have multiple inputs and several outputs. Then each variable or parameter in Eqs. (7-12) and (7-13) becomes a matrix. In the next few sections, however, we focus on single-input, single-output systems.

Another example

The state model

$$\left.\begin{array}{l} \dot{\mathbf{x}} = \mathbf{Ax} + \mathbf{B}u \\ y = \mathbf{Cx} + Du \end{array}\right\} \tag{7-15}$$

precisely defines the op-amp realization described throughout Chap. 6. For instance, Fig. 7.2 shows the system originally derived in Fig. 6.22 for a third order transfer function. In this case, the three $\dot{x}_j$ equations are the summations at the $\dot{x}_j$ nodes:

$$\begin{cases} \dot{x}_1 = -7x_1 + x_2 + 2u \\ \dot{x}_2 = -14x_1 + x_3 + 6u \\ \dot{x}_3 = -8x_1 + 8u \end{cases}$$

and

$$\mathbf{A} = \begin{bmatrix} -7 & 1 & 0 \\ -14 & 0 & 1 \\ -8 & 0 & 0 \end{bmatrix} \quad \mathbf{B} = \begin{bmatrix} 2 \\ 6 \\ 8 \end{bmatrix} \quad \mathbf{C} = [1 \quad 0 \quad 0] \quad D = 0 \tag{7-16}$$

Similarly, for any of the other op-amp realizations of the last chapter, we can write $\mathbf{A}, \mathbf{B}, \mathbf{C}$, and D by inspection. Furthermore, from the state model we can determine

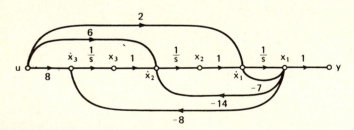

FIG. 7.2 Realization of

$$T(s) = \frac{Y}{U} = \frac{2s^2 + 6s + 8}{s^3 + 7s^2 + 14s + 8}$$

(Similar to Fig. 6.22, except that here we use $+1/s$ branches and the state variables are x_j rather than the z_j of the last chapter.)

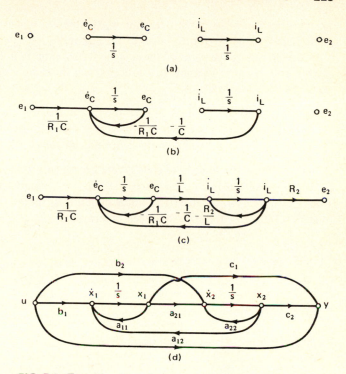

FIG. 7.3 Transition from the state model in terms of first order differential equations to the op-amp or analog realization: (a) Definition of the variables for the state model for Fig. 7.1 (The state variable which we show on the left is immaterial); (b) Inclusion of Eq. (7-6) for $\dot{e}_C$; (c) Complete state model for Fig. 7.1; (d) General signal flow diagram for state model of second order system, Eqs. (7-9)–(7-11); in most cases several of the parameters are zero and the diagram is simpler, as illustrated in (c).

the analog-computer simulation or op-amp realization. Figure 7.3 shows the signal flow diagram for Eq. (7-14) or the network of Fig. 7.1.

State

We have been using the term *state model* and *state variables* so far without any real definition of the concept of *state*. If we try to solve the state equations

$$\begin{aligned} \dot{\mathbf{x}} &= \mathbf{A}\mathbf{x} + \mathbf{B}u \\ y &= \mathbf{C}\mathbf{x} + Du \end{aligned} \Bigg\} \qquad (7\text{-}17)$$

to find y for a given input u, we might use the Laplace transform. If we transform $\dot{x}_j$, we need to know the initial condition $x_j(0)$; these initial conditions (along with the drive u) determine the future values of the response y.

In simple terms, the *state* of the system at any time t_1 is this set of values: $x_1(t_1)$, $x_2(t_1)$, and so on. If we know the state at t_1 and the drive u thereafter, we can find the response. In general terms,

> The state of a system at any time t_0 is the minimum set of numbers $[x_1(t_0), x_2(t_0), \ldots]$ which must be known along with the drive signals for $t \geq t_0$ to determine the system behavior for all $t \geq t_0$.

In these terms, the concept of state is very broad. In a second order electrical network such as that shown in Fig. 7.4, there are two ways energy can be stored. Hence, if we know e_1 for $t \geq 0$, we need two initial conditions; there are two state variables. The system state at $t = 0$ can be described in terms of

$$e_C(0) \quad \text{and} \quad i_L(0)$$

Alternatively, we can use

$$e_2(0) \quad \text{and} \quad \dot{e}_2(0)$$

or, indeed, many other possibilities, such as

$$e_2(0) \quad \text{and} \quad e_C(0)$$

or

$$\dot{e}_2(0) \quad \text{and} \quad \ddot{e}_2(0)$$

The important idea here is that we can only determine $e_2(t)$ for positive time if we know not only the input, but also the state of the system at the start of the solution.

Another example is provided by the system involved when we attempt to predict the population of the United States from 1970 until the year 2000. As a first approach, we might assume that the only inputs are the immigration and emigration rates. The changes in population also depend on the birth and death rates within the country; these in turn depend on the age profiles of the population (if the female population

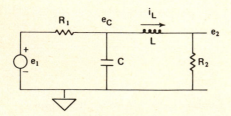

FIG. 7.4 Second order electrical system.

should be predominantly over 50 or under 15, the birth rate would be very low). Hence, the *state* of the system in 1970, at the start of the problem, is measured by the number of women in each age bracket from infancy through the child-bearing years. We might divide this range into five-year intervals and call F_5 the number of femals between the ages of 0 and 5. Then the 1970 state of the system would be described by the ten numbers*

$$F_5, \ F_{10}, \ F_{15}, \ F_{20}, \ \dots, \ F_{50}$$

Final comment

A state model is a set of simultaneous, first order differential equations describing the way the state variables and output depend on the drive and initial conditions (or state).

7.2 SOLUTIONS FOR THE UNDRIVEN SYSTEM

The state model clearly contains much more information than the transfer function. The latter yields only the input-output dynamics; the state model contains this same information plus the dependence of the system behavior on the initial state of the system. In a sense then, we seem to have lost ground in going to the state model. There are, however, important system problems where we must deal with the more detailed system description implicit in the state model.

The total time response consists of two parts: the driven component resulting from u and the free component stemming from the initial state. If we first assume the drive is zero, we can determine the resulting response from

$$\dot{\mathbf{x}} = \mathbf{A}\mathbf{x} \tag{7-18}$$

(The output y is just a linear combination of the x_j's, so it suffices to find the state variables.)

Laplace transform approach

If we proceed formally with the Laplace transform solution, we obtain

$$s\mathbf{X}(s) - \mathbf{x}(0) = \mathbf{A}\mathbf{X}(s) \tag{7-19}$$

Collecting terms in $\mathbf{X}(s)$ on the left side, we have

$$[s\mathbf{1} - \mathbf{A}]\mathbf{X}(s) = \mathbf{x}(0) \tag{7-20}$$

---

*The fact that expected births for the next 20 years or more are largely determined already by the values of F_5, F_{10}, ... is the reason the system is so sluggish in response. Steps introduced today to control population growth are unlikely to have much effect for the next 20 years, unless we can convince the population to reduce radically the average family size.

where we have introduced the unity matrix 1 which is a square matrix of order n with 1's on the main diagonal and 0's elsewhere. Then premultiplication by the inverse of $[s1 - A]$ gives

$$X(s) = [s1 - A]^{-1} x(0) \tag{7-21}$$

Finally, we can return to the time domain and the solution by taking the inverse Laplace transform of *each* component of $[s1 - A]^{-1}$. The resulting matrix of time functions is usually called $\varphi(t)$:

$$x(t) = \varphi(t) x(0) \tag{7-22}$$

These four equations actually represent an enormous amount of work in a system analysis. A simple, second order system suggests the complexity. If the state model is

$$\left. \begin{aligned} \dot{x}_1 &= -8x_2 \\ \dot{x}_2 &= +x_1 - 6x_2 \end{aligned} \right\}$$

the Laplace transform, Eq. (7-19), gives

$$\left. \begin{aligned} sX_1(s) - x_1(0) &= -8X_2(s) \\ sX_2(s) - x_2(0) &= X_1(s) - 6X_2(s) \end{aligned} \right\} \tag{7-23}$$

Collecting terms yields

$$\begin{bmatrix} s & +8 \\ -1 & s+6 \end{bmatrix} \begin{bmatrix} X_1(s) \\ X_2(s) \end{bmatrix} = \begin{bmatrix} x_1(0) \\ x_2(0) \end{bmatrix} \tag{7-24}$$

Next, inversion of the matrix on the left yields

$$\begin{bmatrix} X_1(s) \\ X_2(s) \end{bmatrix} = \begin{bmatrix} \dfrac{s+6}{(s+2)(s+4)} & \dfrac{-8}{(s+2)(s+4)} \\ \dfrac{1}{(s+2)(s+4)} & \dfrac{s}{(s+2)(s+4)} \end{bmatrix} \begin{bmatrix} x_1(0) \\ x_2(0) \end{bmatrix} \tag{7-25}$$

Finally, the inverse transform of each of the four transforms gives

$$\begin{bmatrix} x_1(t) \\ x_2(t) \end{bmatrix} = \underbrace{\begin{bmatrix} 2e^{-2t} - e^{-4t} & -4e^{-2t} + 4e^{-4t} \\ \frac{1}{2}e^{-2t} - \frac{1}{2}e^{-4t} & -e^{-2t} + 2e^{-4t} \end{bmatrix}}_{\phi(t)} \begin{bmatrix} x_1(0) \\ x_2(0) \end{bmatrix} \qquad (7\text{-}26)$$

In other words, the state variable $x_1(t)$ depends on $x_1(0)$ and $x_2(0)$ according to the relation:

$$x_1(t) = (2e^{-2t} - e^{-4t})x_1(0) + (-4e^{-2t} + 4e^{-4t})x_2(0) \qquad (7\text{-}27)$$

and a similar equation describes $x_2(t)$.

The above analysis really represents a *complete* solution for the response of the undriven system. The matrix of time functions, $\varphi(t)$ in Eq. (7-26), describes the way the state variables depend on the initial state. This $\varphi(t)$ is called the *state-transition matrix*, because it defines the transition from the initial state to later states:

State-transition matrix $\qquad \varphi(t) = \mathcal{L}^{-1}[s\mathbf{1} - \mathbf{A}]^{-1} \qquad (7\text{-}28)$

or, in the s domain,

$$\mathbf{\Phi}(s) = [s\mathbf{1} - \mathbf{A}]^{-1} \qquad (7\text{-}29)$$

[where $\mathbf{\Phi}(s)$ is the Laplace transform of the state-transition matrix $\varphi(t)$].

In terms of the signal flow diagram for the system, Fig. 7.5 shows the system used as an example above, with the drive set equal to zero. The initial conditions, $x_1(0)$ and $x_2(0)$, are inserted as shown, and $x_1(t)$ and $x_2(t)$ are measured. In the laboratory, we can *measure* experimentally the four elements of the state-transition matrix $\varphi(t)$ by observing the $x_1(t)$ and $x_2(t)$ which result from $x_1(0)$ and $x_2(0)$ separately. $\varphi(t)$ and its transform $\mathbf{\Phi}(s)$ are, respectively, the four time functions and their transforms which measure the transmission from the initial conditions to the state variables.

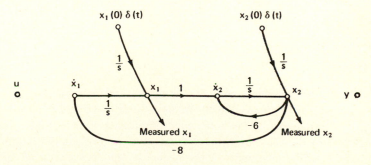

FIG. 7.5 Signal flow diagram (of example in text).

An alternate solution

We can also solve the undriven case without using the Laplace transform. The equations are

$$\dot{x} = Ax \tag{7-30}$$

but if this were a scalar equation in one variable

$$\dot{y} = Ay \tag{7-31}$$

the solution would be

$$y = e^{At} y(0) \tag{7-32}$$

(The validity of this solution can be proved by substituting in the differential equation.) In an exactly analogous manner, the solution of the matrix equation is

$$x = e^{At} x(0) \tag{7-33}$$

where the matrix e^{At} is defined by the power series

$$e^{At} = 1 + At + \frac{A^2}{2!} t^2 + \frac{A^3}{3!} t^3 + \cdots \tag{7-34}$$

We can show that this series does, indeed, converge and that Eq. (7-33) is a solution by substitution in the differential equation (7-30).

In terms of our earlier notation, e^{At} is just the state-transition matrix $\varphi(t)$ [as is clear by a comparison of Eq. (7-33) with (7-22)]. In other words, the state-transition matrix can be found in two ways:

(1)   The Laplace transform:

$$\varphi(t) = \mathcal{L}^{-1}[s1 - A]^{-1} \tag{7-35}$$

(2)   The exponential of the A matrix:

$$\varphi(t) = e^{At} \tag{7-36}$$

In an actual system analysis, we usually use the transform approach if we want to calculate the elements of $\varphi(t)$. We use a version of Fig. 7.5; the transfer function is calculated from each of the initial condition inputs to each of the state variables to obtain $\Phi(s)$. In the usual case, the denominators of all transfer functions are identical (the characteristic polynomial of the system or the determinant of the signal flow graph). The only differences are then in the numerators.

The calculation can be illustrated from Fig. 7.5. As soon as the flow diagram is drawn, we can write immediately the four elements of $\Phi(s)$ from Mason's reduction theorem:

$$
\phi_{11}(s) = \frac{X_1}{x_1(0)} = \frac{(1/s)(1 + 6/s)}{1 + 6/s + 8/s^2} = \frac{s + 6}{(s + 2)(s + 4)}
$$

$$
\phi_{12}(s) = \frac{X_1}{x_2(0)} = \frac{-8/s^2}{1 + 6/s + 8/s^2} = \frac{-8}{(s + 2)(s + 4)}
$$

$$
\phi_{21}(s) = \frac{X_2}{x_1(0)} = \frac{1/s^2}{1 + 6/s + 8/s^2} = \frac{1}{(s + 2)(s + 4)}
$$

$$
\phi_{22}(s) = \frac{X_2}{x_2(0)} = \frac{1/s}{1 + 6/s + 8/s^2} = \frac{s}{(s + 2)(s + 4)}
$$

$$(7\text{-}37)$$

7.3 SOLUTIONS FOR THE DRIVEN SYSTEM

When the system is driven by a single input u, the state variables are determined by

$$\dot{\mathbf{x}} = \mathbf{A}\mathbf{x} + \mathbf{B}u \tag{7-38}$$

(Again, since the output y is just a linear combination of the x_j's and u, we focus first on finding how the x_j's vary with time.) As outlined in the preceding section, we can proceed with the Laplace transform solution.

The first step is transformation of Eq. (7-38):

$$s\mathbf{X}(s) - \mathbf{x}(0) = \mathbf{A}\mathbf{X}(s) + \mathbf{B}U(s) \tag{7-39}$$

which can be rewritten

$$[s\mathbf{1} - \mathbf{A}]\mathbf{X}(s) = \mathbf{x}(0) + \mathbf{B}U(s) \tag{7-40}$$

Solution for $\mathbf{X}(s)$ yields

$$\mathbf{X}(s) = [s\mathbf{1} - \mathbf{A}]^{-1}\mathbf{x}(0) + [s\mathbf{1} - \mathbf{A}]^{-1}\mathbf{B}U(s) \tag{7-41}$$

Introduction of the symbol $\Phi(s)$ for the transform of the state-transition matrix simplfies this equation slightly:

$$\boxed{\mathbf{X}(s) = \Phi(s)\mathbf{x}(0) + \Phi(s)\mathbf{B}U(s)} \tag{7-42}$$

The corresponding form in the time domain is

$$x(t) = \boldsymbol{\varphi}(t)\,x(0) + \boldsymbol{\varphi}(t) * (Bu)$$

(7-43)†

Again, Eqs. (7-42) and (7-43) represent an enormous amount of analytical work in an actual system analysis. The equations do indicate, however, certain important features of the solution:

(1) The solution is the sum of two parts: the response to the initial state $x(0)$, and the response to the drive u.

(2) The undriven component of the response is the same whether u is present or not. In the time domain, this is just the state transition matrix multiplied by the initial state.‡

(3) The driven component of the response is $\boldsymbol{\varphi}(t)$ convolved with $Bu(t)$. In other words, if the system input is viewed as $Bu(t)$, the state transition matrix plays the role of a weighting function or unit impulse response.

A second order example

In order to illustrate the significance of the above equations (which look so deceptively simple), we consider the system depicted in Fig. 7.6. Here there are two state variables, and the system is described by the state model:

$$\left. \begin{aligned} \dot{x}_1 &= -6x_1 + x_2 + 2u \\ \dot{x}_2 &= -8x_1 + 16u \end{aligned} \right\}$$

(7-44)

In the solution for $x_1(t)$ and $x_2(t)$, first we need to find the state-transition matrix or its transform. The $\Phi(s)$ is determined as in the preceding section by finding the

†The symbol * represents the convolution operation.
‡In the preceding section and in this one, we repeatedly use as the initial or starting state $x(0)$; we can, of course, start at any value of t and find the solution thereafter.

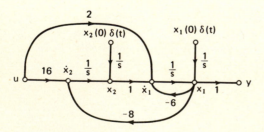

FIG. 7.6 Example of driven system.

transfer functions from $x_1(0)$ and $x_2(0)$ to $x_1(t)$ and $x_2(t)$:

$$\Phi(s) = \begin{bmatrix} \dfrac{s}{(s+2)(s+4)} & \dfrac{1}{(s+2)(s+4)} \\[3mm] \dfrac{-8}{(s+2)(s+4)} & \dfrac{s+6}{(s+2)(s+4)} \end{bmatrix} \tag{7-45}$$

This step is merely four applications of Mason's reduction theorem.

Now the "generalized" drive signal is $\mathbf{B}U(s)$. In our case, Eq. (7-44) shows

$$\mathbf{B} = \begin{bmatrix} 2 \\ 16 \end{bmatrix} \tag{7-46}$$

Hence

$$\mathbf{B}U(s) = \begin{bmatrix} 2U(s) \\ 16U(s) \end{bmatrix} \tag{7-47}$$

We can then write the equation for the transforms of the state variables:

$$\begin{bmatrix} X_1(s) \\[3mm] X_2(s) \end{bmatrix} = \begin{bmatrix} \dfrac{s}{(s+2)(s+4)} & \dfrac{1}{(s+2)(s+4)} \\[3mm] \dfrac{-8}{(s+2)(s+4)} & \dfrac{s+6}{(s+2)(s+4)} \end{bmatrix} \begin{bmatrix} x_1(0) \\[3mm] x_2(0) \end{bmatrix}$$

$$+ \begin{bmatrix} \dfrac{s}{(s+2)(s+4)} & \dfrac{1}{(s+2)(s+4)} \\[3mm] \dfrac{-8}{(s+2)(s+4)} & \dfrac{s+6}{(s+2)(s+4)} \end{bmatrix} \begin{bmatrix} 2U(s) \\[3mm] 16U(s) \end{bmatrix} \tag{7-48}$$

Several comments should be made:

(1) Here we have a *complete* solution for the system behavior. Both state variables, $x_1(t)$ and $x_2(t)$, can be found for any initial state or initial conditions and any drive signal.

(2)   The state-transition matrix is a generalized unit impulse response, its transform a generalized transfer function. For example, if we were interested in only $x_1(t)$ (or y in Fig. 7.6), Eq. (7-48) would reduce to

$$Y(s) \;=\; X_1(s) \;=\; \frac{s}{(s+2)(s+4)}\, x_1(0) \;+\; \frac{1}{(s+2)(s+4)}\, x_2(0)$$

$$+ \;\underbrace{\frac{2s+16}{(s+2)(s+4)}}\; U(s) \tag{7-49}$$

The quantity over the brace is the usual transfer function, $(Y/U)(s)$. The first term on the right includes the transfer function from $x_1(0)$ to y, the second term the transfer function from $x_2(0)$ to y. Essentially, Eq. (7-49) says the output results from three distinct input signals: $u, x_1(0)$, and $x_2(0)$.

We might also rewrite Eq. (7-49) by factoring the transfer function $(Y/U)(s)$:

$$Y(s) \;=\; \frac{2s+16}{(s+2)(s+4)}\left[U(s) + \frac{s}{2s+16}\, x_1(0) + \frac{1}{2s+16}\, x_2(0) \right] \tag{7-50}$$

This form reaffirms the validity of the concept that initial conditions within a system are equivalent to an augmented input driving the inert system.

(3)   Every transfer function in $\Phi(s)$ has the same denominator $(s+2)(s+4)$, which is the system characteristic polynomial.

Solution in the time domain
  Equation (7-43) states that

$$\mathbf{x}(t) \;=\; \boldsymbol{\varphi}(t)\mathbf{x}(0) + \boldsymbol{\varphi}(t) * [\mathbf{B}u(t)] \tag{7-51}$$

In our specific example,

$$\begin{bmatrix} x_1(t) \\ x_2(t) \end{bmatrix} = \begin{bmatrix} \phi_{11}(t) & \phi_{12}(t) \\ \phi_{21}(t) & \phi_{22}(t) \end{bmatrix}\begin{bmatrix} x_1(0) \\ x_2(0) \end{bmatrix} + \begin{bmatrix} \phi_{11}(t) & \phi_{12}(t) \\ \phi_{21}(t) & \phi_{22}(t) \end{bmatrix} * \begin{bmatrix} 2u(t) \\ 16u(t) \end{bmatrix} \tag{7-52}$$

If we are interested in only $y(t)$ or $x_1(t)$, this relation becomes

$$y(t) \;=\; x_1(t) \;=\; \phi_{11}(t)\,x_1(0) + \phi_{12}(t)\,x_2(0) + \underbrace{[2\phi_{11}(t) + 16\phi_{12}(t)]} * u(t) \tag{7-53}$$

Again the system output consists of three parts: the responses to each of the two initial conditions and the response to $u(t)$:

$$y_f(t) = [2\phi_{11}(t) + 16\phi_{12}(t)] * u(t) \tag{7-54}$$

In other words, the unit-impulse response of the inert system is in this case (where $y = x_1$)

$$g(t) = 2\phi_{11}(t) + 16\phi_{12}(t) \tag{7-55}$$

More generally, if

$$y = \mathbf{C}\mathbf{x} + D u \tag{7-56}$$

the unit impulse response of the u-to-y system is

$$g(t) = \mathbf{C}\mathbf{\Phi}(t)\mathbf{B} + D\delta(t) \tag{7-57}$$

The system solution in the time domain, represented by Eq. (7-51), is particularly important when the analysis is carried out on a computer: a possibility which is attractive when the system is only slightly more complex than our second order example, or when the input is a time signal given graphically rather than as a sum of exponentials.

7.4 STATE MODELS FOR SIMPLE CONTROL SYSTEMS

Primary emphasis in the preceding sections is on the mathematical analysis of systems described by state models. In order to see how this formalism is used, we turn to a specific example of a feedback control system (Fig. 7.7).

In this system, the operator hopes to have the output θ follow the input θ_i which he applies at a location remote from the output. The operator turns a dial to adjust θ_i. This angular position is compared to the output θ to generate an error signal e. Whenever the output does not equal the input, an error exists. After amplification, this

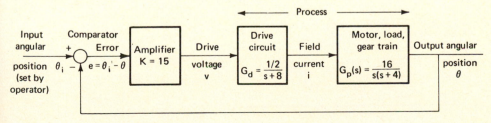

FIG. 7.7 Servomechanism example.

error is applied to the field coil (input) of an electric motor. The field current results in a torque which, through the gear train, changes the output. The complete system is called a servomechanism, a term for a feedback control system in which the output is a displacement (translational or angular).

In this section, we analyze the performance of the system; in the next section, we consider some of the features of the design problem to meet performance specifications.

A state model

Inspection of the block diagram of Fig. 7.7 reveals that the system is third order. What shall we choose as the three state variables?

While there are many possibilities, one choice maintains a close correspondence between the state variables and those signals which we know represent measures of energy storage in the system. In this "natural" definition of state variables, we choose

$$\begin{cases} \theta & \text{Output angular position} \\ \omega & \text{or } d\theta/dt, \text{ the output angular velocity} \\ i & \text{Field current} \end{cases}$$

In terms of these variables, the system diagram can be redrawn as shown in Fig. 7.8.

In the usual state model, we want each state variable to be the output of an integrator. Consequently, we redraw the system diagram, and in the process decompose each first-order transfer function into an integrator with a feedback path (Fig. 7.9). We now have the customary state model, and

$$\begin{cases} \dot{\theta} = \omega & (7\text{-}58) \\ \\ \dot{\omega} = -4\omega + 16i & (7\text{-}59) \\ \\ \dot{i} = -8i - \frac{15}{2}\theta + \frac{15}{2}\theta_i & (7\text{-}60) \end{cases}$$

The transform of the state-transition matrix can be found from the signal flow diagram if we insert initial conditions (on θ, ω, and i). Then by inspection of Fig. 7.9, the system characteristic polynomial is the numerator of

$$1 + \frac{8}{s} + \frac{4}{s} + \frac{32}{s^2} + \frac{120}{s^3}$$

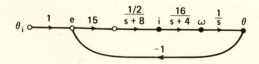

FIG. 7.8 State variables indicated by ● .

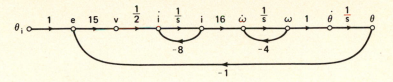

FIG. 7.9 Redrawing Fig. 7.8 in terms of integrators.

or

$$C.P. = s^3 + 12s^2 + 32s + 120 \tag{7-61}$$

The $\Phi(s)$ is, from Fig. 7.10,

$$\Phi(s) = \begin{bmatrix} \dfrac{s^2 + 12s + 32}{CP} & \dfrac{s+8}{CP} & \dfrac{16}{CP} \\[3mm] \dfrac{-120}{CP} & \dfrac{s(s+8)}{CP} & \dfrac{16s}{CP} \\[3mm] \dfrac{-15(s+4)/2}{CP} & \dfrac{-15/2}{CP} & \dfrac{s(s+4)}{CP} \end{bmatrix} \tag{7-62}$$

We can now find the system response to any drive signal and initial state. For example, if we are interested in only the output θ,

$$\theta(s) = \frac{(s^2 + 12s + 32)\,\theta(0) + (s+8)\,\omega(0) + 16i(0) + 120\theta_i(s)}{s^3 + 12s^2 + 32s + 120} \tag{7-63}$$

This expression includes the effects of not only the drive signal θ_i, but also the energy stored in the system at the time θ_i is applied (at $t = 0$).

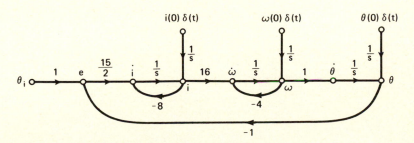

FIG. 7.10 System model with initial conditions inserted.

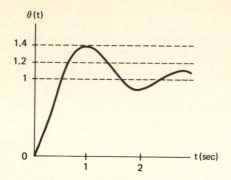

FIG. 7.11 Response of original servomechanism with unit-step input and system inert. [The transfer function is

$$\frac{\theta}{\theta_i} = \frac{120}{(s + 10)(s^2 + 2s + 12)}$$

yielding a resonance with $\omega_n = 3.4$, $\zeta = 0.29$.]

7.5 DESIGN OF THE SERVOMECHANISM

For the servomechanism discussed in the last section, the following typical design problem might be stated.

The basic structure of the system is shown in Fig. 7.7 or the subsequent signal flow diagrams. The system, as designed, tends to oscillate much too violently after a step-function input is applied [Fig. 7.11 shows the response which is measured or calculated from Eq. (7-63) or the θ/θ_i transfer function]. We can change the value of K or add feedback from the state variables ω and i to the comparator. Is it possible to obtain a system with the characteristic polynomial

$$(s + 10)(s^2 + 4s + 12) = s^3 + 14s^2 + 52s + 120 \qquad (7\text{-}64)$$

In other words, can we redesign the system to change the relative damping ratio from 0.29 to slightly more than 0.5? (We are just increasing the damping of the oscillatory mode; the unit-step response would then be as shown in Fig. 7.12.)

The modifications allowed in the system are shown in Fig. 7.13. The characteristic polynomial for this system is the numerator of

$$1 + \frac{4}{s} + \frac{8}{s} + \frac{\alpha K/2}{s} + \frac{32}{s^2} + \frac{2\alpha K}{s^2} + \frac{8\beta K}{s^2} + \frac{8K}{s^3}$$

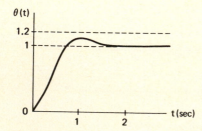

FIG. 7.12 Response desired in redesigned servomechanism.

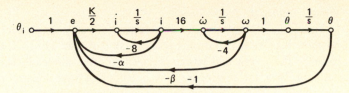

FIG. 7.13 Modified system with feedback of the state variables.

or

$$C.P. = s^3 + \left(12 + \frac{\alpha K}{2}\right)s^2 + (32 + 2\alpha K + 8\beta K)s + 8K \qquad (7\text{-}65)$$

Comparison of Eq. (7-65) with the desired polynomial (7-64) allows us to solve for K, α, and β. The constant term indicates K is again 15. Then the s^2 term gives

$$12 + \frac{15\alpha}{2} = 14 \quad \text{or} \quad \alpha = \frac{4}{15} \qquad (7\text{-}66)$$

and the s term yields

$$32 + 8 + 60\beta = 52 \quad \text{or} \quad \beta = \frac{1}{5} \qquad (7\text{-}67)$$

In other words, by using feedback of not only the output, but also the other state variables, we can control the characteristic polynomial (and the nature of the system response). In building the system of Fig. 7.13, we have to measure the angular velocity ω of the output and the field current i. The error signal is then made to depend on not only θ_i and θ, but also ω and i.

Intuitively, we can see the logic underlying the desirability of feeding back the intermediate state variables ω and i. The reason for the excessive overshoot and oscillation in the original response (Fig. 7.11) is that the error is large just after the step is applied. The error continues positive until the response reaches unity. By that time, however, the motor has acquired so much momentum that the output sails on past unity. The error goes negative and gradually a braking torque is built up to start θ decreasing toward unity.

When we feed back ω and i, we anticipate the future behavior of the output θ. For example, a large, positive angular velocity ω tends to make the error e go negative before θ reaches unity, so that braking can start before the output reaches its desired, final value. Similarly, the field current i anticipates the future changes in velocity.

Final comments

This design example is reasonably typical of many simple problems in control systems engineering. We are confronted with a control task. The motor and gears are chosen to provide the torque and power required to drive the load; then the process in Fig. 7.7 is fixed. Now the control engineer must design the comparator and amplifier to realize an overall system with adequate dynamic characteristics.

We try first a simple amplifier ($K = 15$ in Fig. 7.7) in the hope that the system will be adequate. We find that the amplifier gain cannot be adjusted to yield a response fast enough without excessive overshoot and oscillation. (Reducing K in Fig. 7.7 decreases the overshoot, but also slows the response.)

A next step commonly is to try state-variable feedback. If we can measure the state variables (here ω and i), we add these feedback paths to the comparator and attempt to realize suitable system dynamics (e.g., an appropriate characteristic polynomial).

At this point we should break with the tradition of technical authors and be honest. This procedure frequently does not succeed. (If it would always succeed, there would be little reason for all the graduate courses on linear control system design.) Two problems may occur:

First, it may not be possible or desirable to measure all state variables. If the system is high order, such measurements may be too expensive as far as equipment is concerned, so complicated that poor reliability results, or just too cumbersome. (If there are 120 state variables, the design problem, clearly, is not trivial.)

Second, the basic approach may be unsound because we do not know what to select as a characteristic polynomial, in order to ensure that system response is within the specifications. Unfortunately, the engineer is seldom asked to realize a particular characteristic polynomial. (Often, the person giving the performance specifications does not know what a characteristic polynomial is.) Instead, the engineer may be given a typical input signal, and told that the output should follow this input with an error which exceeds $0.5°$ less than 0.1% of the time. In some of such cases, we can build the system on an analog (or digital) simulator, apply the typical input, and adjust the system parameters until the desired performance is achieved. In such an experimental approach, we really ask our theory primarily such questions as: How might the system be modified to improve performance? What additional feedback paths might it be desirable to insert?

Thus, the concept of system design through state-variable feedback is an exceedingly important technique, a method which is frequently useful, but which by no means solves all design problems.

7.6 CONTROLLABILITY

The state model reveals certain interesting features of the control problem—aspects which may not be apparent from the transfer function approach. In this section, we consider one of these: the controllability of the system. In particular, are there systems where the response can not be controlled, regardless of the choice of input?

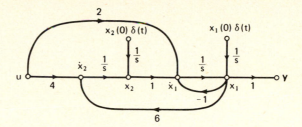

FIG. 7.14 A second order system which is controllable, even though unstable.

The answer is yes; there are such systems. To demonstrate this, we consider a very simple example shown in Fig. 7.14.

For this system, the transfer function from input to output is

$$\frac{Y}{U} = \frac{2(s + 2)}{(s + 3)(s - 2)} \tag{7-68}$$

and the system is unstable. The system is, however, *controllable* in the following sense:

> Given any initial state $[x_1(0)$ and $x_2(0)]$ and *any* desired (later) state, we can find a drive signal $u(t)$ and a time t_1 such that the state at t_1 is the desired value.

In other words, by proper choice of the input, we can "move" the system state from any starting value to any desired final value.

The proof that the system of Fig. 7.14 is controllable in this sense is omitted here since it is not really essential to our development.* We can point out the possibility, however, by showing that the unstable mode in x_1, x_2, and y can be suppressed by a proper choice of $u(t)$. In order to do this, we first find the transform of the state transition matrix by inspection of Fig. 7.14:

$$\Phi(s) = \begin{bmatrix} \dfrac{s}{(s + 3)(s - 2)} & \dfrac{1}{(s + 3)(s - 2)} \\[3mm] \dfrac{6}{(s + 3)(s - 2)} & \dfrac{s + 1}{(s + 3)(s - 2)} \end{bmatrix} \tag{7-69}$$

*A succinct discussion is given in E. Polak and E. Wong, "Notes for a First Course on Linear Systems," Chapter II, Van Nostrand Reinhold, New York, 1970.

Then

$$
\begin{bmatrix} X_1(s) \\[2ex] X_2(s) \end{bmatrix} = \begin{bmatrix} \dfrac{s}{(s+3)(s-2)} & \dfrac{1}{(s+3)(s-2)} \\[3ex] \dfrac{6}{(s+3)(s-2)} & \dfrac{s+1}{(s+3)(s-2)} \end{bmatrix} \begin{bmatrix} x_1(0) \\[2ex] x_2(0) \end{bmatrix}
$$

$$
+ \begin{bmatrix} \dfrac{s}{(s+3)(s-2)} & \dfrac{1}{(s+3)(s-2)} \\[3ex] \dfrac{6}{(s+3)(s-2)} & \dfrac{s+1}{(s+3)(s-2)} \end{bmatrix} \begin{bmatrix} 2 \\[2ex] 4 \end{bmatrix} U(s) \qquad (7\text{-}70)
$$

These two equations for $X_1(s)$ and $X_2(s)$ show that a partial fraction expansion of each would involve poles at -3, $+2$, and the poles of $U(s)$. To observe the unstable mode (pole at $+2$), we consider only those terms. Then the corresponding terms of $x_1(t)$ and $x_2(t)$ are

$$
e^{2t} \text{ components} \atop \text{only} \quad \begin{cases} x_1(t) = \left[\frac{2}{5} x_1(0) + \frac{1}{5} x_2(0) + \frac{8}{5} U(2) \right] e^{2t} \\[3ex] x_2(t) = \left[\frac{6}{5} x_1(0) + \frac{3}{5} x_2(0) + \frac{24}{5} U(2) \right] e^{2t} \end{cases} \qquad (7\text{-}71)
$$

In other words, the x_2 component is just three times the x_1 component. To suppress this mode entirely, we need only choose a drive function $u(t)$ which has a transform having a value at $s = +2$ of

$$
U(2) = -\frac{1}{4} x_1(0) - \frac{1}{8} x_2(0) \qquad (7\text{-}72)
$$

Any $u(t)$ satisfying this condition results in system control with the unstable mode (e^{2t}) never appearing in $x_1(t)$, $x_2(t)$, or $y(t)$.

Equation (7-72) is a generalization of the idea that the unstable Y/U mode can be cancelled by a zero in $U(s)$. If the system is inert $[x_1(0) = x_2(0) = 0]$, the equation shows that $U(2)$ should be zero: $U(s)$ must have a zero at $s = +2$ to cancel the transfer function pole.

An uncontrollable system

A slight modification of Fig. 7.14 results in the system being uncontrollable. Analysis of the system of Fig. 7.15 follows as above (the state-transition matrix is unchanged),

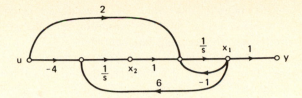

FIG. 7.15 A slight change in the system portrayed in Fig. 7.14 (the +4 input transmittance is changed to -4).

until we reach Eq. (7-70) where the +4 in the final matrix is now -4. Then the two e^{2t} components of x_1 and x_2 are

$$
x_1(t) = \left[\frac{2}{5}x_1(0) + \frac{1}{5}x_2(0) + \frac{2}{5}k_2\right]e^{2t}
$$

$$
x_2(t) = \left[\frac{6}{5}x_1(0) + \frac{3}{5}x_2(0) - \frac{4}{5}k_2\right]e^{2t}
$$

(7-73)

where k_2 is the residue of $U(s)$ in its pole at +2 [$U(s)$ must have a pole at +2 or these terms are zero].

Inspection of Eqs. (7-73) reveals that it is not possible to choose k_2 to make both brackets zero. We can make either the x_1 component or that of x_2 zero, but we cannot make both zero simultaneously. The unstable e^{2t} mode *cannot* be suppressed. If it does not appear in the output (x_1 or y), it will appear in x_2. The unstable mode will not appear only when the system is inert (initial state zero).

The system is uncontrollable.

Meaning of uncontrollability

A retrospective consideration of the above example reveals what is meant by the concept of an uncontrollable system. A system is characterized by certain internally generated modes. If the system is third order, there are three "natural" frequencies—three values of s, each of which leads to terms of the form

$$e^{st}$$

These natural frequencies appear generally in all of the state variables.

By selecting an appropriate drive signal $u(t)$, we can control each of these components of the various state variables. If necessary, we can select $u(t)$ so that a particular natural frequency is killed in every state variable; more generally, we can control the amplitudes of each term in each state variable.

In an uncontrollable system, on the other hand, at least one component (one natural frequency) cannot be controlled in all state variables simultaneously, regardless of the choice of the input.

This uncontrollable component need not be unstable (as it is in the above example). If in Fig. 7.15 the input −4 transmittance is changed to +6, the uncontrollable terms are those in the decaying exponential e^{-3t}.

This discussion is somewhat disconcerting. Are systems often uncontrollable? If so, does this mean that it is then pointless to try to design the system for a desired performance?

Fortunately, the situation is really not so bleak. First, the majority of systems are completely controllable. If a system is not controllable, it means basically that there are modes of oscillation which contribute to the state variables, but which are not under the influence of the input (Fig. 7.16). Even when this situation occurs, the components contributed by the uncontrollable portion may be insignificant (for example, they may decay rapidly to zero). If they are significant, the system can often be readily redesigned to permit control.

Test for controllability

There is a simple test for controllability in terms of the parameters of the state model:

$$\dot{\mathbf{x}} = \mathbf{A}\mathbf{x} + \mathbf{B}u$$
$$y = \mathbf{C}\mathbf{x} + Du$$

(7-74)

The system of order n is controllable if (and only if) the matrix

$$[\,\mathbf{B}\,|\,\mathbf{A}\mathbf{B}\,|\,\mathbf{A}^2\mathbf{B}\,|\,\cdots\,|\,\mathbf{A}^{n-1}\mathbf{B}\,]$$

has a determinant which is not zero. Here the matrix above is formed by using $\mathbf{B}$ as the first column, $\mathbf{A}\mathbf{B}$ as the second column, $\mathbf{A}^2\mathbf{B}$ as the third column, and so on.

To illustrate this theorem, we return to our last example, redrawn in Fig. 7.17 with the lower input transmittance labelled β. (It was stated above that the system is

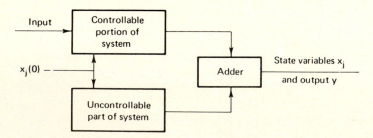

FIG. 7.16 Controllability divides a system into two parts.

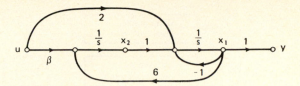

FIG. 7.17 The previous example shown in Fig. 7.15 with the input branch given a transmittance β.

uncontrollable with β equal to -4 or $+6$.) For this system

$$A = \begin{bmatrix} -1 & 1 \\ 6 & 0 \end{bmatrix} \qquad B = \begin{bmatrix} 2 \\ \beta \end{bmatrix} \tag{7-75}$$

Then

$$AB = \begin{bmatrix} -1 & 1 \\ 6 & 0 \end{bmatrix} \begin{bmatrix} 2 \\ \beta \end{bmatrix} = \begin{bmatrix} \beta - 2 \\ 12 \end{bmatrix} \tag{7-76}$$

Since $n = 2$ (there are two state variables), the matrix under Eq. (7-74) reduces to

$$[B \mid AB] = \begin{bmatrix} 2 & \beta - 2 \\ \beta & 12 \end{bmatrix} \tag{7-77}$$

The determinant of this matrix is

$$24 + 2\beta - \beta^2$$

which is zero only when β is -4 or $+6$. Hence, the system is controllable unless β is one of these two values.

This theorem is further illustrated in the problems at the end of the chapter. Here, we have not proved the theorem, but again, this aspect is left for later courses in system analysis.*

7.7 OBSERVABILITY

Figure 7.16 breaks down the system into two parts, one controllable and the other uncontrollable. Analogous to the concept that a portion of the system may be uncontrollable, there is the possibility that part may have no influence on the output. In

*The Polak and Wong Notes (see footnote on page 239) provide a proof.

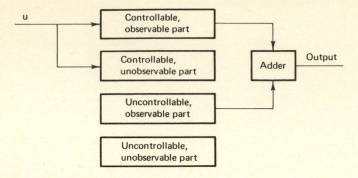

FIG. 7.18 General system decomposition.

other words, observation of the output y over any time interval may not be sufficient to allow determination of the initial values of one or more state variables. Such a system is called *unobservable*.

The situation is depicted in Fig. 7.18, which shows that any system can be broken into four components. Figure 7.19 represents one particularly simple and obvious example of the four parts. In an actual system model, such as that in the last section, the uncontrollable or unobservable portions may not be evident at all (no one state variable is uncontrollable, but rather a particular linear combination of state variables).*

*It is only when we model in normal form (the partial fraction expansion model of Chap. 6) that uncontrollability or unobservability become evident in the sense of Fig. 7.19.

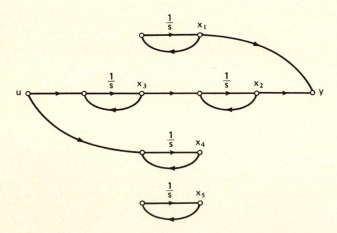

FIG. 7.19 x_1 and x_5 uncontrollable, x_4 and x_5 unobservable.

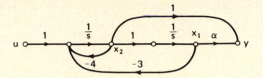

FIG. 7.20 System for example of unobservability.

Test for observability

In terms of the general model for the single-input, single-output linear system,

$$\dot{\mathbf{x}} = \mathbf{A}\mathbf{x} + \mathbf{B}u \left.\right\}$$
$$y = \mathbf{C}\mathbf{x} + \mathbf{D}u \left.\right\}$$

(7-78)

The system of order n is completely observable if and only if

$$[\mathbf{C}^T \mid \mathbf{A}^T \mathbf{C}^T \mid \mathbf{A}^{T2} \mathbf{C}^T \mid \; --- \mid \mathbf{A}^{T(n-1)} \mathbf{C}^T]$$

has a nonzero determinant. (The superscript T indicates the transpose of the matrix: the rows and columns are interchanged.)

As an example, the second order system of Fig. 7.20 is described by the matrices

$$\mathbf{A} = \begin{bmatrix} 0 & 1 \\ -3 & -4 \end{bmatrix} \qquad \mathbf{C} = [\alpha \quad 1]$$

(7-79)

Then

$$\mathbf{C}^T = \begin{bmatrix} \alpha \\ 1 \end{bmatrix} \qquad \mathbf{A}^T \mathbf{C}^T = \begin{bmatrix} 0 & -3 \\ 1 & -4 \end{bmatrix} \begin{bmatrix} \alpha \\ 1 \end{bmatrix} = \begin{bmatrix} -3 \\ \alpha - 4 \end{bmatrix}$$

and

$$[\mathbf{C}^T \mid \mathbf{A}^T \mathbf{C}^T] = \begin{bmatrix} \alpha & -3 \\ 1 & \alpha - 4 \end{bmatrix}$$

(7-80)

which possesses the determinant

$$\alpha^2 - 4\alpha + 3$$

The system is not completely observable if α equals 1 or 3.

Under these conditions, the system output y has no component of the form e^{-t} (if $\alpha = 1$) or e^{-3t} (if $\alpha = 3$). One of the natural frequencies can never be observed at the output. An alternate interpretation is that observation of the output permits measurement of a linear combination of $x_1(0)$ and $x_2(0)$, but not evaluation of the two initial conditions separately.

7.8 CONCLUDING COMMENT

In this chapter, we have introduced the state model in terms of systems which are linear and time-invariant and which have only a single drive signal and a single response. The op-amp realizations of Chap. 6 are state models, with each integrator output a state variable.

The state model is a much more detailed characterization of the system than the transfer function. *The transfer function describes only the controllable, observable portion of the system.* Furthermore, the transfer function strictly yields only the response to a given drive function; the state model also includes the response to the initial state and hence is not restricted to the study of inert systems.

Finally, the importance of the state model also derives from the possibility of extending this description to time-varying and nonlinear systems. In both cases, analysis of any system of even modest complexity usually requires a computer simulation, but the state model (in terms of first order differential equations) is directly useful as a basis for computer programming.

PROBLEMS

7.1   A system is described by the two state equations

$$\frac{dx_1}{dt} = 2x_2 + u$$

$$\frac{dx_2}{dt} = x_1$$

(a) What is the order of the system?
(b) Show a circuit realization.
(c) What initial conditions are required in your circuit of (b)?
(d) Is the system stable?
(e) What is the response when both initial conditions are zero and when $u = 2 \sin t$ for t greater than zero? (The response is the set of two state variables.)
(f) What is the steady state response in (e)?

7.2   Consider the system represented by the differential equation

$$y'''(t) + 7y''(t) + 14y'(t) + 8y(t) = 3u(t)$$

where y is the output and u is the input. [Note $u(t)$ is not a step function.]

(a) Determine four different state-variable realizations for the system. Show a completely labeled signal flow graph for each case. Show the first order differential equation for each state variable and then express the set of equations in the vector differential equation form:

$$\dot{x}(t) = Ax(t) + Bu(t)$$

where x is the state vector. Express the output y as a linear combination of the states.

(b) Determine the initial state vector $x(0)$ as a function of the values of the output and its first two derivatives at $t = 0$.

7.3   A dynamic system is given by the following state equations:

$$\begin{pmatrix} \dot{x}_1 \\ \dot{x}_2 \end{pmatrix} = \begin{pmatrix} 0 & 1 \\ -8 & -6 \end{pmatrix} \begin{pmatrix} x_1 \\ x_2 \end{pmatrix} + \begin{pmatrix} 0 \\ 1 \end{pmatrix} u$$

$$y = (3 \quad 1) \begin{pmatrix} x_1 \\ x_2 \end{pmatrix}$$

$$x(0) = \begin{pmatrix} 1 \\ 1 \end{pmatrix}$$

(a) Draw the signal flow graph including initial conditions.

(b) Find the transfer function Y/U.

(c) The state variables x_1 and x_2 are not accessible to measurement. However, another set of state variables (z_1 and z_2) related to x_1 and x_2 by the equation

$$\begin{pmatrix} z_1 \\ z_2 \end{pmatrix} = \begin{pmatrix} 2 & 1/2 \\ -1 & -1/2 \end{pmatrix} \begin{pmatrix} x_1 \\ x_2 \end{pmatrix}$$

are measurable. Find the state equations for z_1 and z_2.

(d) Draw the signal flow graph including the initial conditions on the state variables z_1 and z_2.

(e) For the original state model (in x_1 and x_2), determine the state transition matrix.

(f) Determine from (e) the output $y(t)$ when $u(t)$ is zero.

7.4 Consider the system represented by the following equation

$$y'''(t) + 7y''(t) + 14y'(t) + 8y(t) = 3u''(t) + 8u'(t) + 15u(t)$$

Repeat part (a) of Prob. 7.2; for one realization, factor $1/(s+1)$ and realize this as a subsystem in tandem with the remainder.

7.5 Consider the system characterized by

$$\frac{Y}{U} = \frac{b_0}{(s+1)(s^2+4s+4)}$$

We want the state model which corresponds to the partial fraction expansion of the transfer function.

(a) Show a completely and clearly labeled signal flow graph.

(b) Show the set of first order differential equations in the state variables.

(c) Express the result from (b) in vector-matrix form: i.e., $\dot{\mathbf{x}} = \mathbf{A}\mathbf{x} + \mathbf{B}u$. Comment on the structure of $\mathbf{A}$. What is the effect of the double pole of the transfer function?

(d) Express the output as a linear construction of the states.

(e) Determine the state transition matrix.

(f) Determine the output y when $b_0 = 4$, $u(t)$ is a unit step function, and the initial conditions are

$$x_1(0) = 1 \quad x_2(0) = x_3(0) = 0$$

where $x_1(t)$ comes from the system natural frequency $s = -1$.

7.6 We are interested in the network shown.

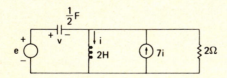

(a) Show that the state-space equations are

$$v'(t) = -v - 12i + e$$

$$i'(t) = -\frac{v}{2} + \frac{e}{2}$$

(b) Find the transfer function $I/E = H(s)$ and the single system differential equation relation e and i.

(c) What kind of terms appear in the response to $2e^{-3t}$? Is the system stable?

(d) If the state variables are v and i, and the output is the resistor voltage v_R, determine the expanded form of the equation $\dot{x} = Ax + Be$. Determine the matrix output equation for v_R.

(e) Draw the signal flow graph for the state model.

(f) Draw an analog-computer simulation using Rs, Cs, and ideal amplifiers.

7.7 (a) Construct the signal flow graph for the state equations

$$\begin{pmatrix} \dot{x}_1 \\ \dot{x}_2 \end{pmatrix} = \begin{pmatrix} 0 & 1 \\ -5 & -4 \end{pmatrix} \begin{pmatrix} x_1 \\ x_2 \end{pmatrix} + \begin{pmatrix} 0 \\ 1 \end{pmatrix} u$$

with the initial conditions

$$\begin{pmatrix} x_1(0) \\ x_2(0) \end{pmatrix} = \begin{pmatrix} 1 \\ 1 \end{pmatrix}$$

Apply Mason's rule to find the transfer function from u to x_1.

(b) It is desired to construct a signal flow graph with new state variables

$$\begin{pmatrix} z_1 \\ z_2 \end{pmatrix} = \begin{pmatrix} 1 & 1 \\ 0 & 1 \end{pmatrix} \begin{pmatrix} x_1 \\ x_2 \end{pmatrix}$$

Construct this signal flow graph including the initial conditions for z. Now find the transfer function to z_1.

7.8 (a) Find two distinct signal flow graphs for the transfer function (one must be the partial fraction expansion)

$$G(s) = \frac{s^2 + 3s + 2}{s(s + 1)^3}$$

Do not cancel the common factor in numerator and denominator.

(b) How many state variables are there in each of your realizations?

(c) On the basis of (b), can you comment critically on these realizations?

7.9 Write the state equations for the system shown, where x is the input and y the output. Is the system controllable? Is it observable from the output? We might also use as output the voltage across C or across L. What would be the possible applications of such a system?

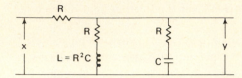

7.10 A unicycle suspension system is shown.

(a) We can first assume that the mass of the body of the car is the only significant mass and the spring shown is the only significant spring coefficient. Consider only vertical motion. The road contour determines the excitation, and the vertical velocity of the body is the desired response. Find a state model with the state variables chosen to be the energy-storage variables.

(b) In a more complete model, consider the unsprung mass, the springiness of the tire, and the tire friction. The last may be represented by a second ideal damper element having a coefficient B_1. The source is now the displacement below this spring. Determine the state model.

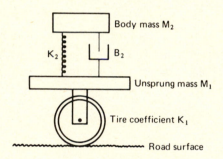

7.11 Consider a three-car commuter train. We want to focus attention on the forward-and-backward motion of the individual cars. In other words, initially we neglect the sideways motion or any pitching or rolling or vertical motion. Construct and discuss a state model which might indicate quantitatively the manner in which motion occurs. Include in your discussion consideration of the sources or inputs which are relevant.

If we neglect all damping and consider the cars as simple masses which are coupled by pure, ideal springs, what is the order of the system? How many different modes of sinusoidal motion are possible? Discuss for such a system the manner in which the frequencies of oscillation depend upon the parameters. Do the frequencies depend on the particular drive signals which are important?

7.12 The system shown represents a "force simulator" inserted in an aircraft to develop for the pilot a force from the stick onto the pilot's hand. In the older low-speed aircraft, the pilot and joy stick are directly coupled mechanically to the control surfaces (e.g., the elevator). In modern high-speed aircraft, however, the force required to move the control surface is so great that power amplification must be provided, usually through a hydraulic amplifier. When the amplifier is included, the joy stick is decoupled from the control surface, and the pilot has no feel of the forces acting on the surface. In order to create an artificial feel, a mechanical network is added at the stick. The figure shows one possibility, simplified for analysis by orientation of the elements so that all motion is in the vertical direction.

(a) Write the equations of motion (define carefully all velocities required).

(b) Determine a state model.

(c) If v_a is a constant, what force f_s must the pilot apply to hold v_s equal to zero (to hold the stick fixed in position)?

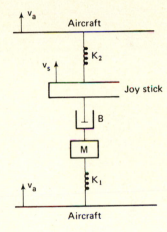

8 MULTIDIMENSIONAL SYSTEMS

Driving an automobile down one lane of a highway serves as a familiar systems problem. These are the input signals: the curvature of the road, and the position and velocity of other vehicles just ahead of our car. The driver observes these inputs and responds by adjusting the accelerator pedal (or brake) and the angular position of the steering wheel—both signals then, serving as inputs to the vehicle.

There are also several output signals from the system: signals which the system is supposed to control. The position and velocity of the car along the lane are certainly two primary outputs, but we are also interested in the sideways position and velocity, the acceleration and its derivative (called the jerk) since they determine riding comfort for the passengers, and the vertical vibration induced as the car passes over the rough road surface.

Thus, analysis of this system requires consideration of several inputs and outputs, in contrast to the examples of the last two chapters which were restricted to a single input and a single output. While we might try to view the car system as a set of separate or decoupled systems (e.g., one determining vertical motion), there is often interaction or coupling among these parts, and we have to treat the system as *multidimensional* (more than one input and/or output).

In this chapter, we want to consider how the concepts of the preceding chapters are modified or extended when the system is multidimensional.

8.1 SIMPLE MULTIDIMENSIONAL DESIGN PROBLEM

Before considering the properties of multidimensional systems, we consider one very simple design problem, shown in Fig. 8.1. The process we are trying to control has two outputs (y_1 and y_2) and two control signals (m_1 and m_2), the four signals interrelated by the four P transfer functions. There are two system inputs (u_1 and u_2). Can we insert feedback, represented by the dashed lines and the H transfer functions, to achieve adequate control?

In the system, u_1 tends to affect both y_1 and y_2; likewise, u_2 influences both y_1 and y_2. If we want to achieve a simple control scheme, we might logically ask that u_1 influence y_1 only, and that u_2 influence y_2. Can we select the four H functions so that this *noninteracting* control is achieved, and also so that the two resulting transfer functions

$$T_1 = \frac{Y_1}{U_1} \quad \text{and} \quad T_2 = \frac{Y_2}{U_2}$$

have desired performance characteristics?

A matrix formulation

Matrices are a real help in multidimensional problems. If we try to use Mason's reduction theorem to find the Y_2/U_1 and Y_1/U_2 transfer functions (and the conditions the Hs must satisfy to make these both zero), we are rapidly entrenched in a maze of algebraic equations. If the system were 3×3 (three inputs and three outputs), the analysis would be totally hopeless.

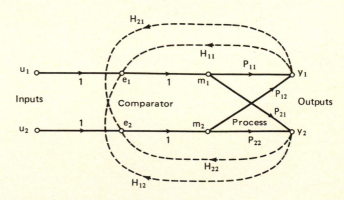

FIG. 8.1 A two-dimensional system.

Fortunately, matrix notation allows us to derive very simply the conditions on the Hs. In matrix terms, the outputs y_1 and y_2 are related to m_1 and m_2 by the transform relations:

$$\mathbf{Y} = \mathbf{PM} \tag{8-1}$$

where

$$\mathbf{Y} = \begin{bmatrix} Y_1(s) \\ Y_2(s) \end{bmatrix} \qquad \mathbf{M} = \begin{bmatrix} M_1(s) \\ M_2(s) \end{bmatrix} \qquad \mathbf{P} = \begin{bmatrix} P_{11} & P_{12} \\ P_{21} & P_{22} \end{bmatrix} \tag{8-2}$$

Similarly, the remainder of the system is described by

$$\mathbf{M} = \mathbf{1E} \tag{8-3}$$

$$\mathbf{E} = \mathbf{U} + \mathbf{HY} \tag{8-4}$$

where $\mathbf{H}$ is the matrix

$$\begin{bmatrix} H_{11} & H_{12} \\ H_{21} & H_{22} \end{bmatrix}$$

We can now combine the above equations to eliminate the variables $\mathbf{M}$ and $\mathbf{E}$. Equations (8-3) and (8-1) yield

$$\mathbf{Y} = \mathbf{PE} \tag{8-5}$$

and with Eq. (8-4), we have

$$\mathbf{Y} = \mathbf{P}[\mathbf{U} + \mathbf{HY}] \tag{8-6}$$

Before bringing the terms in $\mathbf{Y}$ to the left side, we can premultiply both sides by $\mathbf{P}^{-1}$:

$$[\mathbf{P}^{-1} - \mathbf{H}]\mathbf{Y} = \mathbf{U} \tag{8-7}$$

or

$$\mathbf{Y} = [\mathbf{P}^{-1} - \mathbf{H}]^{-1}\mathbf{U} \tag{8-8}$$

This equation shows immediately how to select the four H feedback transfer functions. In terms of the $\mathbf{P}$ and $\mathbf{H}$ components defined above, Eq. (8-7) can be written

$$
\begin{bmatrix}
\dfrac{P_{22}}{\Delta_P} - H_{11} & -\dfrac{P_{12}}{\Delta_P} - H_{12} \\[4mm]
-\dfrac{P_{21}}{\Delta_P} - H_{21} & \dfrac{P_{11}}{\Delta_P} - H_{22}
\end{bmatrix}
\begin{bmatrix} Y_1 \\[4mm] Y_2 \end{bmatrix}
=
\begin{bmatrix} U_1 \\[4mm] U_2 \end{bmatrix}
\tag{8-9}
$$

where Δ_P is the determinant of P or $P_{11}P_{22} - P_{12}P_{21}$. Then we have

Conditions for noninteracting control:

$$
H_{12} = -\frac{P_{12}}{\Delta_P} \tag{8-10}
$$

$$
H_{21} = -\frac{P_{21}}{\Delta_P} \tag{8-11}
$$

As soon as these conditions are satisfied and, if T_1 and T_2 are the desired transfer functions, H_{11} and H_{22} are found from:

$$
T_1 = \frac{Y_1}{U_1} = \frac{1}{P_{22}/\Delta_P - H_{11}} \tag{8-12}
$$

$$
T_2 = \frac{Y_2}{U_2} = \frac{1}{P_{11}/\Delta_P - H_{22}} \tag{8-13}
$$

Equations (8-10)–(8-13) are amazing. They tell us that, given a multidimensional system, we can use feedback to achieve both

(1) noninteracting control, and
(2) any desired transfer functions (here T_1 and T_2).

Of course, we must be sure that the H transfer functions which come from these equations can be built. If Eq. (8-10) should give

$$
H_{12} = s^3 + s^2 + s + 1 \tag{8-14}
$$

for example, we would hesitate to try to build the third-derivative term (the noise would dominate the output). Except for this practical aspect, the analysis above provides a total design.

An example

As an illustration, we consider control of the process described by

$$P_{11}(s) = \frac{s}{(s + 1)(s + 2)} \qquad P_{12}(s) = \frac{-2}{(s + 1)(s + 2)}$$
$$P_{21}(s) = \frac{1}{(s + 1)(s + 2)} \qquad P_{22}(s) = \frac{s + 3}{(s + 1)(s + 2)} \qquad (8\text{-}15)$$

Then the **P** determinant is

$$\Delta_P = \frac{1}{(s + 1)(s + 2)} \qquad (8\text{-}16)^*$$

For a noninteracting system, we should choose [Eqs. (8-10) and (8-11)]:

$$H_{12} = -\frac{P_{12}}{\Delta_P} = 2$$
$$H_{21} = -\frac{P_{21}}{\Delta_P} = -1 \qquad (8\text{-}17)$$

Regardless of the choice of H_{11} and H_{22}, these two values give noninteraction: y_1 depends only on u_1, y_2 on u_2.

To select H_{11} and H_{22}, we consider Eqs. (8-12) and (8-13), which become

$$T_1 = \frac{1}{s + 3 - H_{11}}$$
$$T_2 = \frac{1}{s - H_{22}} \qquad (8\text{-}18)$$

or

$$H_{11} = s + 3 - \frac{1}{T_1}$$
$$H_{22} = s - \frac{1}{T_2} \qquad (8\text{-}19)$$

---

*In general, Δ_P has double poles at the simple poles of each P function. In many cases, however, the numerator of Δ_P cancels one of these denominator factors.

We must now select T_1 and T_2 to meet performance specifications and also to yield realizable H_{11} and H_{22} functions. We might, for example, select T_1 and T_2 each as

$$\frac{1}{s + 8}$$

to achieve a time constant of 1/8 second. Then

$$H_{11} = -5 \qquad H_{22} = -8 \tag{8-20}$$

and all four feedback elements (the Hs) are just constants.

In this design, by imposing the condition for noninteracting controls, we have effectively decomposed the 2×2, multidimensional problem into two one-input, one-output problems. Obviously, this decomposition is a convenience in design, but it is not always desirable. We certainly sacrifice some freedom, possibly some speed of response, and some performance capability, by forcing the total system to behave as a set of simpler, isolated systems.

Furthermore, to employ the above design approach, we have to use four feedback paths, from each y to each e. In a system of higher dimensions, we would have a multitude of feedback paths to insert (in the 4×4 case, 16 paths).

In the next two sections, we turn attention to a different facet of multidimensional systems: how to synthesize such systems, or build a simulation—essentially the extension of the concepts of Chap. 6 to the system with more than one input and/or output.

8.2 SIMULATION OF MULTI-INPUT OR MULTI-OUTPUT SYSTEMS

If there are several inputs and only one output, the system is described by an equation of the form

$$Y(s) = T_1(s) U_1(s) + T_2(s) U_2(s) + T_3(s) U_3(s) \tag{8-21}$$

where, to be specific, we consider three inputs. The corresponding block diagram is shown in Fig. 8.2. If the three transfer functions (T_1, T_2 and T_3) possess entirely different sets of poles (or natural frequencies), the system is realized by the addition of the three components of $y(t)$ at the output (Fig. 8.3).

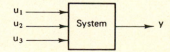

FIG. 8.2 Three-input, one-output system.

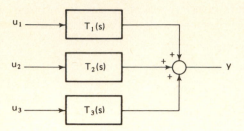

FIG. 8.3 Appropriate realization when T_1, T_2, and T_3 are distinct systems.

The interesting case occurs when the poles of all three transfer functions are identical— the situation which often exists when a single process is involved, and u_1, u_2, and u_3 represent three different ways in which the one system can be controlled. In such a case, the model should involve, if possible, only a single realization of each of the poles.

The problem (and the solution as well) can be illustrated in terms of the specific example

$$Y(s) = \frac{12}{s^3 + 8s^2 + 19s + 12} U_1 + \frac{6(s + 2)}{s^3 + 8s^2 + 19s + 12} U_2$$

$$+ \frac{s^2 + 7s + 10}{s^3 + 8s^2 + 19s + 12} U_3 \qquad (8\text{-}22)$$

The denominator polynomial can be realized by many different feedback configurations; the one of particular interest with multiple inputs has all feedback from x_1 (to $\dot{x}_1, \dot{x}_2$, and $\dot{x}_3$). The numerators are then realized by the feed-forward branches from u_1, u_2, and u_3 into the $\dot{x}_j$ terminals (Fig. 8.4).

The solution indicates clearly why the configuration with feedback from x_1 is selected when the system possesses several input signals, each related to the output by a transfer function with a different numerator: each of the separate numerators is realized by appropriate branches from u_j into the basic feedback structure. Figure 8.4 also demonstrates that the system of Eq. (8-22) is fundamentally third order: only three integrators are required for the simulation.

More than one output

The analogous problem with a single input and several outputs is readily handled by the flow diagram with all feedback to $\dot{x}_n$. The separate numerators are then realized by appropriate transmittances from the state variables to the various outputs y_1, y_2, ..., y_m. The example of Fig. 8.5 shows the approach.

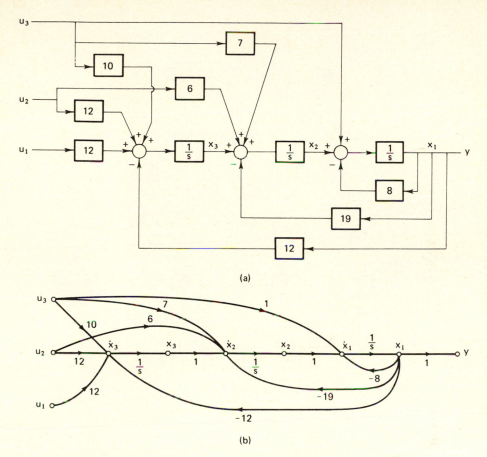

FIG. 8.4 Realization of Eq. (8-22): (a) Block diagram; (b) Signal flow diagram.

8.3   MULTIPLE INPUTS AND OUTPUTS

When the system possesses both multiple inputs and multiple outputs, determination of a model is more difficult. We can no longer utilize one of the simple diagrams derived for the case of one input and one output.

A very simple problem

As a very simple example with two inputs u_1 and u_2, two outputs y_1 and y_2, we consider the system described by the equations:

$$Y_1 = \frac{3}{s+2} U_1 + \frac{2}{s+2} U_2 \left.\right\}$$

$$Y_2 = \frac{9}{s+2} U_1 + \frac{6}{s+2} U_2 \left.\right\}$$

(8-23)

One correct realization is shown in Fig. 8.6.

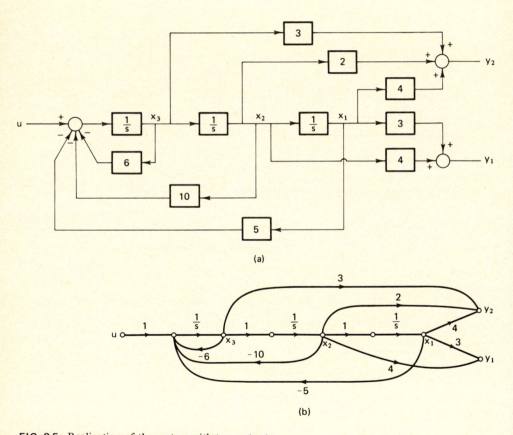

(a)

(b)

FIG. 8.5 Realization of the system with two outputs:

$$\left\{ \begin{array}{l} Y_1 = \dfrac{4s+3}{s^3 + 6s^2 + 10s + 5} U \\[2ex] Y_2 = \dfrac{3s^2 + 2s + 4}{s^3 + 6s^2 + 10s + 5} U \end{array} \right.$$

(a) Block diagram; (b) Signal flow diagram.

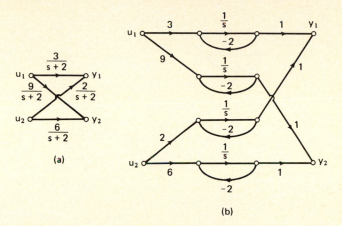

FIG. 8.6 Representation and simulation of 2×2 system example: (a) Diagram for Eqs. (8-23); (b) Realization of (a) with integrators.

Do we actually need four integrators, as depicted in part (b) of the figure? Are there really four initial conditions which must be specified independently? In other words, is the system really fourth order?

Since each of the four transfer functions has a pole at -2, we might be optimistic and hope we can do the job with only one integrator (Fig. 8.7). Can we add branches from

$$\left. \begin{array}{l} u_1 \text{ to } \dot{x}_1 \\ u_2 \text{ to } \dot{x}_1 \end{array} \right\}$$

$$\left. \begin{array}{l} x_1 \text{ to } y_1 \\ x_1 \text{ to } y_2 \end{array} \right\}$$

so that all four of the desired transfer functions are realized?

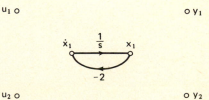

FIG. 8.7 An optimistic start in the realization of Eqs. (8-23).

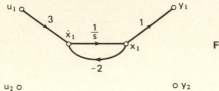

FIG. 8.8 Realization of

$$Y_1 = \frac{3}{s+2} U_1.$$

Boldly, we move ahead. The u_1 to y_1 path should have a transfer function

$$\frac{3}{s+2}$$

so we insert a 3 from u_1 to $\dot{x}_1$, a 1 from x_1 to y_1. (It makes no difference how we split up the 3 factor into two parts.) We now have Fig. 8.8.

From u_2 to y_1, we would like

$$\frac{2}{s+2}$$

This can be realized by a branch of gain 2 from u_2 to $\dot{x}_1$, with Fig. 8.9 the result.

Now we turn to the second of Eqs. (8-23):

$$Y_2 = \frac{9}{s+2} U_1 + \frac{6}{s+2} U_2 \qquad (8\text{-}24)$$

The first term specifies the transmission from u_1 to y_2. This term can be realized by a branch from x_1 to y_2 with gain 3 (Fig. 8.10). Now we have inserted all possible branches; we still have to realize the correct transmission from u_2 to y_2:

$$\frac{6}{s+2}$$

If we inspect Fig. 8.10, we find that we have already realized that term exactly (the product of 2 from u_2 to $\dot{x}_1$ and 3 from x_1 to y_2).

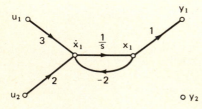

FIG. 8.9 System for

$$Y_1 = \frac{3}{s+2} U_1 + \frac{2}{s+2} U_2$$

(the first of the two equations to be realized).

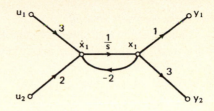

FIG. 8.10 Simulation for

$$Y_1 = \frac{3}{s+2} U_1 + \frac{2}{s+2} U_2$$

$$Y_2 = \frac{9}{s+2} U_1 + \frac{6}{s+2} U_2$$

Obviously we were lucky;* If the two equations had been

$$Y_1 = \frac{3}{s+2} U_1 + \frac{2}{s+2} U_2 \Bigg)$$

$$Y_2 = \frac{9}{s+2} U_1 + \frac{4}{s+2} U_2 \Bigg\}$$ (8-25)

we still would have proceeded in exactly the same way to realize the first three residues (3, 2, and 9). The result would be the two equations of Fig. 8.10. Then to realize Eqs. (8-25), we would need to add a component to Y_2:

$$-\frac{2}{s+2} U_2$$

and the complete simulation would be as shown in Fig. 8.11.

Figures 8.10 and 8.11 show that the original 2×2 system can be realized by two integrators at most, and under certain conditions, by one integrator. The system certainly need not be fourth order, as we had originally postulated.

*It was not really luck; our good fortune occurred because the determinant of the residues

$$\begin{vmatrix} 3 & 2 \\ 9 & 6 \end{vmatrix}$$

is zero, as we point out near the end of this section.

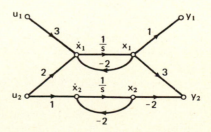

FIG. 8.11 Simulation for

$$Y_1 = \frac{3}{s+2} U_1 + \frac{2}{s+2} U_2 \Bigg)$$

$$Y_2 = \frac{9}{s+2} U_1 + \frac{4}{s+2} U_2 \Bigg\}$$

A more complete example

This minimum realization (i.e., simulation of minimum order) can be more adequately illustrated by the following two-dimensional example:

$$
\left.
\begin{aligned}
Y_1 &= \frac{3s^2 + 12s + 11}{s^3 + 6s^2 + 11s + 6} U_1 + \frac{5s^2 + 19s + 16}{s^3 + 6s^2 + 11s + 6} U_2 \\[2mm]
Y_2 &= \frac{4s + 9}{s^2 + 5s + 6} U_1 + \frac{9s + 21}{s^2 + 5s + 6} U_2
\end{aligned}
\right\}
\qquad (8\text{-}26)
$$

In such a case, we ordinarily describe the system by the matrix of transfer functions, $H(s)$, where the matrix equation

$$
\mathbf{Y} = \mathbf{HU} \qquad (8\text{-}27)
$$

represents the equations

$$
\begin{bmatrix} Y_1 \\ Y_2 \end{bmatrix} = \begin{bmatrix} H_{11} & H_{12} \\ H_{21} & H_{22} \end{bmatrix} \begin{bmatrix} U_1 \\ U_2 \end{bmatrix} \qquad (8\text{-}28)
$$

Thus, in our example

$$
\mathbf{H}(s) = \begin{bmatrix} \dfrac{3s^2 + 12s + 11}{(s+1)(s+2)(s+3)} & \dfrac{5s^2 + 19s + 16}{(s+1)(s+2)(s+3)} \\[4mm] \dfrac{4s + 9}{(s+2)(s+3)} & \dfrac{9s + 21}{(s+2)(s+3)} \end{bmatrix} \qquad (8\text{-}29)
$$

The problem of how to simulate (or model) such a set of transfer functions is not trivial. A first approach, albeit naive, might be the system shown in Fig. 8.12, but the engineer recognizes at once that the system is probably not tenth order, as implied by this simulation. With each of the various poles appearing in so many of the four transfer functions, we should be able to find a realization utilizing fewer than ten integrators.

A systematic procedure does exist for achieving a simulation of minimum complexity when the four characteristic polynomials can be factored. First we make a partial fraction expansion of the $\mathbf{H}(s)$ matrix (a partial fraction expansion of each function, with the terms for identical poles collected). In the present example, the four transfer

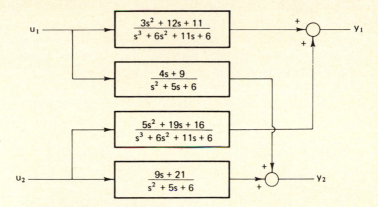

FIG. 8.12 System for approach to Eqs. (8-26).

functions possess three different poles (at $-1, -2$, and -3). Hence, if we make a partial fraction expansion of *each* transfer function, we can write the $H(s)$ matrix as

$$\mathbf{H}(s) = \frac{1}{s+1}\begin{bmatrix} 1 & 1 \\ 0 & 0 \end{bmatrix} + \frac{1}{s+2}\begin{bmatrix} 1 & 2 \\ 1 & 3 \end{bmatrix} + \frac{1}{s+3}\begin{bmatrix} 1 & 2 \\ 3 & 6 \end{bmatrix} \tag{8-30}$$

Here the matrix multiplying $1/(s+2)$ simply represents the residues of the four transfer functions in the pole at -2; if $H_{jk}(s)$ has no pole, the residue is zero.

The matrices in Eq. (8-30) are termed the *residue matrices* for the two-dimensional system:

$$\mathbf{k}_{-1} = \begin{bmatrix} 1 & 1 \\ 0 & 0 \end{bmatrix} \qquad \mathbf{k}_{-2} = \begin{bmatrix} 1 & 2 \\ 1 & 3 \end{bmatrix} \qquad \mathbf{k}_{-3} = \begin{bmatrix} 1 & 2 \\ 3 & 6 \end{bmatrix} \tag{8-31}$$

We determine next the *rank* of each of these residue matrices. The rank is the maximum order of a nonzero determinant. For example, the determinant of $\mathbf{k}_{-2}$ has the value 1; since this value is not zero, the rank of $\mathbf{k}_{-2}$ is two. On the other hand, both $\mathbf{k}_{-1}$ and $\mathbf{k}_{-3}$ have determinants which are zero; in both cases, a one-by-one determinant derived from these matrices (by using the element in the first row and column) is not zero. Therefore, both $\mathbf{k}_{-1}$ and $\mathbf{k}_{-3}$ have rank one. Hence, the three ranks are

$$r_{-1} = 1 \qquad r_{-2} = 2 \qquad r_{-3} = 1 \tag{8-32}$$

The original set of transfer functions [or the $H(s)$ of Eq. (8-30)] can now be simulated *with each pole realized a number of times equal to the corresponding rank* of Eq.

(8-32). Thus, we require only one realization of the poles at both -1 and -3, but two realizations of the pole at -2.

The basic dynamical structure of the canonic model can now be drawn (Fig. 8.13). The only problem remaining is to find those branches connecting u_1 and u_2 to the loops (and the loops to y_1 and y_2) in such a way as to realize the four specified transfer functions.

For the final step, we work with Eq. (8-30), which is repeated here:

$$\mathbf{H}(s) = \frac{1}{s+1}\begin{bmatrix} 1 & 1 \\ 0 & 0 \end{bmatrix} + \frac{1}{s+2}\begin{bmatrix} 1 & 2 \\ 1 & 3 \end{bmatrix} + \frac{1}{s+3}\begin{bmatrix} 1 & 2 \\ 3 & 6 \end{bmatrix} \tag{8-33}$$

The first term states that Y_1 (the top row) contains the factors $[1/(s+1)]\,U_1$ and $[1/(s+1)]\,U_2$; hence we feed from both u_1 and u_2 to $\dot{x}_1$ through unity branches, then from x_1 to y_1 through a unity branch. This now completes the realization of the first term in Eq. (8-33)—the pole at -1.

When we turn now to the second term of Eq. (8-33)—the pole at $-2-$, the rank here is two and we have two loops available in Fig. 8.13. We realize first the dependence of y_1 on u_1 and u_2. For this, we use the top row again and

$$Y_1 = \frac{1}{s+2}\,U_1 + \frac{2}{s+2}\,U_2 \tag{8-34}$$

We insert a unity branch from x_2 to y_1, a branch of unity gain from u_1 to $\dot{x}_2$, and a branch of gain 2 from u_2 to $\dot{x}_2$. The $\dot{x}_3 - x_3$ loop can be reserved for realization of the second row of the $\mathbf{k}_{-2}$ matrix.

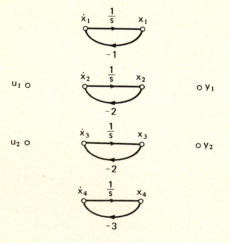

FIG. 8.13 Structure of the feedback loops. (We still need to insert the branches connecting to u_1, u_2, y_1, and y_2.)

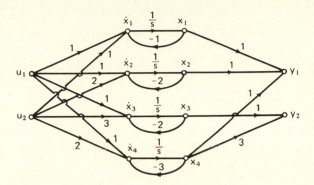

FIG. 8.14 Signal flow diagram for Eqs. (8-26).

The final term of Eq. (8-33) is simulated by branches from u_1 to u_2 to $\dot{x}_4$ and branches from x_4 to both y_1 and y_2. The $u_1 \rightarrow \dot{x}_4$, $x_4 \rightarrow y_1$ branches realize the 1 entry in $\mathbf{k}_{-3}$; the $x_4 \rightarrow y_2$ branch then realizes the 3 entry; the $u_2 \rightarrow \dot{x}_4$ branch finally gives the entire second column of $\mathbf{k}_{-3}$. The complete simulation model is shown in Fig. 8.14.

The result of this problem is remarkable! Even though solution requires factoring of the characteristic polynomial of each of the given transfer functions (we are essentially deriving a partial-fraction state model), the result is of major significance. First, the solution demonstrates that the original system, which might be of order ten, *is actually of order four*: only four integrators are required for the simulation, and only four state variables suffice (or only four initial conditions are required to determine the response from specified input signals). Thus, the solution leads step-by-step to a simulation model of minimum possible complexity; the method provides an important building block in both analog and digital simulation work.

In addition, the model shows clearly the state variables which must be measured in order to determine system response. The system order, revealed by the flow diagram, indicates the complexity of the stability analysis problem* and the nature of the problem of optimization—the selection of $u_1(t)$ and $u_2(t)$ in order to optimize the system performance. In other words, the basic model reveals the fundamental nature of the process described by the original set of transfer functions (or differential equations or measured process characteristics).

*For example, if we insert feedback around the process by adding branches from y_1 to u_1 and y_2 to u_2 (branches with constant transmittances), the characteristic polynomial for the closed-loop system is fourth order in s. The determinant of the $\mathbf{H}(s)$ matrix of transfer functions possesses only four poles (simple poles at -1 and -3, a double pole at -2) after the common factors in numerator and denominator are canceled.

8.4 CONTROLLABILITY AND OBSERVABILITY

The concept of controllability and observability discussed in Secs. 7.6 and 7.7 can be extended to multidimensional systems. In this case, the general state model is

$$
\begin{aligned}
\dot{\mathbf{x}} &= \mathbf{A}\mathbf{x} + \mathbf{B}\mathbf{u} \\
\mathbf{y} &= \mathbf{C}\mathbf{x} + \mathbf{D}\mathbf{u}
\end{aligned} \tag{8-35}
$$

where $\mathbf{u}$ and $\mathbf{y}$ are now vectors (column matrices) rather than single (scalar) signals.

In Sec. 7.6, we stated that the system of order n is completely controllable if the matrix

$$[\,\mathbf{B}\,|\,\mathbf{A}\mathbf{B}\,|\,\mathbf{A}^2\mathbf{B}\,|\ \cdots\ |\,\mathbf{A}^{n-1}\,\mathbf{B}\,]$$

has a nonzero determinant. In the one-dimensional case, this matrix is $n \times n$ (n rows and n columns); we need consider only the determinant of the matrix.

For the multidimensional system, this same matrix is in general $n \times nr$ (n rows and r columns), where r is the number of different inputs. The system is controllable if, and only if this matrix is of rank n (i.e., there is an $n \times n$ submatrix which has a nonzero determinant).

Example

The second order system of Fig. 8.15(a) serves as an illustration (the same system as in Fig. 8.11). Here

$$
\begin{aligned}
\dot{x}_1 &= -2x_1 + 3u_1 + 2u_2 \\
\dot{x}_2 &= -2x_2 + u_2 \\
y_1 &= x_1 \\
y_2 &= 3x_1 - 2x_2
\end{aligned} \tag{8-36}
$$

The matrices of Eqs. (8-35) are

$$
\mathbf{A} = \begin{bmatrix} -2 & 0 \\ 0 & -2 \end{bmatrix} \qquad
\mathbf{B} = \begin{bmatrix} 3 & 2 \\ 0 & 1 \end{bmatrix} \tag{8-37}
$$

The expanded matrix is

$$
[\,\mathbf{B}\,|\,\mathbf{A}\mathbf{B}\,] = \begin{bmatrix} \underbrace{3 \quad\ \ 2}_{\mathbf{B}} & \underbrace{-6 \quad -4}_{\mathbf{A}\mathbf{B}} \\ 0 \quad\ \ 1 & 0 \quad\ -2 \end{bmatrix} \tag{8-38}
$$

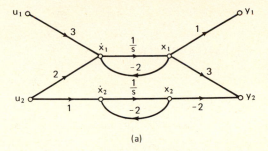

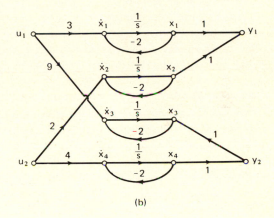

(a)

FIG. 8.15 Realizations of

$$Y_1 = \frac{3}{s+2} U_1 + \frac{2}{s+2} U_2$$

$$Y_2 = \frac{9}{s+2} U_1 + \frac{4}{s+2} U_2$$

(a) Second order system (a minimum realization); (b) Brute-force realization with four integrators

$$\dot{x}_1 = -2x_1 + 3u_1$$
$$\dot{x}_2 = -2x_2 + 2u_2$$
$$\dot{x}_3 = -2x_3 + 9u_1$$
$$\dot{x}_4 = -2x_4 + 4u_2$$

(b)

which does have rank 2 (the first two columns, for example, have a determinant of 3). Hence, the second order system realized in Fig. 8.15(a) is completely controllable.

In contrast, the brute-force simulation with four integrators shown in Fig. 8.15(b) is described by

$$\mathbf{A} = \begin{bmatrix} -2 & 0 & 0 & 0 \\ 0 & -2 & 0 & 0 \\ 0 & 0 & -2 & 0 \\ 0 & 0 & 0 & -2 \end{bmatrix} \qquad \mathbf{B} = \begin{bmatrix} 3 & 0 \\ 0 & 2 \\ 9 & 0 \\ 0 & 4 \end{bmatrix} \qquad (8\text{-}39)$$

and the expanded, test matrix is

$$[\,B\,|\,AB\,|\,A^2B\,|\,A^3B\,] \;=\; \begin{bmatrix} 3 & 0 & -6 & 0 & 12 & 0 & -24 & 0 \\ 0 & 2 & 0 & -4 & 0 & 8 & 0 & -16 \\ 9 & 0 & -18 & 0 & 36 & 0 & -72 & 0 \\ 0 & 4 & 0 & -8 & 0 & 16 & 0 & -32 \end{bmatrix}$$

$$(8\text{-}40)$$

This matrix is of rank only 2, not 4 as required if the fourth order system is to be controllable. Hence, the realization of (b) is not controllable—a fact which is intuitively obvious since the two state variables, x_1 and x_3, are controlled in the same way by u_1. We cannot specify future values of x_1 and x_3 independently and expect to find a $u_1(t)$ which will move the system to this specified state.

Observability

The extension of the concept of observability to multidimensional systems follows a similar pattern. The condition for complete observability of the system states is that the expanded, $n \times nm$ matrix (where m is the number of outputs)

$$[\,C^T\,|\,A^T C^T\,|\,A^{T2}C^T\,|\;\cdots\;|\,A^{T(n-1)}C^T\,]$$

has rank n.

If we were to analyze the two simulations of Fig. 8.15, we would find that the minimal realization is completely observable.

Significance of the transfer function

These concepts of controllability and observability lead to an understanding of the relation between the transfer function model and the state model. The transfer function (or the matrix of transfer functions in a multidimensional system) is equivalent to a controllable, observable state model.

If the actual, physical system we are studying is completely controllable and observable, the two models are equivalent. If the system is not controllable or not observable, the transfer function gives no indication how to insert the additional state variables (beyond those of the minimum realization). It is in this sense that the state model is a more complete system description than the transfer function.

Even if the actual system is known to be controllable and observable, the state model is sometimes advantageous in system analysis and design. If the state model is constructed from an analysis of the actual system, we can often choose as state variables those signals which physically represent energy storage, or those signals which can be readily measured to evaluate the initial state. If we start from the transfer function, we can select an enormous variety of different state variables; often it is not clear which choice leads to the simplest evaluation of the initial state.

8.5 A FINAL, SIMPLE EXAMPLE

The preceding sections of this chapter focused on the extensions of the earlier theory to multidimensional systems. We should not lose sight of the fact that many systems problems do not require an arsenal of theoretical weapons for solution. Rather, these problems demand only basic understanding of system models and the ability to make intelligent deductions about dynamic performance, stability, and so on.

Throughout engineering, very often we find this somewhat intangible relationship between theory and practice. Very few systems are ever designed using advanced or complex theory. The most important function of the theory is to give the human designer a better understanding of the phenomena which may appear in system performance, a better feeling for the parameters which should be studied in depth (often in an analog or digital simulation), and a better creative imagination for the ways in which the system might be modified to improve performance. Successful engineering is still very much an art.

The following simple problem hopefully illustrates some of these ideas.

The system

A blending system is one in which we mix two fluids (A and B) to obtain an output which has the proper concentration by weight of fluid A. A basic scheme for controlling the blending operation is shown in Fig. 8.16.

Two outputs are to be controlled: the flow rate q_o of the mixture in pounds/hour, and the concentration c_o, or the percentage of weight of liquid A. There are also two inputs: the desired values of flow and concentration, q_d and c_d.

The error in flow is used to adjust the flow of liquid A, the error in concentration to adjust the flow of B. This separation of the two control functions simplifies the design. (Clearly, we could have interchanged roles here, with the concentration error controlling q_A.) Each comparator and valve control combination is described by a transfer

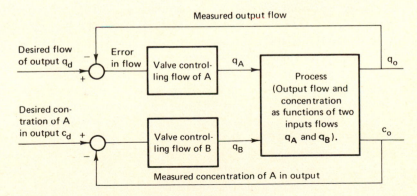

FIG. 8.16 General operation of the blending system.

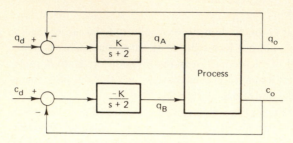

FIG. 8.17 Block diagram before the process is described quantitatively.

function

$$\frac{K}{s + 2}$$

(There is a time constant of 1/2 second in the response.) Thus, the pictorial diagram of Fig. 8.16 leads to the block diagram of Fig. 8.17.

Now we need to describe quantitatively the operation of the process. In a simple blending operation, fluid A flows in at a rate q_A, fluid B at the rate q_B, and the two fluids are added. Hence

$$q_o = q_A + q_B \tag{8-41}$$

The concentration of A in the output is

$$c_o = \frac{q_A}{q_A + q_B} \tag{8-42}$$

These two equations describe the process: how q_o and c_o are determined from q_A and q_B. While Eq. (8-41) is linear, Eq. (8-42) is nonlinear. We might go to a digital simulation, represent these two equations plus the relationships shown in Fig. 8.17, and evaluate the response for various input signals. Such computer analysis indicates accurately system performance; unless we test with a very large range of signals, however, we are always worried that we may have missed exactly those signals which yield particularly important system performance characteristics. For example, the system may be unstable within a range of operating conditions, yet our test signals may never put us in this range.

Simulation studies are, of course, essential. Before undertaking these, however, we can carry the analytical study slightly further by considering the system behavior when we make small changes around an operating point. In other words, we *linearize* the process in the vicinity of an operating point Q_A and Q_B.

This type of incremental linearization means that we consider

$$q_A \to Q_A + q_A$$

$$q_B \to Q_B + q_B$$

Instead of q_A, we use the normal value Q_A plus a very small change called q_A. Then the two process equations, (8-41) and (8-42), become

$$\text{Output flow} = Q_A + Q_B + q_A + q_B \qquad (8\text{-}43)$$

$$\text{Output concentration} = \frac{Q_A + q_A}{Q_A + Q_B + q_A + q_B} \qquad (8\text{-}44)$$

Equation (8-43) states that the output flow is a constant value $(Q_A + Q_B)$ plus a small variation $(q_A + q_B)$ subject to control by the process input. If we let this output flow be represented by $Q_o + q_o$, where the capital letter stands for the constant value, we have

$$q_o = q_A + q_B \qquad (8\text{-}45)$$

Equation (8-44) can be manipulated to give

$$C_o + c_o = \frac{Q_A(1 + q_A/Q_A)}{(Q_A + Q_B)[1 + (q_A + q_B)/(Q_A + Q_B)]} \qquad (8\text{-}46)$$

If q_A and q_B are both small compared with Q_A or Q_B, the right side of this relation is of the form

$$\frac{Q_A}{Q_A + Q_B} \qquad \frac{1 + x}{1 + y}$$

where x and y are much less than unity. This can be written

$$\frac{Q_A}{Q_A + Q_B} \qquad (1 + x)(1 - y)$$

or approximately

$$\frac{Q_A}{Q_A + Q_B} \qquad (1 + x - y)$$

where we have thrown away all terms in x^2, y^2, xy, or higher powers of the small quantities x and y. Then Eq. (8-46) becomes

$$C_o + c_o = \frac{Q_A}{Q_A + Q_B}\left(1 + \frac{q_A}{Q_A} - \frac{q_A + q_B}{Q_A + Q_B}\right) \tag{8-47}$$

Elimination of the constant values gives the linearized, incremental relationship

$$c_o = \frac{Q_B}{(Q_A + Q_B)^2} q_A - \frac{Q_A}{(Q_A + Q_B)^2} q_B \tag{8-48}$$

which can also be written in terms of Q_o and C_o:

$$c_o = \frac{1 - C_o}{Q_o} q_A - \frac{C_o}{Q_o} q_B \tag{8-49}*$$

Equations (8-45) and (8-49) allow us to draw the signal flow diagram of Fig. 8.18 for the linearized system. All signals here are changes from the normal operating values.

*Equation (8-49) was derived entirely algebraically. We could also have worked with Eq. (8-44) by noticing that we want

$$c_o = \left[\frac{\partial}{\partial q_A}\left(\frac{Q_A + q_A}{Q_A + Q_B + q_A + q_B}\right)\right]_{q_A = q_B = 0} q_A + \left[\frac{\partial}{\partial q_B}\left(\frac{Q_A + q_A}{Q_A + Q_B + q_A + q_B}\right)\right]_{q_A = q_B = 0} q_B \tag{8-50}$$

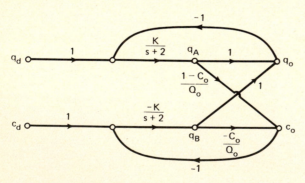

FIG. 8.18 Linearized two-dimensional blending system.

Dynamics

The dynamics of the system can be investigated rather easily (as one example of the studies which can now be carried out). The determinant of the signal flow graph is

$$\Delta = 1 \underbrace{- \frac{K}{s+2}\left(1 + \frac{C_o}{Q_o}\right) + \frac{1 - C_o}{Q_o}\left(\frac{K}{s+2}\right)^2}_{\sum T_l} + \underbrace{\frac{C_o}{Q_o}\left(\frac{K}{s+2}\right)^2}_{\sum' T_l T_j} \tag{8-51}$$

If we consider the specific operating point

$$C_o = \tfrac{1}{4} \qquad Q_o = 1$$

algebraic manipulation of Eq. (8-51) yields a system characteristic polynomial (the numerator of Δ) which is

$$C.P. = s^2 + \left(4 + \tfrac{5}{4}K\right)s + \left(4 + \tfrac{5}{2}K + K^2\right) \tag{8-52}$$

We can now find the system natural frequencies (the zeros of the characteristic polynomial) as K assumes various values. (This very simple system, with only two poles, is never unstable for any positive K.)

Final comment

This example of a blending operation illustrates a broad class of simple multidimensional systems. In the more detailed design of the system, we would want to investigate the way the incremental linear model changes with the operating point (C_o, Q_o), the possible advantages of interchanging the roles of q_A and q_B, and the effects of additional time lags in the two valve operations (e.g., second order rather than first order transfer functions).

Perhaps the primary impression, from the fleeting glimpse in this chapter at multidimensional systems, is that such problems are basically no different from the corresponding one-dimensional problem of the same total order.

PROBLEMS

8.1 One of the most familiar examples of a multi-input, multi-output system is the steering of a car on a modern, multilane expressway. For this man-machine system, construct a block diagram which shows the various input and output signals, also the various signals through which the driver influences the car motion.

Discuss briefly how each of the transfer functions involved in the system model might be measured experimentally.

8.2 The transfer function of a system is

$$T(s) = \frac{s^2 + 4}{s^3 + 3s^2 + 2s + 1}$$

(a) Draw the signal flow graph of a possible realization of this system.

(b) Redraw this graph using $-1/s$ as integrators. Show that no more than three sign inverters are needed in this realization.

(c) Actually, only two sign changers are needed. How can this be done?

(d) Suppose that the system has two inputs, one a signal x and the other a noise n. The output y is formed from x and n according to

$$Y = \frac{s^2 + 4}{s^3 + 3s^2 + 2s + 1} X + \frac{s + 2}{s^3 + 3s^2 + 2s + 1} N$$

Supplement the signal flow graphs drawn for (a) and (b) to show the addition of the second input.

8.3 An industrial control is described by the transfer functions

$$Y_1 = \frac{4s + 12}{s^2 + 6s + 8} X_1 + \frac{5s + 14}{s^2 + 6s + 8} X_2$$

$$Y_2 = \frac{3s + 8}{s^2 + 6s + 8} X_1 + \frac{4s + 10}{s^2 + 6s + 8} X_2$$

where X_1 and X_2 are inputs, Y_1 and Y_2 are outputs. Determine an analog simulation using Rs, Cs, and op-amps.

8.4 Determine an analog computer simulation of minimal complexity for the set of transfer functions:

$$\mathbf{T}(s) = \begin{bmatrix} \dfrac{2}{s+1} + \dfrac{1}{s+2} & \dfrac{1}{s+1} + \dfrac{2}{s+2} \\ \dfrac{6}{s+1} & \dfrac{3}{s+1} + \dfrac{6}{s+2} \end{bmatrix}$$

8.5 The single-loop feedback system is described by the two transfer function matrices:

$$G(s) = \begin{bmatrix} \dfrac{1}{s} & \dfrac{1}{s+2} \\ 5 & \dfrac{1}{s+1} \end{bmatrix} \qquad H(s) = \begin{bmatrix} 1 & 0 \\ 0 & 1 \end{bmatrix}$$

Here G describes the relations between the two outputs and the two error signals, H describes the relations between the two feedback signals to the comparators and the two outputs. There are also two inputs. Determine the matrix of closed-loop transfer functions for this system.

8.6 A system is described by the matrix of transfer functions

$$T(s) = \begin{bmatrix} \dfrac{1}{s^2} & \dfrac{1}{s^2(s+1)} \\ \dfrac{2}{s(s+2)} & \dfrac{1}{s} \end{bmatrix}$$

Determine a minimal-complexity analog simulation. Write the state equations for the system. (This and the preceding problem are based on problems in the second edition of the excellent text by B. C. Kuo, "Automatic Control Systems," Prentice-Hall, Englewood Cliffs, N. J., 1967.)

8.7 A system is described by the three state equations

$$\dot{x}_1 = -2x_1 + x_2$$
$$\dot{x}_2 = -6x_1 + x_3 + 4u$$
$$\dot{x}_3 = -12x_1 + u$$

and the outputs y and z, where

$$y = x_1 \qquad z = x_2$$

If the initial state is described by

$$[x_1(0), x_2(0), x_3(0)]$$

determine the matrix of transfer functions relating the four inputs (u and the initial values of the state variables) to the two outputs. What is the order of this system?

Initial conditions can always be handled this way in a linear system—treating the initial conditions as independent inputs.

9 OP-AMP APPLICATIONS

Operational amplifiers, resistors, and capacitors—the building blocks of network synthesis and analog simulation! Whether we are working with one-input, one-output systems (Chap. 6) or more complex problems (Chap. 8), we can pass logically and simply from a transfer function description to a realization suitable for laboratory construction.

When we are confronted with a problem in which the system is actually to be built, we should take advantage of the flexibility inherent in the procedures described in the preceding chapters. That is, we should look at the alternate solutions and select that solution which is the easiest to construct.

This chapter completes our study of linear op-amp circuits, by considering briefly some of the range of alternative possibilities which comprise the engineer's arsenal of design weapons.

9.1 AN ESSENTIAL CHARACTERISTIC IN SIMULATION

Analog simulation or analog computer studies are used for the adjustment of design parameters to ensure adequate or optimum performance. For example, we are designing a feedback control system. The process to be controlled has been chosen, but we can adjust two parameters, A and B, each anywhere within the range from unity to ten.

The transfer function of the total system is

$$T(s) = \frac{20(s + 1)}{s^4 + 4s^3 + (8 + A)s^2 + 12s + B} \tag{9-1}$$

We might simply substitute a large number of combinations of A and B values (e.g., $A = 1$, $B = 1$; $A = 1$, $B = 3$; and so on). In each case, we determine system stability or plot the unit-step response or make any appropriate evaluation of system dynamics. After all this calculation, we study the results and select specific values for A and B.

While this brute-force approach is perhaps good for the employment of engineers, the work is certainly uninspiring. A far better approach is to build the system in the laboratory and vary experimentally the parameters A and B while observing the response on an oscilloscope or paper recorder. This approach to the design is the most important purpose of analog simulation.*

If we build a system with the $T(s)$ of Eq. (9-1), there is one additional requirement which is imposed on the realization: if possible, a *single* element (R or C) in the simulation should correspond to A, and a *separate single* element to B. If this characteristic can be achieved, it is simple to vary A and B. If, on the other hand, one resistance in the simulation were $9/(8 + A)$ and another $(30 - A)$, the engineer has to decide on a value of A, then calculate each R, then adjust them separately, and so on.

Thus, *each variable parameter in the actual system should be determined by a single parameter in the simulation.* Preferably, any parameter A should be determined by a simulator parameter (R or C or $1/R$ or $1/C$) which is just equal to A.

System parameter is a coefficient of $T(s)$

Equation (9-1) is an example when the system parameter is a coefficient in the transfer function. Here we want each parameter (A or B) in

$$T(s) = \frac{20(s + 1)}{s^4 + 4s^3 + (8 + A)s^2 + 12s + B} \tag{9-2}$$

to be realized by a simulator parameter.

The solution is straightforward with the techniques of Chap. 6. Division by s^4 gives

$$T = \frac{20/s^3 + 20/s^4}{1 + 4/s + 8/s^2 + A/s^2 + 12/s^3 + B/s^4} \tag{9-3}$$

One possible realization is given in Fig. 9.1.

Success in building a simulation with each system parameter represented by a single gain constant may require considerable ingenuity (and indeed may not be feasible if

*Digital simulation can also be used (Chap. 14).

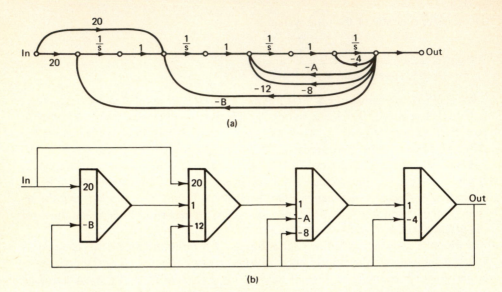

(a)

(b)

FIG. 9.1 Simulation of $T(s)$ of Eq. (9-2) with each parameter represented by a single gain constant. (a) Signal flow diagram with both A and B as gains of single branches; (b) simulation using integrators (no concern is given to the simplification of signs).

the parameter A appears in several different ways within the transfer function). For instance, the simple transfer function

$$T(s) = \frac{3A}{s + A/(1 + B)} \tag{9-4}$$

has the two parameters, A and B. $T(s)$ can be rewritten

$$T(s) = \frac{3(A/s)}{1 + [1/(1 + B)](A/s)} \tag{9-5}$$

which is represented by Fig. 9.2.

Now we can focus on B, which we would like to realize as the gain of a single branch. We recognize that $1/(1 + B)$ is itself the transfer function of a single-loop feedback

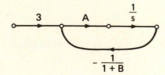

FIG. 9.2 Initial realization of $T(s)$.

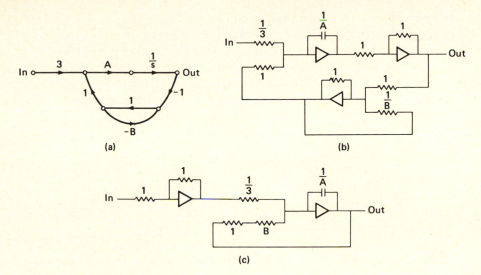

FIG. 9.3 Simulation of

$$T(s) = \frac{3A}{s + A/(1 + B)}$$

(a) Signal flow diagram; (b) Circuit realization; (c) An alternate realization based on Fig. 9.2.

system with $-B$ the feedback gain, and Fig. 9.3 results. Actually, the form $1/(1 + B)$ in Fig. 9.2 might also suggest the system of Fig. 9.3(c), since the total integrator input resistance on the feedback path is just $(1 + B)$.

The absence of any cut-and-dried, straightforward procedure for going from the transfer function to the simulation makes these problems interesting and represents the inherent value of the human designer (instead of automated, computer design).

System parameter is defined in the circuit

Figure 9.4 shows a mechanical vibration damper. An instrument is to be isolated from the vibrations of the table on which it "sits." The instrument is first mounted on a platform through spring K_2 and damper B_2; the platform in turn is connected to the table through K_1 and B_1. The input signal is the velocity v_t of the table, the response the velocity v_i of the instrument. If we use d'Alembert's principle (summing the forces on each mass, including the inertial force), we obtain the same equations as the node equations for the electrical analog of Fig. 9.4(b). Thus, in this simple case, we can use either the mechanical or the electrical system as a basis for the simulation.

To design such a system, we want to select the parameters (for example, M_p, K_1, K_2, B_1, and B_2) in such a way that, with typical table velocities v_t, the instrument motion stays within allowable limits. If we build an analog simulator, we want each system parameter to correspond to an adjustable simulator parameter.

As a simulation in this sense, we can use the electrical analog of Fig. 9.4(b). Each system parameter is represented by an electrical parameter. In many cases, the RLC network is inconvenient or uneconomical, and we prefer an op-amp simulation.

As a first step in the derivation of the op-amp system, we write the state equations for either system of Fig. 9.4. The state variables are the four energy-storage measures: f_1, f_2, v_p, and v_i. The two equations in $\dot{f}_1$ and $\dot{f}_2$ can be written immediately in terms of the v's:

$$\frac{1}{K_1} \dot{f}_1 = v_t - v_p \tag{9-6}$$

$$\frac{1}{K_2} \dot{f}_2 = v_p - v_i \tag{9-7}$$

For the $\dot{v}_p$ and $\dot{v}_i$ equations, the two node equations are used:

$$M_p \dot{v}_p = f_1 - f_2 + B_1(v_t - v_p) - B_2(v_p - v_i) \tag{9-8}$$

$$M\dot{v}_i = f_2 + B_2(v_p - v_i) \tag{9-9}$$

Equations (9-6)–(9-9) constitute the state model. In the transition from these to the op-amp system, we want the parameters B_1 and B_2 to appear only once; hence, we form the quantities $(v_t - v_p)$ and $(v_p - v_i)$ within the system, before multiplying by B_1 and B_2, respectively (Fig. 9.5).

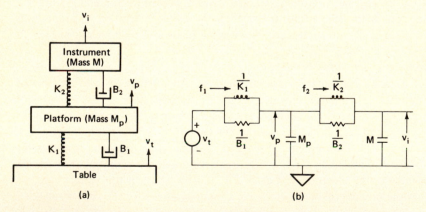

FIG. 9.4 Mechanical system and an electrical network with the same dynamics: (a) A diagrammatic representation of the system for vibration isolation; (b) Electrical analog of (a). Values of parameters are in ohms, farads, and henrys or a consistent set of units. v_t, v_p, and v_i are voltages, f_1 and f_2 currents.

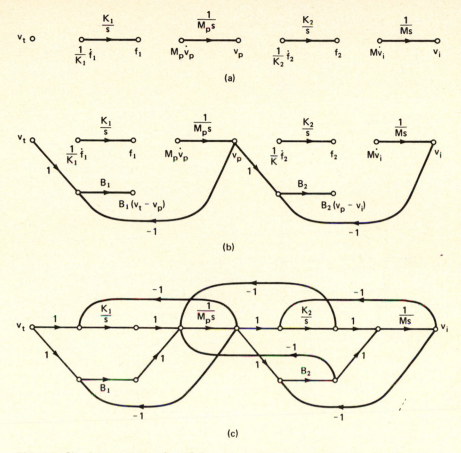

FIG. 9.5 Simulator programming of the state model of Eqs. (9-6)–(9-9): (a) First step in constructing the signal flow graph for Eqs. (9-6)–(9-9); (b) Second step in simulation; each term in Eqs. (9-6)–(9-9) is now available; (c) Final signal flow diagram.

The op-amp realization can then be constructed directly from the signal flow diagram. Figure 9.6 shows one of several possibilities in terms of ideal amplifiers, resistors, and capacitors.

Final comment

The theory or science of simulation is elementary. Proceeding from a transfer function or a state model to an op-amp realization is straightforward (Chap. 6). When we impose the additional requirement that each system parameter is to be represented by one, and only one, simulator parameter, we move from a science to an art. The system parameters can appear in so many different ways within the transfer function that successful simulation may require considerable ingenuity on the part of the engineer.

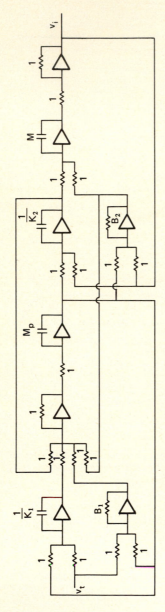

FIG. 9.6 Simulation of the vibration isolator (no attempt has been made to minimize the number of amplifiers).

9.2   INITIAL CONDITIONS

Another aspect of simulation was slighted in Chaps. 6 and 8: the necessity of inserting proper initial conditions for each of the integrator outputs. In terms of the state-model concepts of Chap. 7, these initial conditions constitute the *initial state*. In terms of the mathematical viewpoint of integration, these are the constants of integration. (If the integrator is started at $t = 0$, the output is the definite integral

$$e_{out} = -\frac{1}{RC} \int_0^t e_{in}\, dt \qquad (9\text{-}10)$$

where e_{in} is the input signal.)

Methods to insert initial conditions

The signal flow diagram for the basic two-input integrator is shown in Fig. 9.7(a). In terms of this diagram, the initial value of e_o can be inserted in either of two ways:

(1)   A step function of value $e_o(0)$ can be added to e_o, part (c) of the figure.
(2)   An impulse of value $-Ce_o(0)$ can be added to the integrator input as in (d).

Circuit realization of initial conditions

While we might add impulses (in practice, very short and large pulses) as in Fig. 9.7(d), the integrator-output initial condition is usually realized by starting the solution with

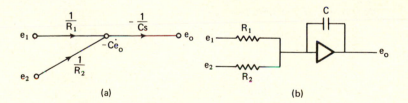

(a)   (b)

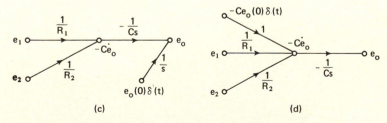

(c)   (d)

FIG. 9.7   Inserting the initial condition in the signal flow diagram: (a) Signal flow diagram for an integrator with two inputs [part (b)]; (b) Circuit for (a); (c) Step added to e_o; (d) Impulse added to integrator input.

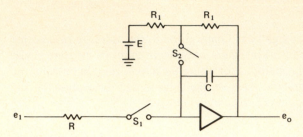

FIG. 9.8 Circuit for realizing the desired initial condition and a hold condition.

the capacitor C charged to the desired value—essentially a step function is added to the output. Figure 9.8 shows a common circuit where the operation is controlled by two switches, S_1 and S_2 (which are normally electronic rather than mechanical switches).

When S_1 is opened and S_2 closed (before $t = 0$ or the start of the simulator operation), capacitor C is charged through the circuit of Fig. 9.9. The resulting output voltage is $-E$; hence E in Fig. 9.8 should be selected as the negative of the desired initial value of the output.

$$-E \ = \ e_o(0) \tag{9-11}$$

At $t = 0$, S_1 is closed and S_2 opened; the circuit starts integrating from its initial value $e_o(0)$.

At the end of the desired integration interval (or when we want to stop the solution to observe or record the value of various variables), S_1 is opened. Thereafter, no current flows through C (except for the very small amplifier-input current and the capacitor leakage), and e_o holds at a constant value.

Required initial conditions

In order to simulate a system, we must know the initial values of the integrator outputs: the initial state, in terms of the particular set of state variables represented by our simulation. If the simulation is developed directly from the state equations written from the system (as in the vibration isolator of Sec. 9.1), the state variables may be

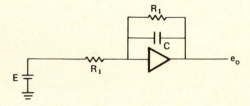

FIG. 9.9 Charging circuit (S_2 closed, S_1 open in Fig. 9.8).

the energy-storage variables of the system. Then the initial state is simply the initial inductor currents and capacitor voltages in an electrical system, or the corresponding variables in nonelectrical systems (e.g., velocity of a mass and displacement of a spring in a mechanical network).

In many cases, however, the simulation is derived from the transfer function. For example, Fig. 9.10 shows one possible simulation for a given $T(s)$. In such a problem, the three initial conditions often given are

$$y(0) \qquad \dot{y}(0) \qquad \ddot{y}(0)$$

Building the simulator in the laboratory requires knowledge of the initial state $x(0)$, or

$$x_1(0) \qquad x_2(0) \qquad x_3(0)$$

The determination of $x(0)$ can be illustrated by this particular example. The state equations are written from the signal flow diagram

$$\dot{x}_1 = -2x_1 + x_2 \tag{9-12}$$

$$\dot{x}_2 = -10x_1 + x_3 + u \tag{9-13}$$

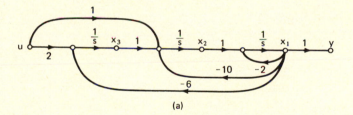

(a)

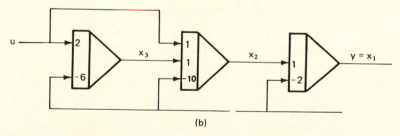

(b)

FIG. 9.10 Simulation of

$$T(s) = \frac{Y}{U} = \frac{s + 2}{s^3 + 2s^2 + 10s + 6}$$

(a) Signal flow diagram; (b) Integrator interconnection.

$$\dot{x}_3 = -6x_1 + 2u \tag{9-14}$$

and the model is completed by the relation for the output y; here

$$y = x_1 \tag{9-15}$$

These four equations are valid at all times, hence for $t = 0 +$. Then Eq. (9-15) states that

$$\boxed{x_1(0) = y(0)} \tag{9-16}$$

Differentiating Eq. (9-15) once and substituting $\dot{y}$ for $\dot{x}_1$, y for x_1 yields a revised Eq. (9-12):

$$\dot{y} = -2y + x_2 \tag{9-17}$$

or

$$\boxed{x_2(0) = \dot{y}(0) + 2y(0)} \tag{9-18}$$

Finally a differentiation of Eq. (9-17) gives $\dot{x}_2$ in terms of $\ddot{y}$ and $\dot{y}$; substitution in Eq. (9-13) yields:

$$\ddot{y} + 2\dot{y} = -10 y + x_3 + u \tag{9-19}$$

or

$$\boxed{x_3(0) = \ddot{y}(0) + 2\dot{y}(0) + 10y(0) - u(0)} \tag{9-20}$$

The three boxed equations above give the initial state in terms of the three initial conditions on the output: $y(0)$, $\dot{y}(0)$, and $\ddot{y}(0)$.

In other words, the state model is essentially solved for x_1, x_2, and x_3 in terms of y and its derivatives. The resulting equations are valid at 0+, and serve to establish the required initial conditions $x(0)$ from $y(0)$, $\dot{y}(0)$, and so on.

This process of "inverting" the state equation is closely related to the concept of observability, first encountered in Sec. 7.7.* If the system is observable, we can solve the state equations for the initial state $x(0)$ in terms of the initial conditions $y(0)$, $\dot{y}(0)$,

---

*Indeed, the mathematical condition for observability is derived on this basis.

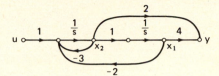

FIG. 9.11 Unobservable system.

etc. In other words, if the system is observable, measurements on the output can give the initial state.

If, on the other hand, the system is unobservable, we cannot solve the state equation for $x(0)$. Figure 9.11 shows such a system, for which

$$
\begin{aligned}
\dot{x}_1 &= x_2 \\
\dot{x}_2 &= -2x_1 - 3x_2 + u \\
y &= 4x_1 + 2x_2
\end{aligned}
\tag{9-21}
$$

Differentiation of the last equation gives

$$
\dot{y} = 4\dot{x}_1 + 2\dot{x}_2 = 4(x_2) + 2(-2x_1 - 3x_2 + u)
\tag{9-22}
$$

or

$$
\left.
\begin{aligned}
\dot{y} &= -4x_1 - 2x_2 + 2u \\
y &= 4x_1 + 2x_2
\end{aligned}
\right\}
\tag{9-23}
$$

These two relations show that $x_1(0)$ and $x_2(0)$ can not be determined from $y(0)$ and $\dot{y}(0)$. Indeed, in this system, $y(0)$ and $\dot{y}(0)$ can not be specified independently. From the standpoint of initial conditions, the system is really first order: As soon as $y(0)$ and $u(0)$ are given, $\dot{y}(0)$ is determined.

The problems at the end of the chapter illustrate the algebra required to move from given values of $y(0)$, $\dot{y}(0)$, and the like to the initial state $[x_1(0), x_2(0)$, etc.] for various common forms of simulation. When the simulation has all feedback from x_1, the algebra is simple. When the feedback is entirely to $\dot{x}_n$, the algebra is straightforward, but more tedious (a set of n simultaneous equations must be solved).

9.3 OP-AMPS WITH TWO INPUTS

Thus far, we have considered ideal amplifiers which possess only a single input: the amplifier output (Fig. 9.12) is

$$
e_o = Ke_i
\tag{9-24}
$$

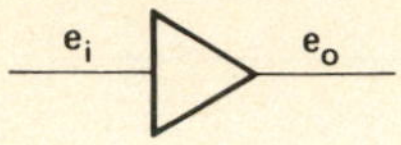

FIG. 9.12 Ideal amplifier with single input.

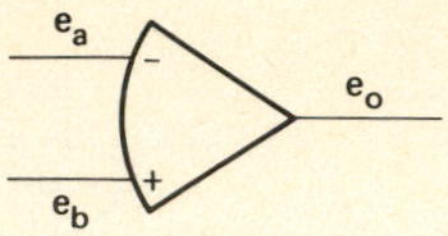

FIG. 9.13 Op-amp with + and − inputs.

where K is a large, negative value. Many commerically available amplifiers have both inverting and noninverting input terminals (Fig. 9.13). For this device,

$$e_o = A(e_b - e_a) \tag{9-25}$$

where A is positive; in other words, the output is the gain times the difference of the two input signals.

Use as a single-input amplifier

The amplifier of Fig. 9.13 can be used in the mode described in Chaps. 6–8 as a single-input device similar to Fig. 9.12. We apply a signal only to the e_a input and connect the e_b terminal to ground (usually through a resistor, sized according to the manufacturer's recommendation). Figure 9.14 shows the form of the single integrator with two inputs under these circumstances.

If the goal of the circuit design is an integrator, there is no particular advantage of the noninverting (+) input of the amplifier. The techniques of Chaps. 6 and 8 are used directly for the realization of a given transfer function or state model.

Differential amplifier

In instrumentation applications, we often want to amplify the difference between two signals. For example, when using a strain gage to measure displacement or strain or a temperature-sensitive resistance to measure temperature, we often work on a comparison basis. The voltage e_1 across the sensitive element is compared to the voltage e_2 across a constant reference element.

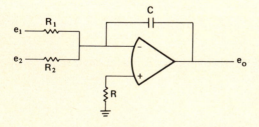

FIG. 9.14 Integrating op-amp with two input signals.

In the thermometry example, a rise in temperature results in an increase in e_1 compared to e_2. $e_1 - e_2$ measures the temperature. An increase in the power supply voltage for the system, in contrast, causes both e_1 and e_2 to rise and is primarily reflected in an increase in $e_1 + e_2$.

A circuit which amplifies the difference of two inputs (and rejects the sum) is called a *differential amplifier* (or an instrumentation amplifier). The ideal differential amplifier thus yields an output e_o which is related to the two inputs e_1 and e_2 by

$$e_o = A(e_1 - e_2) \tag{9-26}$$

with

$$\frac{e_o}{e_1 + e_2} = 0 \tag{9-27}$$

The simplest differential amplifier is shown in Fig. 9.15. If we follow the analysis approach used in Chap. 6 and assume that the gain of the amplifying element is so large that the input must be essentially zero, we start from the assumption

$$e_a - e_b = 0 \tag{9-28}$$

Inspection of the circuit reveals that

$$e_b = \frac{R}{R + R_2} e_2 \tag{9-29}$$

since both inputs of the amplifying element draw no current (each Y_{in} is zero). In addition

$$e_a = \frac{R_o}{R_o + R_1} e_1 + \frac{R_1}{R_o + R_1} e_o \tag{9-30}$$

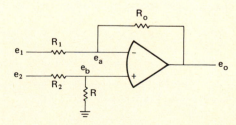

FIG. 9.15 Basic differential amplifier.

Substitution of the last two equations in (9-28) yields the relation for e_o in terms of e_1 and e_2

$$e_o = -\frac{R_o}{R_1}\left(e_1 - \frac{1 + R_1/R_o}{1 + R_2/R}\, e_2\right) \tag{9-31}$$

Equation (9-31) shows that we can realize the ideal differential amplifier if we select

$$\frac{R_1}{R_o} = \frac{R_2}{R} \tag{9-32}$$

in Fig. 9.15. With the careful choice of resistances so that this condition is satisfied, the circuit has the output

$$e_o = -\frac{R_o}{R_1}(e_1 - e_2) \tag{9-33}$$

and any desired gain can be obtained by choice of the ratio R_o/R_1.

The configuration of Fig. 9.15 is only one of many possible forms for the differential amplifier.* In the style of Chap. 6, we can achieve the same result using two amplifiers which have only the inverting input, as shown in Fig. 9.16. Here, the rejection of $(e_1 + e_2)$ in the output depends again on the matching of resistance values: in this case, the equality of the R values and of the R_1 values.

9.4 MORE GENERAL OP-AMPS

Since the beginning of Chap. 6, we have discussed operational amplifiers, Rs and Cs, as the basic building blocks in the realization of given transfer functions or state models. Throughout this development, attention was focused exclusively on component networks which are single elements. For instance, in Fig. 9.17, we have always

*Several other circuits of practical importance are discussed in the article by Larry L. Schick, Linear Circuit Applications of Operational Amplifiers, *IEEE Spectrum*, April 1971, pp. 36–50.

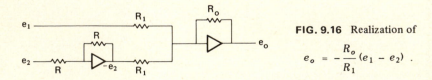

FIG. 9.16 Realization of

$$e_o = -\frac{R_o}{R_1}(e_1 - e_2).$$

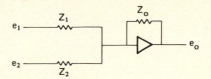

FIG. 9.17 Basic op-amp.

considered Z_1 and Z_2 each to be a resistance; Z_o has been either a resistance or a capacitive impedance. Even when we considered amplifiers with two inputs in the last section, we made the associated networks single elements.

There is no real reason why each of the Z elements in Fig. 9.17 should not be a two-port network, as shown for the one-input system in Fig. 9.18. Again, we can find E_b/E_a by recognizing that e_{in} is essentially zero. Consequently, Network A is working into a short circuit to ground, and i_α is just the short-circuit output current (Fig. 9.19).

Electrical engineers use the symbol $y_{21}(s)$ for the transfer function

<u>Short - circuit output current</u>

　　　Input voltage

for a two-port network. Hence

$$I_\alpha = y_{21A} E_a \tag{9-34}$$

where y_{21A} is the short-circuit transfer admittance for network A. Similarly, i_β is given by

$$I_\beta = y_{21B} E_b \tag{9-35}$$

Since the amplifying element draws no current, the node equation at e_{in} in Fig. 9.18 is

$$y_{21B} E_b + y_{21A} E_a = 0 \tag{9-36}$$

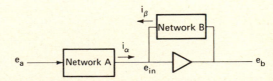

FIG. 9.18 Generalized op-amp.

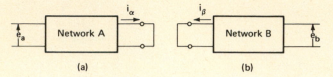

FIG. 9.19 Analysis of Fig. 9.18 requires equations for i_α and i_β: (a) Calculation of i_α in Fig. 9.18; (b) Calculation of i_β.

or

$$\boxed{\frac{E_b}{E_a} = -\frac{y_{21A}}{y_{21B}}}$$

(9-37)

Equation (9-37) gives the transfer function of the generalized op-amp.

Conventional integrator

Equation (9-37) must be valid for the simple integrator (Fig. 9.20). Here y_{21A} is the transform of the short-circuit output current when $E_a = 1$, or

$$y_{21A} = \frac{1}{R}$$

Similarly

$$y_{21B} = Cs$$

and Eq. (9-37) becomes the familiar

$$\frac{E_b}{E_a} = -\frac{1}{RCs}$$

(9-38)

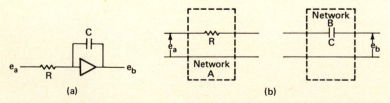

FIG. 9.20 Integrator as example of Eq. (9-37): (a) Op-amp; (b) Component networks.

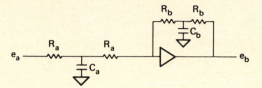

FIG. 9.21 Op-amp using simple *RC* ladders.

Op-amp with simple two-port networks

What can we achieve with the op-amp of Fig. 9.21?

Calculation of y_{21} for each network* reveals that

$$y_{21A} = \frac{1}{R_a{}^2 C_a} \frac{1}{s + 2/R_a C_a} \tag{9-39}$$

or that

$$\frac{E_b}{E_a} = \frac{R_b{}^2 C_b}{R_a{}^2 C_a} \frac{s + 2/R_b C_b}{s + 2/R_a C_a} \tag{9-40}$$

If we are given a transfer function of this form to be synthesized, we can determine appropriate values of the parameters. As an example,

$$T(s) = 200 \frac{s + 2}{s + 20} \tag{9-41}$$

Then

$$\left.\begin{array}{l} R_a C_a = \tfrac{1}{10} \ \text{realizes the pole at } -20 \\[2ex] R_b C_b = 1 \ \ \text{gives the zero at } -2 \\[2ex] \dfrac{R_b}{R_a} = 20 \ \text{yields the coefficient 200} \end{array}\right\} \tag{9-42}$$

A possible system is shown in Fig. 9.22 (C_a is arbitrarily chosen as unity; then R_a is found from the first equation, R_b from the third, and C_b from the second).

---

*In each case y_{21} is determined by short-circuiting the output terminals, then calculating the transfer function: current in this short circuit divided by the input voltage (each in terms of its Laplace transform).

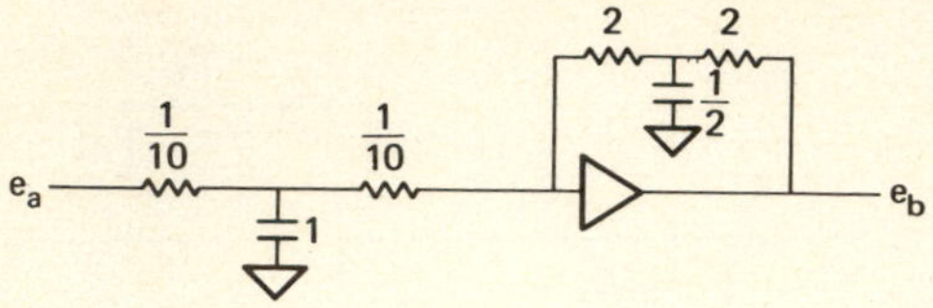

FIG. 9.22 System for

$$\frac{E_b}{E_a} = 200\,\frac{s+2}{s+20}\,.$$

How useful is this generalized op-amp?

The above solution is initially very attractive. We have realized a given transfer function with a single amplifying element and associated Rs and Cs—in contrast to the simulations of Chap. 6, where a system of order n required $(n + 2)$ amplifiers. The example above, however, is hardly general (the transfer function has only one pole and one zero); can this approach be extended to higher order systems? If so, how usefully? Several comments can be made.

(1) First, the approach is quite general. If the desired transfer function is given as the ratio of two polynomials in s,

$$T(s) \;=\; \frac{p(s)}{q(s)} \;=\; -\,\frac{y_{21A}}{y_{21B}} \tag{9-43}$$

we can find appropriate functions y_{21A} and y_{21B}, from which the two RC networks can be synthesized.* As the degree of each polynomial increases, however, the network becomes far more complicated, and the synthesis procedure becomes tedious analytically. Indeed, the network rapidly becomes so complex that, if the degree of p or q is greater than 2, the system is simpler if we use more than one amplifier—that is, if we revert to the approach of Chap. 6.

(2) If the transfer function is known in factored form, we are always better off if we realize the given $T(s)$ as the tandem connection of simpler networks. (This fact is also true if the approach of Chap. 6 is used.) For example, if

$$T(s) \;=\; \frac{K}{(s^2 + \alpha s + \beta)(s^2 + \gamma s + \delta)} \tag{9-44}$$

the best realization is that shown in Fig. 9.23.

*The procedures for going from a given $y_{21}(s)$ to an RC, two-port network are discussed in courses on network synthesis and are beyond the scope of this book. It turns out that we can always find a network when y_{21} has only simple negative-real poles and no zero on the positive real axis.

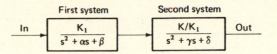

FIG. 9.23 Realization of a fourth order transfer function.

If this decomposition of the fourth order problem into two second order problems is used, the adjustment of circuit parameters is greatly simplified. In the form of Fig. 9.23, the parameters of the "first system" are adjusted to realize precisely the coefficients α and β; then γ and δ are obtained by adjustment (or "tuning") of the second system. If a *single* system is used for the original $T(s)$, tuning for the desired α and β affects the values of γ and δ.

(3) Even if we break down the given transfer function into separate factors, the generalized op-amp leads to difficulties. We have not achieved the desirable 1:1 identification between system parameters and transfer function parameters—the correspondence discussed in Sec. 9.1. For example, in the system of Fig. 9.22, adjustment of the two resistances R_a affects both the pole and the constant multiplier of $T(s)$.

Thus, the generalized op-amp (using two-port input and feedback networks) offers the possibility of reducing radically the required number of amplifiers. We pay for this in terms of

(a) many more Rs and Cs (when integrated circuits are used, the result may be a more expensive system);

(b) design complexity;

(c) tuning difficulties.

As a consequence, the simulation approach of Chap. 6 is often preferable to the generalized op-amp.

9.5 DIFFERENTIATORS

We conclude this chapter with one other topic regarding the use of op-amps in system realization and simulation. Throughout these four chapters on simulation, our consideration has been essentially restricted to systems in which the dynamics are realized by integrators. We could try to use differentiators, and indeed there are applications (as in prediction) when differentiation must be attempted.

The simplest differentiator conceptually is shown in Fig. 9.24: the input element is a capacitor, the output a resistor, and the transfer function as shown. What difficulties do we encounter with this system?

In the first place, as mentioned in Sec. 4.1 where differentiators were first discussed, we often do not really want a transfer function αs. Such a characteristic has a gain which increases with frequency. In actual applications, the signal to be differentiated

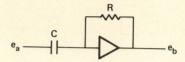

FIG. 9.24 Basic differentiator, with

$$\frac{E_b}{E_a} = -RCs \, .$$

has a defined bandwidth, and it is only over this band that we want to differentiate. At higher frequencies, where noise dominates, we would like low gain.

There is another difficulty with Fig. 9.24. An actual amplifier always has a gain which falls at high frequency. As a specific example, we might have an amplifier gain of the form

$$K = -\frac{K_o}{(1 + s/10^4)^2} \tag{9-45}$$

(The gain is $-K_o$ at low frequencies, but falls off in the vicinity of 10^4 rad/sec.) With this amplifier gain, what is the transfer function of the total system of Fig. 9.24?

Since the amplifier gain is no longer essentially infinite at all frequencies, the amplifier input is not zero, and we must analyze the circuit of Fig. 9.25. Calculation of the transfer function yields, when $RC = 10^{-3}$,

$$\frac{E_b}{E_a} = -10^{-3}s \, \frac{1}{1 + (10^{-3}s + 1)(10^{-4}s + 1)^2/K_o} \tag{9-46}$$

If the characteristic polynomial of this transfer function is inspected, we find that the system is unstable for

$$K_o > 24.2$$

Since we commonly want a low-frequency gain of at least 10^4, the differentiator circuit is unsatisfactory. (We have assumed K falls off as $1/s^2$ at high frequencies; if the amplifier is more carefully designed to fall at $1/s$, the system is stable, but described by a pair of conjugate complex poles with intolerably low ζ.)

Thus, quite aside from the noise problem, the ideal differentiator is unsatisfactory because of the stability problem with a practical amplifier. To circumvent this difficulty (also the noise problem), we follow the same procedure of Sec. 4.1 and use the system of Fig. 9.26.

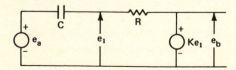

FIG. 9.25 Circuit representation of differentiating op-amp.

FIG. 9.26 Practical, differentiating op-amp with

$$\frac{E_b}{E_a} = -R_b C_a s \frac{1}{(R_b C_b s + 1)(R_a C_a s + 1)}.$$

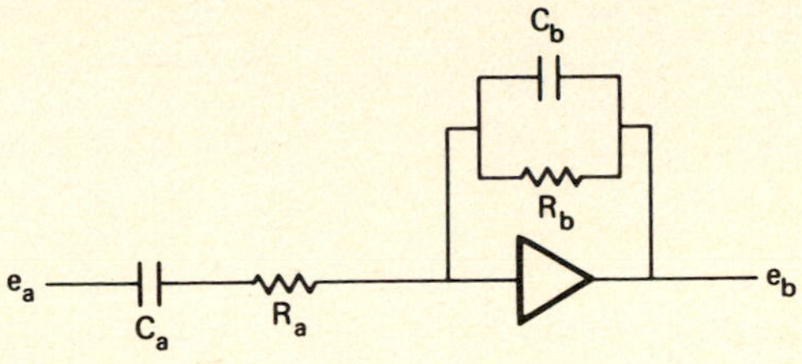

9.6   FINAL COMMENT

Chapters 6–9 constitute an introduction to the closely related topics of state models, simulations, and op-amp realizations for linear systems. Chapter 6 showed that any given transfer function can be realized simply by RC op-amps. In Chap. 7, we considered in more detail the analytical properties of the corresponding state models. In Chap. 8, these ideas were extended to multidimensional systems. Finally, in this chapter, we considered a few important features that arise in practice.

To a very great extent, modern electrical system design is built on the principles outlined in these four chapters, whether we are considering networks (e.g., filters) or feedback control systems.

APPENDIX—GENERALIZED OP-AMPS USING A DIFFERENCE AMPLIFIER

Section 9.4 discusses generalized op-amps when the active element is an amplifier without a noninverting input. This restriction is particularly important when field-effect transistors are used, since these elements tend to have poor linearity when used as difference amplifiers.

We can also discuss the generalization idea, however, with difference amplifiers, as an extension of Sec. 9.3. In other words, how do we design for a specified transfer function when the single active element is to be a difference amplifier. In this appendix, we consider one particular example, not as a complete discussion of this topic, but rather to suggest the range of synthesis possibilities which are important in the design of op-amp circuits.

A specific example to illustrate the problem

Using one difference amplifier and *RC* networks, we wish to realize the transfer function

$$T(s) = \frac{E_2}{E_1} = \frac{K(s + 10)}{s^2 + 2s + 100} \tag{A9-1}$$

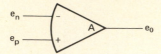

FIG. A9.1 Symbol for difference amplifier.

The value of K is immaterial; the important characteristics are the complex poles (resonant frequency of 10, Q of 5) and the zero at -10.

The difference amplifier: The difference amplifier (Fig. A9.1) gives an output

$$e_0 = A(e_p - e_n) \tag{A9-2}$$

We assume A is very large (the gain is effectively infinite), so that the amplifier forces

$$e_p = e_n \tag{A9-3}$$

Furthermore, the output impedance is zero and both input admittances are zero.

Basic configuration: For the realization, we choose the system configuration shown in Fig. A9.2(a), which can be simplified to the flow diagram of Fig. A9.2(b). Our task is then to select the two transfer functions $G(s)$ and $L(s)$: these must be realizable by RC networks.*

Naturally, we have no reason as yet to know that this configuration (which was chosen arbitrarily) will allow us to realize the given $T(s)$. Also, there are certainly many other configurations which are possible—some of which may give much simpler or cheaper systems.

Determination of $G(s)/L(s)$: For the system of Fig. A9.2(b),

$$T = \frac{G/L}{1 - G/L} \tag{A9-4}$$

Since we know T and want G/L, we rewrite this equation as

$$\frac{G}{L} = \frac{T}{1 + T} \tag{A9-5}$$

$$\frac{G}{L} = \frac{K[(s + 10)/(s^2 + 2s + 100)]}{1 + K[(s + 10)/(s^2 + 2s + 100)]} \tag{A9-6}$$

$$\frac{G}{L} = \frac{K(s + 10)}{s^2 + (2 + K)s + (100 + 10K)} \tag{A9-7}$$

*Furthermore, we shall see later that the network for $G(s)$ should start with a series R in order to accomplish the adding required in

$$e_s = e_1 + e_2 .$$

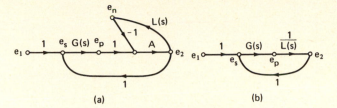

FIG. A9.2 Configuration arbitrarily chosen for the design: (a) Configuration chosen for realizing $T(s)$. Here

$$E_2 = A(E_p - E_n)$$
$$= A(E_p - LE_2)$$

If $A \gg 1$, $E_p = LE_2$; (b) Configuration equivalent to (a). Since the gain A is large,

$$E_2 = \frac{1}{L} E_p .$$

Equation (A9-6) reveals that a root locus plot (Fig. A9.3) would show the poles of G/L as a function of K. If K is large enough, the poles of G/L are real and can be realized as poles of an RC network for $G(s)$.

The denominator of Eq. (A9-7) has real factors for $K > 45$. The actual choice of K is arbitrary, so we might as well select a value which gives easy numbers: e.g., a K of 58. Then

$$\frac{G}{L} = \frac{58(s + 10)}{s^2 + 60s + 680} = \frac{58(s + 10)}{(s + 15.2)(s + 44.8)} \tag{A9-8}$$

Where are we now? If we find a $G(s)$ and $L(s)$ such that G/L is as given in Eq. (A9-8), the system of Fig. A9.2 realizes a

$$T(s) = \frac{58(s + 10)}{s^2 + 2s + 100} \tag{A9-9}$$

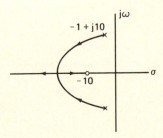

FIG. A9.3 Root locus plot to find poles of G/L.

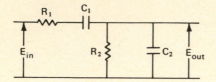

FIG. A9.4 Network for $G(s)$ of Eq. (A9-10).

$G(s)$ network: We know G/L; we can arbitrarily decide which part to realize in G, which part in $1/L$. We might decide, for example, to make

$$G(s) = \frac{As}{(s + 15.2)(s + 44.8)} = \frac{As}{s^2 + 60s + 680} \qquad \text{(A9-10)}$$

This $G(s)$ can be realized by a ladder network of the form shown in Fig. A9.4. The ladder transfer function is second order (two energy storage elements); the two zeros at $s = 0$ and $s = \infty$ are realized, respectively, by the series C_1 and the shunt C_2.

To find appropriate element values, we calculate E_{out}/E_{in} for the ladder. We assume

$$E_{out} = 1$$

Then

$$I_{in} = C_2 s + \frac{1}{R_2} = \frac{R_2 C_2 s + 1}{R_2}$$

and

$$E_{in} = 1 + \frac{R_1 C_1 s + 1}{C_1 s} \frac{R_2 C_2 s + 1}{R_2}$$

or

$$\frac{E_{out}}{E_{in}} = \frac{(1/R_1 C_2)s}{s^2 + (1/R_1 C_1 + 1/R_2 C_2 + 1/R_1 C_2)s + (1/R_1 C_1)(1/R_2 C_2)} \qquad \text{(A9-11)}$$

Comparison of Eqs. (A9-11) and (A9-8) shows that again we have freedom. An easy choice is

$$\frac{1}{R_1 C_1} = 26 \qquad \frac{1}{R_2 C_2} = 26$$

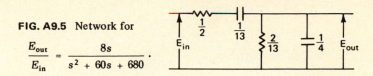

FIG. A9.5 Network for

$$\frac{E_{out}}{E_{in}} = \frac{8s}{s^2 + 60s + 680}.$$

Then the constant term in the denominator is 680. The s coefficient in the denominator is 60 if we choose

$$\frac{1}{R_1 C_2} = 8$$

These three equations can be satisfied in many ways; one is

$$R_1 = \frac{1}{2} \text{ (arbitrarily chosen)}$$

$$C_1 = \frac{1}{13} \text{ (from first equation)}$$

$$C_2 = \frac{1}{4} \text{ (from third equation)}$$

$$R_2 = \frac{2}{13} \text{ (from second equation)}$$

The result is shown in Fig. A9.5.

Figure A9.2(b) shows that we want

$$E_p = G(s)(E_1 + E_2) \tag{A9-12}$$

We might as well perform the adding operation in the network for $G(s)$. Hence, we come to the system of Fig. A9.6 for the realization of Eq. (A9-12), with

$$G(s) = \frac{4s}{s^2 + 60s + 680} \tag{A9-13}$$

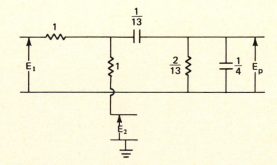

FIG. A9.6 Realization of
$E_p = G(s)(E_1 + E_2)$ with

$$G(s) = \frac{4s}{s^2 + 60s + 680}.$$

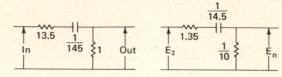

FIG. A9.7 Network for Eq. (A9-15).

FIG. A9.8 Network for $L(s)$ in Fig. A9.2(a).

$L(s)$ *network:* We have $G(s)$ given by Eq. (A9-13) and G/L by Eq. (A9-8). Hence we want

$$L(s) = \frac{s}{(29/2)(s + 10)} \tag{A9-14}$$

which can be written as a voltage-divider ratio

$$L(s) = \frac{1}{29/2 + 145/s} = \frac{1}{1 + 13.5 + 145/s} \tag{A9-15}$$

Figure A9.7 shows the network. Element values (R and C) closer to those in Fig. A9.6 are obtained if we divide all impedances in Fig. A9.7 by 10 (the Out/In voltage ratio is unchanged). Thus, the required $L(s)$ is realized by Fig. A9.8.

Final system and comments

Figure A9.9 shows the final system, the realization of the specified $T(s)$ by an active RC network, using a single difference amplifier as the active element.

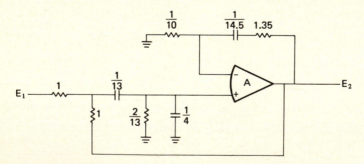

FIG. A9.9 System for

$$\frac{E_2}{E_1} = \frac{58(s + 10)}{s^2 + 2s + 100} \ .$$

Several comments are in order:

(1)   Figure A9.9 is only one of *many* possible solutions. We arbitrarily chose:

   (a)   the configuration of Fig. A9.2(a) and the "1" in various branch transmittances;
   (b)   the value of K determining the real poles of G/L;
   (c)   the way G/L was split between G and $1/L$;
   (d)   the network for G;
   (e)   the particular choice for R_1, which then determined the other parameters in the G network;
   (f)   the network for L.

The problem can be reworked with a different decision at any of these points; a different final system then results.

(2)   The system of Fig. A9.9 has zero output impedance, so such systems can be cascaded. To realize the overall transfer function

$$\frac{s(s + 1)}{(s^2 + as + b)(s^2 + cs + d)}$$

we would use two tandem networks, separately realizing

$$\frac{s}{s^2 + as + b} \quad \text{and} \quad \frac{s + 1}{s^2 + cs + d}$$

(3)   This section demonstrates an alternative solution to the use of operational amplifiers (and the analog-computer format in a multiloop configuration).

(4)   We assumed above that the amplifier gain A was high, so that we force the equality of E_p and E_n. This is not necessary, and we can use low-gain amplifiers in this "positive-feedback" configuration. The only difference above if A is not large comes in Eq. (A9-8). Instead of G/L, we have

$$G\frac{A}{1 + AL} = \frac{58(s + 10)}{(s + 15.2)(s + 44.8)} \tag{A9-16}$$

Now we must select G and L so that each is RC (each has negative real, finite poles only) and each can be realized by a simple network. One easy solution is to select

$$G = K_1 \frac{s + 10}{(s + 15.2)(s + 44.8)} \tag{A9-17}$$

realized by the network of Fig. A9.10. Then element values are chosen as before and

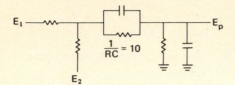

FIG. A9.10 Network for G in form of Eq. (A9-17).

K_1 is found. Then we return to Eq. (A9-16) and find L from

$$\frac{A}{1 + AL} = \frac{58}{K_1} \tag{A9-18}$$

In this case, L is just a constant realized by a resistive voltage divider (L is positive if $A > 58/K_1$).

PROBLEMS

9.1 (a) Using ideal high-gain amplifiers, resistors, and capacitors, find a realization of the transfer function

$$\frac{Y(s)}{X(s)} = T(s) = \frac{s^2 + 3s + 12}{s^4 + 3s^3 + 8s^2 + 2s + 12}$$

(b) Redraw your solution to (a) so that the system responds eight times as fast; indicate all R and C values.
(c) In the answer to (b), what is the new transfer function?
(d) In the answer to (a), call the outputs of the four integrators z_1, z_2, z_3, and z_4, enumerating from left to right. Redraw the circuit with the four integrator outputs $4z_1, z_2, 8z_3$, and z_4. In other words, scale the amplitudes.
(e) What is the transfer function for the solution to (d)?
(f) In your solution to (a), we want to put in initial conditions of

$$y(0) = 2 \qquad \qquad y''(0) = 3$$

$$y'(0) = -2 \qquad \qquad y'''(0) = -1$$

Write the equations to be solved to find the initial state of the system.

9.2 A system is described by the transfer function

$$\frac{Y(s)}{X(s)} = \frac{2s + 6}{s^3 + 4s^2 + 10s + 6}$$

where x is the input and y the output. The system is initially inert. Using high-gain ideal voltage amplifiers, resistors, and capacitors, determine a circuit realization for the system. Use batteries and switches to show how the initial conditions would be inserted if they were not zero. Determine the initial state in terms of initial values of y and its derivatives.

9.3   The figure shows two flow diagrams for the same transfer function. In both cases, the state variables are indicated. For each case, determine the initial values of the state variables in terms of the initial values of y and its derivatives.

How can the state variables be measured if only y is observable—that is, we cannot get into the system, but we can measure y and its derivatives. Show how initial conditions can be represented on the flow diagram (use either diagram) by either impulse or step inputs.

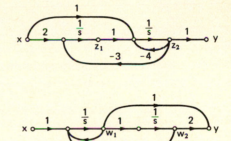

9.4   For the solutions of Prob. 6.4, write the equations from which we can determine initial capacitor voltages from given initial conditions on the output and its derivatives.

9.5   What is the transfer function of the system shown in the figure?

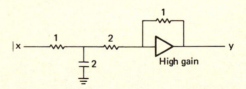

9.6   An analog computer operational-amplifier circuit is shown in the diagram.
(a)   Determine the transfer function relating input and output.
(b)   What operation does this device perform at low frequencies?
(c)   What operation at high frequencies?
(d)   What does the circuit reduce to at low frequencies? At high frequencies? Do these answers agree with (b) and (c)?

(e) Determine the new transfer function if a resistance R were added between ground and the connection point of R and $10C$.

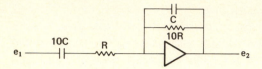

9.7   The servomechanism described in Sec. 7.4 is to be resimulated with the following parameters:

$$G_p = \frac{16}{s^2} \quad \text{(corresponding to no damping)}$$

K replaced by the transfer function $\dfrac{9s}{s + a}$

We want a potentiometer available which can be varied to represent an adjustment of a. We intend to build the circuit in the laboratory and experimentally adjust a to obtain satisfactory system performance. In addition, each other parameter should be represented by a single resistance or capacitance.

Design the system.

9.8   The circuit shown is used for temperature measurement. The silicon diode has a temperature coefficient of about -2 mv/°C. As long as the current is constant, this coefficient is reasonably constant over a wide temperature range. The constant current is provided by the current-limiting diode. (From *Control Engineering*, October 1969, p. 108.)

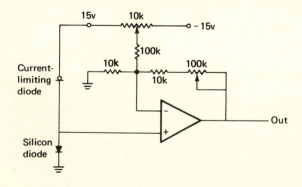

To calibrate the device, first we place the silicon diode in a cold environment corresponding to a known temperature and zero output voltage; we adjust the $10k$ pot. Then a known high-temperature environment is used with the $100k$ pot adjusted for the desired output.

(a) Does the calibration require iterative adjustments of the two pots? In other words, must we go back and forth, with each adjustment affecting the other?

(b) If the temperature range covered is to be $32°F$ to $212°F$, and the output can vary to as much as 5 v, describe in detail the circuit operation at each end of the range.

9.9   The circuit shown is used as a D/A converter (a device to convert a digital signal—here four binary digits—to analog form). Explain the operation of the system in detail.

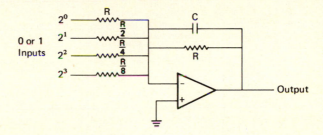

9.10 The figure shows a constant current source (taken from *IEEE Spectrum*, Sept. 1970, p. 48). Explain the operation of the system.

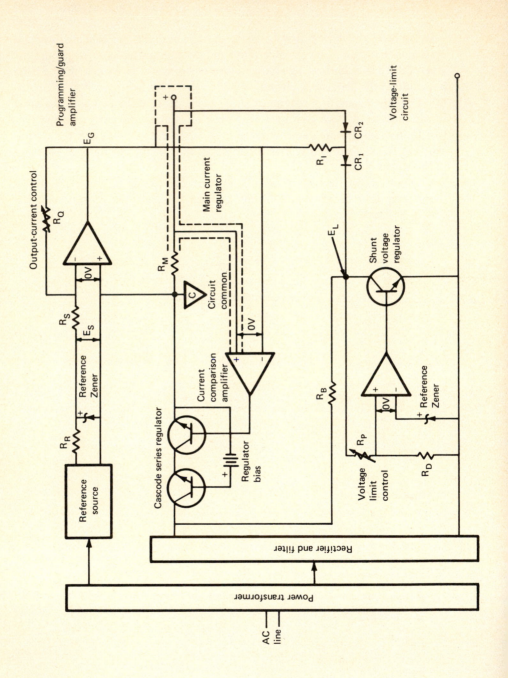

10 NONLINEAR SIMULATIONS

To a considerable extent, the electrical engineer has been unique among engineers because of the availability of inexpensive electrical components which are *linear*: the R, L, and C passive components and the range of active elements from vacuum tubes through transistors to integrated circuits. The electrical engineer's ability to construct exceedingly complex systems in a logical way from the interconnection of a multitude of basic elements depends on linearity.* The preceding chapters have presented some of the more important techniques in the analysis and the design of such linear systems; in the following chapters, we will consider other aspects of linear systems.

As the electrical engineer moves into system engineering, however, he often faces the task of controlling a process which is inherently nonlinear. Figure 10.1 depicts an example of a system currently being developed by electronic engineers (which was actually being tested at the time this material was written). An expressway has two lanes in each direction; traffic is to enter Lane 1 at point H, after accelerating on the

---

*In recent years, the scope of the electrical engineer has been extended to a new breadth by the availability of simple logic elements, which also can be interconnected in large numbers in a straightforward way to design digital circuitry.

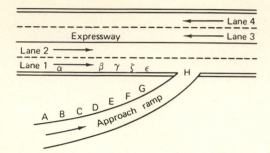

FIG. 10.1 Automated entry system for an expressway.

approach ramp. We desire a system which will guide the driver at A and bring him at normal speed into Lane 1 at H when there is a suitable break in the traffic flow along Lane 1.

The system consists of the following elements. There are traffic detectors (α, β, . . .) all along Lane 1 to detect appropriate gaps in the vehicular flow. Simultaneously, a car waiting to enter the expressway is held by a red light at A. When a suitable gap is found, the light at A turns green and, as the car accelerates, the lights at B, C, D, and so on turn green in sequence to pace the car as it picks up speed. If the driver keeps up with the green lights, he is brought to H just as the traffic gap appears and at expressway speed, so that he can move directly into the vehicular flow.

In this simple description, the concept at first sounds great. The driver desiring to enter the expressway needs only to follow the directions conveyed by the lights along his ramp. Drivers on the expressway can be confident that no car will try to merge unless there is an adequate traffic gap.

In actual operation, there are several problems. If the driver starting at A does not accelerate fast enough, he will reach H too late to hit the gap; sensors must detect this lag well before H and signal the driver to stop, rather than try to enter the traffic stream. Furthermore, along the expressway the system must detect a car moving from Lane 2 to Lane 1 just before H—a motion which is possible because there is a vehicular gap in Lane 1—or prevent such transfer by a divider between the two lanes.

From our standpoint, the interesting feature of the system is that there are nonlinearities which have dominant effects on performance. For example, the acceleration achievable by the entering car on the ramp is limited (by the mechanical capabilities of the vehicle and, most importantly, by the driver's comfort and judgment of safety). If the entering car must be stopped just before the merger, deceleration is limited. As a final example of nonlinearity, the importance of the error with which the car is brought into the center of the vehicular gap is a nonlinear function of that error: when the error is very small, it is unimportant; errors greater than a given size are catastrophic.

Thus, the system is inherently nonlinear; the engineer cannot design solely on the basis of linear system components.

Unfortunately, nonlinearity appears in a wide variety of forms. There is no straight-forward method of analysis or design when the system is nonlinear. For this reason, simulation is particularly important. In this chapter, we want to consider a few of the methods which have proved particularly useful in simulating nonlinear systems.

10.1   SYSTEM REDUCTION AROUND A NONLINEARITY

The systems of interest often possess a *single* nonlinear element, with all other components linear. As an example, Fig. 10.2 portrays a two-loop feedback system with a single nonlinear element: the amplifier which saturates for large values of its input signal m. (If the system is a servomechanism controlling the output position c, G_2 represents the dynamics of the load; the amplifier produces a force or torque which causes output motion; and the saturation means that the motor or amplifier is limited in the maximum force it can develop.)

When nonlinearity occurs in only a single element, we can reduce the linear system around the nonlinearity with the techniques useful for any linear system analysis. In Fig. 10.2, essentially we stand straddling the amplifier and look out at the remainder of the system. The signal m is simply

$$M = G_1 R - (H + \beta G_1)C \tag{10-1}$$

while the output c is given by

$$C = G_2 Y \tag{10-2}$$

In other words, the system can be reduced to the single-loop form shown in Fig. 10.3.

The possibility of system instability arises because of the closed-loop nature of the system. Thus, the stability question in this example reduces to the consideration of the single loop containing the transfer function $G_2(H + \beta G_1)$ in tandem with the non-linearity. If the nonlinear amplifier could be represented by a gain N, the stability

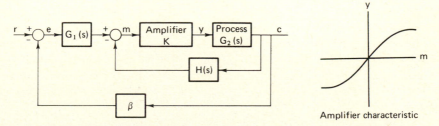

FIG. 10.2 Two-loop feedback control system.

Amplifier characteristic

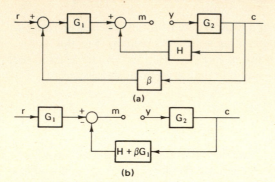

(a)

(b)

FIG. 10.3 Reduction of the linear system around the nonlinearity: (a) Linear system seen as we look outward from the nonlinearity; (b) Equivalent of (a).

would depend on the characteristic polynomial, which is the numerator of

$$1 + NG_2(H + \beta G_1)$$

The reduction shown in Fig. 10.3 is comparatively trivial. In other cases, when the original system is far more complex, the simplification may be major. The important concept here is that we can manipulate the linear portions around the nonlinearity.

A circuit example

This concept can be illustrated further by the circuit example of Fig. 10.4: a network containing four linear resistors and one nonlinear resistive element. We wish to generate a plot of the output voltage e_2 as a function of the input e_1.

The nonlinear element is described by the given characteristic for $v(i)$: the voltage v as a function of the current i. First we reduce the circuit around the nonlinear element in an attempt to find the current i for each value of system input e_1. Once i is found, v can be read from Fig. 10.4(b), and e_2 then calculated.

Straddling the nonlinear element and looking outward, we see that the network can be replaced by Thevenin's theorem (Fig. 10.5, where the terminals of n are viewed as the system output port).

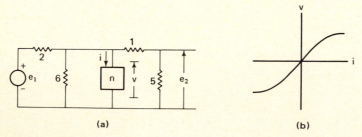

(a)

(b)

FIG. 10.4 Nonlinear system without energy storage: (a) Circuit with the nonlinear element n; (b) Characteristic of n.

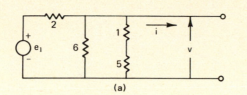
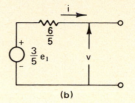

(a)   (b)

FIG. 10.5 Reduction of the linear part of the system: (a) Circuit seen from n; (b) Equivalent of (a).

Figure 10.5(b) shows that all linear elements taken together result in one relationship between i and v:

$$v = \frac{3}{5}e_1 - \frac{6}{5}i \tag{10-3}$$

The nonlinear element imposes a second constraint, represented by Fig. 10.4(b). These two constraints allow evaluation of the two unknowns, i and v. For example, for any selected e_1, we can draw Eq. (10-3) on the nonlinear characteristic; the intersection of the two curves gives the solution for i and v for this e_1 (Fig. 10.6). Variation of e_1 simply moves the straight line vertically so that the v intercept is always $\frac{3}{5}e_1$.

For the particular value $e_1 = 4.5$, Fig. 10.6 gives a v of 2. The original circuit shows that e_2 is just $\frac{5}{6}v$, or in this case $\frac{10}{6}$. Repetition of the graphical analysis for a range of values of e_1 permits point-by-point determination of the curve of e_2 versus e_1 for the total nonlinear system.

This particular circuit example is almost trivial, but it does illustrate two important features:

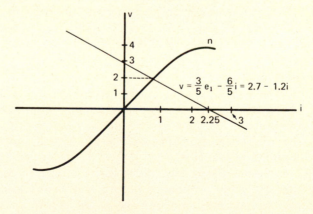

FIG. 10.6 Graphical determination of i and v when $e_1 = 4.5$.

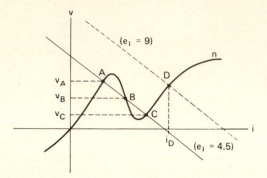

FIG. 10.7 Graphical analysis when n is a more complex characteristic.

(1) The value of reduction of the linear portion of the system—a reduction which will be essential in stability analysis in Chap. 16.

(2) The existence of more than one possible solution for a nonlinear problem. If we change the characteristic of the nonlinear element from the saturating curve of Fig. 10.6 to the undulating curve of Fig. 10.7, the same graphical analysis yields three distinct solutions (v_A, v_B, and v_C)—three possible *equilibrium points* where we can satisfy the two constraints [the linear equation (10-3) and the nonlinear characteristic]. The correct solution now depends on the past history of the system. For instance, if e_1 is gradually increased from 0 to 4.5, the current i builds up from 0 and, when e_1 reaches 4.5, we are at v_A. If e_1 is initially 9 (i at i_D) and is reduced to 4.5, the system moves toward C in Fig. 10.7; when $e_1 = 4.5$, v is v_C.

Thus, even this relatively simple circuit illustrates some of the troublesome properties of nonlinear systems.

10.2 LINEARIZATION

The power of analysis and design methods for linear systems is so great that, whenever it is possible, we try to *linearize* the system: i.e., to approximate the actual characteristics by a linear model. Such linearization may be accomplished in several ways.

Graphical linearization

If the system contains a single nonlinear element described by a graphical characteristic, we can often find a linear approximation directly from the graph. The first step is a guess of the probable range of variation of the input signal, m in Fig. 10.8. The appropriate approximation depends critically on this particular operating range (as shown in Fig. 10.9).

In deciding which linearization to use, we are, of course, faced with an impossible task. The range of variation of m is only known after the system is analyzed; the analysis is possible only after we have linearized. We can only make the most intelligent possible guess as to the range of m, linearize, analyze the system, then repeat the

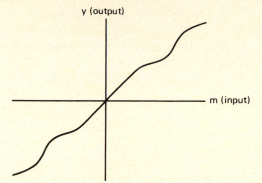

FIG. 10.8 A nonlinear characteristic.

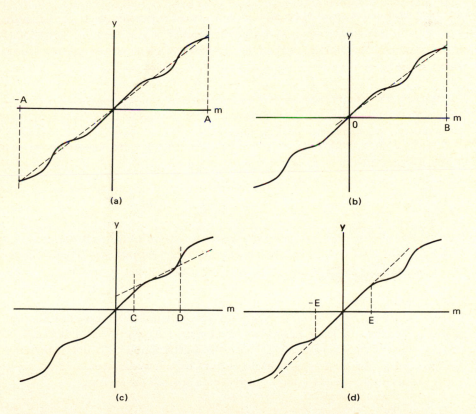

FIG. 10.9 Four possible linearizations depending on expected variation of input m: (a) Expected variation of m from $-A$ to A; (b) Expected variation of m from 0 to B; (c) Expected variation of m from C to D; (d) Expected variation of m from $-E$ to E.

process with an improved estimate of the m variation. The problem arises in feedback systems, where the input m depends on the output y; fortunately, in feedback studies we often find that the system output is relatively insensitive to particular parameter values.*

Two other comments should be made on Fig. 10.9:

(1) When the linear approximation passes through the origin,

$$y = \alpha m \tag{10-4}$$

The approximation is simply a gain α. When the straight line intercepts the y axis at y_0,

$$y = \alpha m + y_0 \tag{10-5}$$

and the nonlinearity is described by a gain α plus an offset (constant output).

(2) The approximation of (d) is an *incremental linearization,* valid in the vicinity of the origin. In other words, this representation is valid for *small signals, m*. We might also make an incremental linearization in the vicinity of any operating point (other than the origin). The linearized gain is then the slope of the nonlinear characteristic at the operating point.

Nonlinearity described in three variables

When three variables are involved in the description of the nonlinearity, the graphical linearization is more complicated. For example, the flow q through a valve depends on the displacement x of the valve, also the pressure drop p across the orifice. Typical graphical characteristics are shown in Fig. 10.10.

The linear approximation must have the form

$$q = \alpha x + \beta p + \gamma \tag{10-6}$$

which graphically represents *parallel, equally spaced* straight lines. α is the slope; γ is value of q with $x = p = 0$; and β is measured as the change in q divided by the change in p with x constant (as we move vertically from one line to another).

Analytic linearization

If the nonlinearity is described analytically (by equations), we can plot the characteristic, then linearize graphically. If we are interested in an incremental linearization in the vicinity of a known operating point, it is usually simpler to work analytically. As illustrated by the mixing system at the end of Chap. 9, we can replace each variable (e.g., y) by its normal value Y_0 plus an incremental or small signal change y_1:

$$y \to Y_0 + y_1$$

---

*Indeed, such insensitivity is a primary motivation for using feedback, as we shall see in Chapter 12.

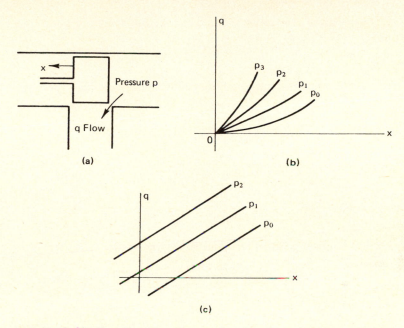

FIG. 10.10 Nonlinearity with output q a function of two inputs, x and p: (a) Valve: when x is positive, the orifice opens and flow q increases; (b) Nonlinear characteristics. As pressure increases (from p_0 to p_1, etc.), flow increases for a constant x; (c) Linear approximation.

In the equations, we then discard all nonlinear terms (powers of y_1 higher than the first), under the assumption that the signals are small enough that these higher order terms are negligible.

As an example, an amplifier with saturation might be described by the relation

$$y = 2\left(m - \frac{1}{16}m^3\right) \tag{10-7}$$

where m is the input, y the output. We wish to linearize for operation in the immediate vicinity of $m = 2$ (point O in Fig. 10.11).

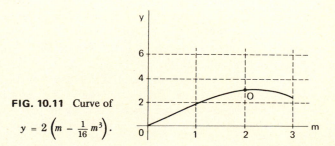

FIG. 10.11 Curve of

$$y = 2\left(m - \frac{1}{16}m^3\right).$$

The linearized description is found by replacing m by

$$2 + m_1$$

where m_1 is the signal measured from 2. Substitution in Eq. (10-7) gives

$$y = 2\left[(2 + m_1) - \frac{1}{16}(2 + m_1)^3\right] \tag{10-8}$$

or

$$y = 3 + \frac{1}{2}m_1 - \frac{3}{4}m_1^2 - \frac{1}{8}m_1^3 \tag{10-9}$$

Since the total y is just $3 + y_1$, neglecting terms in m_1^2 and m_1^3 gives the incremental model

$$y_1 = \frac{1}{2}m_1 \tag{10-10}$$

Equation (10-10) can also be found by a Taylor series expansion of $y(m)$ around the operating point $(m = 2, y = 3)$; the slope is

$$\left.\frac{dy}{dm}\right|_{m=2} = 2 - \left.\frac{6}{16}m^2\right|_{m=2} = \frac{1}{2} \tag{10-11}$$

or the gain of the incremental linear approximation.

This analytical approach can be extended directly to nonlinearities in which the output is a function of any number of inputs.

10.3 SIMULATION OF NONLINEARITIES

While linearization is a commendable goal, to do so often requires using a relatively crude model of the actual components. Very frequently the nonlinearity plays an essential role in determining system performance. In such circumstances, simulation becomes a most important factor in design, just because of the difficulty inherent in the analysis of a nonlinear system.

The power of the analog simulator is illustrated by Fig. 10.12—a position control system for a radar antenna. The radar measures the error between the target position and the direction in which the antenna is pointed. This error signal is amplified, then excites the motor which moves the antenna in a direction to reduce the error. As the

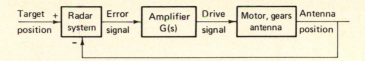

FIG. 10.12 Block diagram of antenna control system.

target moves, the antenna automatically is driven to follow. Design of such a system involves selection of the motor and gears, and choice of the amplifier transfer function $G(s)$.

In such a system, ordinarily the radar and amplifier operate as linear devices. The motor-gears-load, however, may be quite nonlinear because of saturation of the motor, backlash in the gear train, and so forth. Because the motor, gears, and antenna are expensive components easily damaged, it is not convenient to design $G(s)$ by building the system and then trying different $G(s)$ functions. On the other hand, because of the nonlinearities, usually we cannot solve analytically for the response to various input signals.

Faced with these difficulties, we turn to simulation: to build an electronic analog simulator which has the same dynamic characteristics as the actual components. In this simulator, $G(s)$ can be varied until we find the best overall system.*

Thus, design requires an ability to simulate a given nonlinear system. In the following paragraphs, we consider the expansion of our RC op-amp circuits to include non-linearities, although the types of nonlinearities are restricted to those which can be represented by ideal diodes in relatively simple circuits. Fortunately, this restriction still allows simulation of many of the nonlinearities which occur in real systems.

Ideal diode

The ideal diode is an approximation to actual electronic diodes; it has the characteristic that, in terms of the notation of Fig. 10.13:

Device is a short-circuit when $i > 0$

Device is an open-circuit when $e < 0$

In other words, the ideal diode is a switch which turns on and off according to the applied signal (e or i). The voltage e cannot be positive; the current i cannot be negative.

*Indeed, we may also wish to drive different motors, various gear ratios, and so forth.

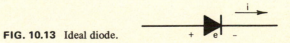

FIG. 10.13 Ideal diode.

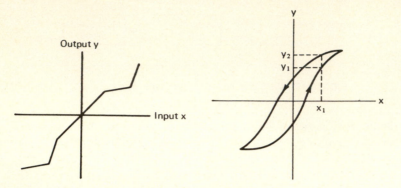

FIG. 10.14 Typical characteristic.   **FIG. 10.15** Characteristic with memory.

If the diode is conducting $(i > 0)$, as i decreases the diode shuts off when i reaches zero and tries to go negative. The diode then remains off (open with e negative) until the voltage e rises to zero and tries to go positive, at which time the diode conducts.

With ideal diodes, resistors, capacitors, and ideal high-gain voltage amplifiers, we can simulate many of the nonlinearities common in physical systems.

10.4 SIMULATION OF PIECEWISE LINEAR INPUT-OUTPUT CHARACTERISTIC

In this section, we wish to realize (using operational amplifiers, diodes, resistors, and batteries) a specified piecewise linear input-output characteristic (as shown in Fig. 10.14). We assume that the desired characteristic is

(a) Single-valued [both $y(x)$ and $x(y)$] —that is, each value of x corresponds to only one y, and vice versa.

(b) Monotonic—that is, the slope is always positive (or it can be always negative).

Constraint (a) means that the device is zero-memory. In other words, the output at any time depends only on the input signal at that time, and not on the past history of the input. Characteristics such as depicted in Fig. 10.15 are not included. Here, the input x_1 causes two possible outputs y_1 and y_2; the output is y_1 if the earlier input was less than x_1, y_2 if greater.

Components to be used

We wish to use ideal voltage amplifiers of high-gain, ideal diodes, resistors, and batteries, as shown in Fig. 10.16.

FIG. 10.16 Components used to realize Fig. 10.14 characteristic.

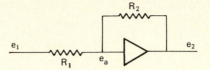

FIG. 10.17   Basic op-amp circuit.

The ideal diode has the characteristic that it is a short circuit when current flows in the direction of the arrow, an open circuit when the voltage across the diode is positive on the right (in Fig. 10.16).

Basic principle of synthesis

The gain of the amplifier is sufficiently large so that the amplifier input voltage is negligible in the circuit of Fig. 10.17. With this condition satisfied, the $e_1 - e_2$ relation can be determined by one node equation written at e_a:

$$\frac{e_1}{R_1} + \frac{e_2}{R_2} = 0 \quad \text{or} \quad e_2 = -\frac{R_2}{R_1} e_1 \qquad (10\text{-}12)^*$$

Thus, any desired slope for e_2/e_1 can be realized by choosing the ratio R_2/R_1. If e_2/e_1 is to be positive, we need two op-amp circuits in tandem, one with a gain of (-1).

A change in slope can be realized by changing R_2 or R_1. If R_2 is to be reduced, we use the circuit shown in Fig. 10.18: the diode and battery are inserted so that, over the desired range of e_2, the diode conducts and R_3 is placed in parallel with R_2. If the $|e_2/e_1| = R_2/R_1$ is to be increased, R_1 is decreased by a similar circuit in parallel with R_1.

*The correct node equation is

$$\frac{e_1 - e_a}{R_1} + \frac{e_2 - e_a}{R_2} - i_{in} = 0$$

where i_{in} is the amplifier input current. If i_{in} is negligible and if $|e_a| \ll e_1$ or e_2, we obtain Eq. (10-12).

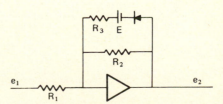

FIG. 10.18   Circuit with one change in slope.

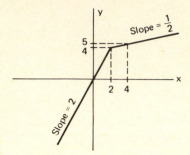

FIG. 10.19 Desired characteristic.

Thus, each change in slope is achieved by addition of a diode-battery-R circuit in parallel with R_2 or R_1. If the slope decreases, the circuit parallels R_2; if the slope increases, R_1.

Simplest example

The characteristic shown in Fig. 10.19, with only one change in slope, is the simplest example. The system is synthesized in the following steps.

(1) First we realize the slope desired as the characteristic passes through the origin—a slope of +2 (Fig. 10.20). The first op-amp realizes an output $z = -2x$ (any convenient resistances can be used as long as the ratio is $R_2/R_1 = 2$; here we show $100K$ and $200K$). The second circuit restores the plus sign, so that the total system yields

$$y = 2x \tag{10-13}$$

(2) We now move out from the origin of Fig. 10.19 in either direction. In this example, the only desired change in slope is the one which occurs at

$$x = 2, \ y = 4$$

The slope is reduced at this point, so we add a circuit in parallel with R_2. We must determine three things for this circuit: the size of R, the direction of the diode, and the value of the battery voltage:

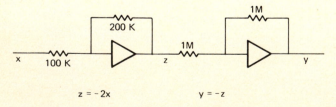

FIG. 10.20 Characteristic desired near origin.

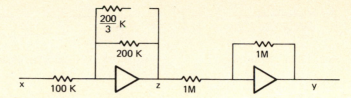

FIG. 10.21 Circuit after step (a).

(a) Size of R_3 in parallel with R_2. Since the new slope is to be 1/2, the total "R_2" must be $50K$—so that "R_2"$/R_1 = 1/2$. Hence R_3 in parallel with $200K$ must be $50K$, or

$$\frac{1}{R_3} + \frac{1}{200} = \frac{1}{50} \tag{10-14}$$

$$\frac{1}{R_3} = \frac{3}{200} \quad \text{or} \quad R_3 = \frac{200}{3} K \tag{10-15}$$

(b) Direction of diode. We now want to insert a diode which conducts when

$$x > 2 \qquad y > 4$$

from Fig. 10.19. Since we are working on the feedback (or R_2) branch, the diode behavior depends on z or $-y$, not on x. Hence we want diode conduction when

$$-y < -4 \quad \text{or} \quad z < -4 \tag{10-16}$$

We want conduction whenever z is *less* than a certain number; *hence the diode must have the arrow directed to the right,* so that smaller and smaller values of z lead to continued conduction (Fig. 10.22).

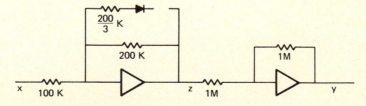

FIG. 10.22 Circuit after step (b).

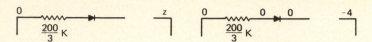

FIG. 10.23 Diode circuit.   FIG. 10.24 Voltages when diode just conducts.

(c) Battery. Finally we must determine the battery voltage and polarity. If the diode circuit is redrawn (Fig. 10.23), at the moment the diode just conducts, the voltage on the left is 0 (since this is the amplifier input); then the voltage between $R = (200/3)K$ and the diode is zero. With the diode just conducting, the voltage to the right of the diode is zero. At this time, Eq. (10-16) tells us z should be -4. Hence the voltages are as shown in Fig. 10.24. Thus, the battery must be inserted as shown in Fig. 10.25, with a magnitude of 4 volts, and positive to the left. The complete circuit is shown in Fig. 10.26.

Final example

The desired characteristic is shown in Fig. 10.27 with the slopes indicated for each linear segment. Since there are three changes in slope, three diodes and batteries are required.

(a) The slope of $+1/2$ through the origin is realized by Fig. 10.28.

(b) The break at

$$x = -2, \qquad y = -1, \qquad z = +1$$

is realized. Here the slope increases to 3 so the total input R is decreased to $1/3$. The R to be added (Fig. 10.29) is

$$\frac{1}{R} + \frac{1}{2} = \frac{1}{1/3} \qquad R = \frac{2}{5}$$

Since this R is to be inserted for $x < -2$, the diode must have the arrow to the left (Fig. 10.30). At the break, $x = -2$, so the voltage on the right side of the diode must be -2. Hence the circuit is given in Fig. 10.31.

(c) The same type of analysis realizes the break at $x = -3$, $y = -4$ by addition of a circuit in parallel with the feedback resistor R_2. Since the input resistance is now $\frac{1}{3}M$ and the slope is to be 1, the total feedback resistance must be $\frac{1}{3}M$. Hence the added

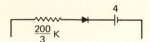

FIG. 10.25 Battery required to fit voltages in Fig. 10.24.

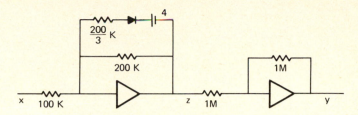

FIG. 10.26 Complete circuit.

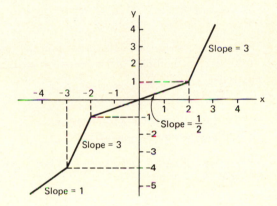

FIG. 10.27 Desired characteristic for second example.

FIG. 10.28 Linear behavior near origin (all Rs in megohms).

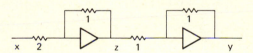

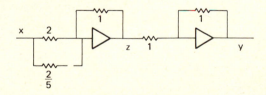

FIG. 10.29 Resistance required to increase slope to 3.

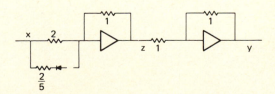

FIG. 10.30 Diode added.

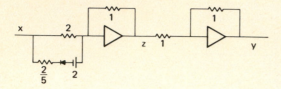

FIG. 10.31 First break in slope realized.

R must be

$$\frac{1}{R} + \frac{1}{1} = 3 \qquad R = \frac{1}{2} M$$

The diode and battery are determined as in the first example, with the result Fig. 10.32.

(d) Finally, the break at $x = 2$, $y = 1$ is synthesized, and the final circuit is shown in Fig. 10.33.

10.5 SIMULATION OF EVEN, ZERO-MEMORY NONLINEARITY

The same approach can be used to realize an even, zero-memory nonlinearity—for example

$$y = ax^2 \tag{10-17}$$

where x is the input, y the output. The synthesis is accomplished in the following steps:

(1) The given nonlinearity is approximated by a piecewise-linear characteristic (such as shown in Fig. 10.34, where the slopes are indicated).

(2) We next generate the signal y which is

$$y = \begin{cases} x & \text{when} \quad x > 0 \\ -x & \text{when} \quad x < 0 \end{cases} \tag{10-18}$$

by a circuit with two diodes (Fig. 10.35).

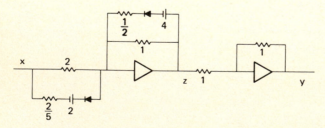

FIG. 10.32 Second break included.

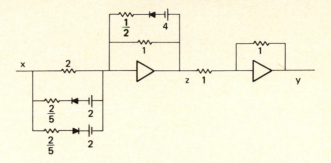

FIG. 10.33 Simulation of piecewise-linear characteristic of Fig. 10.27.

(3)   Next the break at

$$x = 2, \quad y = 2 \tag{10-19}$$

to a slope of +4 is realized by a diode circuit in parallel with the upper input resistor of Fig. 10.35 as shown in Fig. 10.36.

(4)   Finally, the break at $(-2, +2)$ is realized, and the final circuit is shown in Fig. 10.37.

The two signals, $\begin{cases} -x \\ 0 \end{cases}$ and $\begin{cases} 0 \\ x \end{cases}$, can be added in an op-amp circuit so that one diode suffices to realize both breaks in slope, but then additional amplifiers are required.

The circuit is readily modified if y is not an even function of x—that is, if $y(-x) \neq y(x)$—but if the slope is always negative for x negative, positive for x positive. The changes are merely in the slopes realized for each segment; each of the four slopes is realized in Fig. 10.37 by a definite set of elements.

10.6   SIMULATION OF A NONLINEARITY PLUS A TRANSFER FUNCTION

We conclude this discussion of simulation by considering the example of a simple electric motor with torque saturation and a dynamical load.

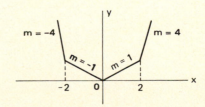

FIG. 10.34 Desired piecewise-linear characteristic; we assume $y(x_1) = y(-x_1)$.

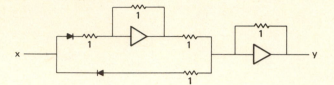

FIG. 10.35 Realization of $y = |x|$.

Problem description

An electric motor accepts an electric input signal (voltage or current) and yields a mechanical output (torque or angular motion, if the output is rotational).* In the type of motor to be considered here, the torque developed by the motor is proportional to the applied voltage.† This torque turns the motor shaft and the load according to the relationship

$$\ell_d = J_m \frac{d^2\theta}{dt^2} + B_m \frac{d\theta}{dt} + \ell_L \tag{10-20}$$

In other words, the developed torque ℓ_d divides into three components: part $J_m \theta''$ is used to accelerate the rotor of inertia J_m; part $B_m \theta'$ is used to overcome the friction of the rotor; and the rest ℓ_L is the torque delivered to drive the load (which might be a rotating antenna, an output dial, or any mechanical system). Thus, in Eq. (10-20), the symbols represent:

ℓ_d	developed torque
θ	angular position of motor shaft

*Motors may also yield a translational output—force or displacement. In these paragraphs, we restrict consideration to rotational systems.

†In a motor, the applied signals create rotating magnetic fields. The torque is developed because the rotor (the rotational part) tries to turn toward a position where the stored magnetic energy is minimized. In other words, the system seeks a condition of minimum energy. The magnetic field continually rotates, however, so that the rotor is turned continually. When the applied signal reverses in polarity, the field rotates in the opposite direction, and the torque turning the rotor changes direction as well.

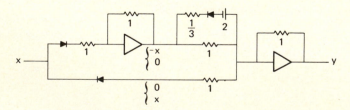

FIG. 10.36 Break at $x = 2$ added.

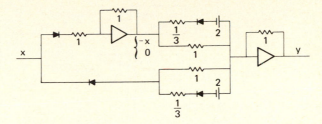

FIG. 10.37 A circuit to realize the characteristic of Fig. 10.34:
$\begin{cases} -x \\ 0 \end{cases}$ means the signal is $-x$ when $x > 0$, 0 when $x < 0$.

J_m	moment of inertia of rotor
B_m	damping or friction coefficient for rotor
ℓ_L	torque to load

For the specific problem considered here, we make three additional assumptions:

(a) The load is simply an inertia J_L which is on the motor shaft. Therefore

$$\ell_L = J_L \theta''$$

and the developed torque is related to the output θ by the equation:

$$\ell_d = (J_m + J_L)\theta'' + B_m \theta' \quad \text{or} \quad \frac{L_d}{\theta} = \frac{1}{(J_m + J_L)s^2 + B_m s} \tag{10-21}$$

(b) The developed torque depends on the electrical voltage applied to the motor according to the graph of Fig. 10.38.

(c) The applied voltage is the output of an amplifier for which the transfer function is $G(s)$. This amplifier is driven by an error signal e which is derived as shown in Fig. 10.39.

The system depicted in Fig. 10.39 is a *servomechanism*—a feedback system in which the output angular position θ is to be controlled by an input voltage e_{in}. In order to achieve this control, θ is measured electrically in the device called a *sensor*: the sensor

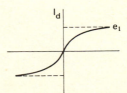

FIG. 10.38 Developed torque as a function of applied voltage.

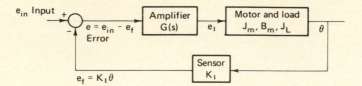

FIG. 10.39 System in which motor is used.

output is a voltage e_f which is proportional to θ. This $e_f = K_1\theta$ is compared to e_{in} to generate the error signal e.

To understand system operation, we first assume all signals are zero for $t < 0$. At the start $(t = 0)$, e_{in} is abruptly moved to +5 and fixed there. The output θ does not change instantaneously, so e_f stays at 0. Hence, the error e jumps to 5. This error is amplified in G and then drives the motor. As a result, θ starts to increase. Then e_f starts to increase and the error e is reduced. When $K_1\theta$ reaches 5, the error is zero, the drive is removed from the motor, and the system stops with

$$e_{in} = K_1\theta \qquad \text{or} \qquad \theta = \frac{1}{K_1} e_{in} \qquad\qquad (10\text{-}22)$$

In other words, the system attempts to maintain Eq. (10-22)—the output proportional to the input.

Of course, our explanation is greatly oversimplified. An obvious problem arises because of the inertia of the rotor and load. If, when θ reaches e_{in}/K_1, the output has any velocity, the momentum will cause θ to continue to increase beyond e_{in}/K_1; the error then goes negative and the system is driven "backward" to bring θ back to the desired e_{in}/K_1. [The behavior is shown in curve (b) of Fig. 10.40.] While response (b) may be satisfactory, it is also possible to have so much overshoot that we reach the behavior shown in (c): here the system is actually unstable, and θ never settles down to its desired value.

The exact form of the response depends on the specific values for the different parameters of Fig. 10.39, and can be calculated from the set of differential equations which describes the system. Because of the nonlinearity of Fig. 10.38, however, calculation is extremely tedious, and use of an analog simulation is highly preferable. We build an electronic circuit (an analog simulator) which has the same characteristics as

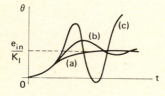

FIG. 10.40 Possible responses of system in Fig. 10.39 with step input.

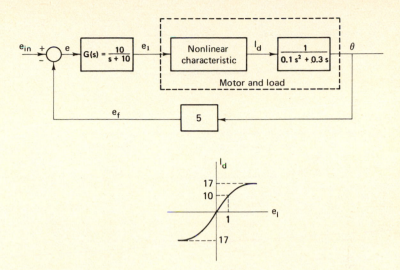

FIG. 10.41 System to be simulated.

the system of Fig. 10.39; we then drive this simulator electrically and observe the output signal.

Specific system to be simulated

In order to work with a numerical example, we use the parameters shown in Fig. 10.41. (Since both electrical and mechanical quantities are involved, we must be sure to use a consistent set of units. While MKS units can always be used, they are not necessary.) A division of the motor and load into two parts is shown: the applied voltage e_1 causes a developed torque ℓ_d according to the nonlinear characteristic shown; this ℓ_d is related to θ by the transfer function of Eq. (10-21), where we have chosen the parameter values

$$J_m + J_L = 0.1 \qquad\qquad B_m = 0.3 \qquad\qquad (10\text{-}23)$$

Hence, our task is to simulate the system of Fig. 10.41.

The signal flow diagram

We construct the signal flow diagram of Fig. 10.42 to represent the elements of the block diagram of Fig. 10.41. In this signal flow diagram, the nonlinear element is shown by a dashed line.

If we decide to use only Rs, Cs, and ideal amplifiers for the linear portion of the system, we have to convert the signal flow diagram into a form in which all integrations and adders involve a minus sign. If we use commercial op-amps with flexibility on sign changing in each op-amp circuit, we can stop with part (b) of Fig. 10.42.

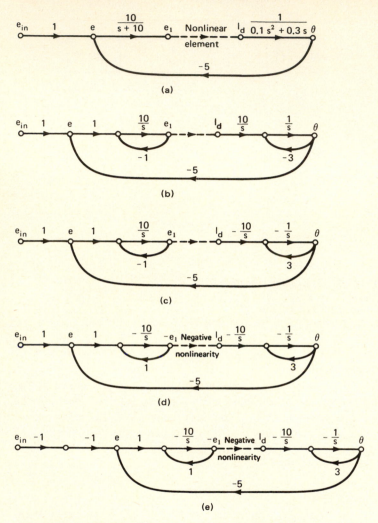

FIG. 10.42 Preparation of signal flow diagram for circuit design: (a) Direct replacement of Fig. 10.41; (b) All s dependence realized by integrators; (c) Last 2 integrators realized with minus signs; (d) First integrator and non-linearity realized with negative signs; (e) Final signal flow diagram.

As soon as Fig. 10.42 is drawn, we can go directly to the circuit of Fig. 10.43, which shows the linear portion of the system. Completion of the simulation now requires only realization of the nonlinear characteristic, redrawn in Fig. 10.44(a), with one possible piecewise-linear approximation shown. The simulation with two breaks in slope is shown in Fig. 10.44(b), and the total system realization in Fig. 10.45.

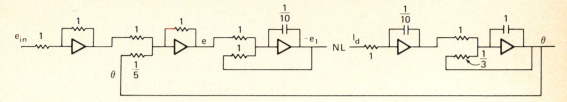

FIG. 10.43 Simulation of the linear portions of the system.

Final comment

The example emphasizes that a nonlinear system involving frequency-dependent elements is simulated by separating the linear, frequency-dependent portions from the nonlinear, zero-memory portion. Such separation is not always possible, but many practical nonlinearities can be handled in this way. Then we can use the standard techniques for each portion of the system.

Incidentally, it is perhaps interesting to note that this system is actually unstable. A step input causes the output θ to oscillate indefinitely at a constant amplitude (the oscillation does not grow without bound because the motor torque is limited). Experimental tests indicate that the system can be stabilized if the resistance of 1/5 is increased to 1/3; in other words, K_1 in the original system is decreased from 5 to a value less than 3. In Chap. 16, we will see how this oscillation and the remedy can be predicted.

10.7 CONCLUSION

This chapter is a minimal introduction to the simulation and study of nonlinear systems. Indeed, we have focused almost exclusively on linearization or piecewise linearization of the characteristics of the nonlinear element. There are many nonlinear systems where neither approach is particularly valuable.

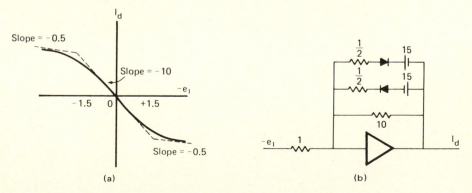

FIG. 10.44 Nonlinear element: (a) Desired nonlinear characteristic; (b) Simulation of piecewise-linear characteristic (assuming 1/2 in parallel with 10 is about 1/2).

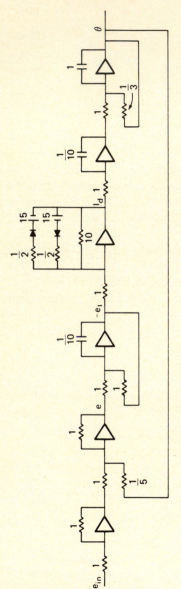

FIG. 10.45 Simulation of the total servomechanism.

Three familiar examples illustrate the far greater generality of the subject of non-linear systems. A very important nonlinearity has an output y which is the product of two inputs u and x:

$$y = Kux \tag{10-24}$$

Such a device is used in modulation—to change the band of frequencies in which a signal contains energy. For instance, Fig. 10.46 shows a scheme for transmitting, over a telephone wire, heart and breathing sounds which can be diagnosed by a physician at a location remote from the patient. The nonlinearity operates on the basis that the "modulated signal" is the product of the patient's signal and the 2000-Hz sinusoid. (It can be shown trigonometrically that this modulated signal has energy from 1000–3000 Hz, hence can be transmitted through the telephone system.)

Because of the two multipliers, the signal received by the doctor is

$$\text{Output} = \text{Input} \times [\sin(2\pi \times 2000t)] \times [\sin(2\pi \times 2000t)] \tag{10-25}$$

A theorem in trigonometry states that

$$\sin^2 \theta = \frac{1}{2} - \frac{1}{2} \cos 2\theta \tag{10-26}$$

Hence, Eq. (10-25) becomes

$$\text{Output} = \text{Input} \times \frac{1}{2} - \text{Input} \times \frac{1}{2} \cos(2\pi \times 4000t) \tag{10-27}$$

The first term carries the information; the second term contains only high-frequency components (above 3000 Hz) and can be removed by filtering. Thus, with this system, the information originally in the band from 0 to 1000 Hz can be transmitted over a communication system which operates with a 300–3000 bandwidth.

A second type of nonlinearity, beyond the scope of the discussion here, is exemplified by the static friction opposing relative motion of two bodies. When one block sits on a

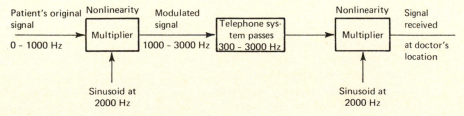

FIG. 10.46 Use of multiplicative nonlinearity.

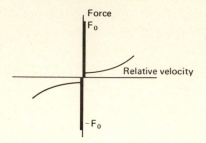

FIG. 10.47 Frictional force.

surface for a period of time, the force required to initiate relative motion increases as the two surfaces in contact tend to merge into one another (or set). The force-versus-velocity characteristic has the form of Fig. 10.47, with the amplitude of the pulse F_0 dependent on the duration of the state of zero relative velocity.

Finally, we often encounter nonlinearity of the hysteresis type (e.g., in magnetization curves). The flux or output builds up as the current or input is increased. Reduction of the input to zero does not bring the output back to zero.

Thus, the very simple piecewise-linear nonlinearities considered in this chapter constitute only one part of the entire subject of nonlinear systems. Fortunately, however, many important nonlinearities can be adequately modeled by a piecewise-linear characteristic. If not, analysis and system design usually require numerical analysis or digital simulation, which is introduced in Chap. 14.

PROBLEMS

10.1 Determine a circuit using ideal amplifiers, Rs, Cs, batteries, and ideal diodes which realizes

$$y = 4x^3$$

over the range $-2 < x < 2$. Use only two diodes.

10.2 Design a circuit to approximate

$$y = 4x^3$$

over the range $-10 < x < 10$ with as much accuracy as possible, using no more than four diodes.

10.3 Many elements possess both a *dead zone* and *saturation*, as shown in the figure. The dead zone means that the input magnitude $|x|$ has to exceed unity before there is any significant response; saturation is represented by the fact that $|x|$ increasing beyond 2 does not cause a significant change in y.

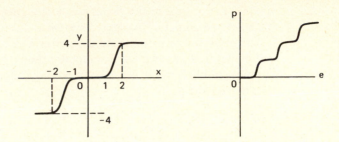

(a) Cite one or more examples of devices that have this property.

(b) The figure of p as a function of e shows a plot of human sensation of pain versus the intensity of the cause. How would you represent such a characteristic as the sum of characteristics of the form of dead zone and saturation?

(c) Simulate y versus x, as given, by ideal amplifiers and electrical components.

10.4 (a) Design a circuit realizing the $y(x)$ characteristic shown in the figure.

(b) If we apply a signal x varying randomly but never exceeding 2, what is the gain of the device?

(c) If we apply a signal $x = 5 \sin 70t$, we might try to linearize the device by representing it by an equivalent gain. What gain would you use (assume the characteristic is odd)?

(d) If x is a very large sinusoid, say amplitude greater than 20, and $y(x)$ is again odd, what equivalent gain would you use?

(e) Sketch the equivalent gain versus the amplitude of x if x is sinusoidal. This "equivalent gain" is very nearly the *describing function* discussed at the end of Chap. 16. We can test the stability of a feedback system, for example, over the range of gains and find out if there is any signal amplitude at which the system is unstable.

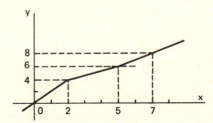

10.5 We have op-amps and diodes available and wish to generate "infinitely clipped speech"—that is, a signal which has the same zero crossings as the original speech signal, but between zeros is either +10 or −10 volts, according to whether the original speech signal was + or -, respectively. The figure shows the input and output of the desired system.

Design a system to approximate this performance. Any number of amplifiers and diodes may be used, but no amplifier should have an output greater than 10 volts. In other words, we wish to amplify, then clip, then amplify again, then clip off the peaks, etc.

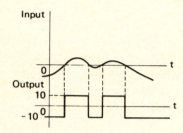

10.6   Using ideal diodes, ideal amplifiers, resistors, and batteries, determine a circuit realizing the input-output characteristic shown.

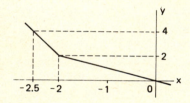

10.7   One type of nonlinearity which arises in practical systems is

$$y = x_1 x_2$$

The response y is the product of the two inputs. To realize this, we can use many different approaches, with two being illustrated in this problem.

(a) We can realize the blocks shown to obtain the squares of the sum and difference of the two input signals. Design a simple, approximate realization, using four or fewer diodes in each squarer circuit.

(b) We can use logarithmic amplifiers, as shown. Again, realize a rough simulation. Over what range of x_1 and x_2 does your system operate?

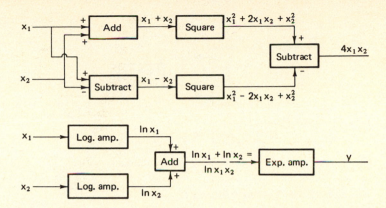

10.8   Design a circuit using ideal diodes and op-amps to realize the characteristic

$$y = |2x|$$

Attempt to minimize the required number of diodes.

10.9   Using only two diodes, find a circuit to realize the characteristic shown.

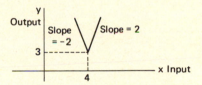

10.10   The circuit shows a quarter-squaring operation to obtain the product of two input signals, x and y. Operation is based on the fact that

$$(x + y)^2 - (x - y)^2 = 4xy$$

The blocks labeled "squaring module" are realized by diode-shaping networks. Explain how the output $-xy/10$ is obtained.

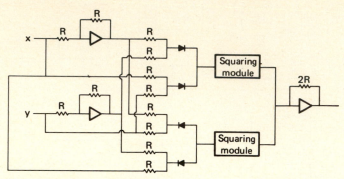

10.11 Once a multiplier is available (Prob. 10.10), we can obtain the square root of an input signal. Explain how the circuit shown yields an output $-\sqrt{10z}$, where z is the input signal.

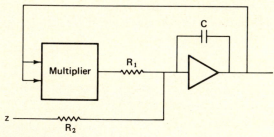

11 GAIN AND PHASE PLOTS

The normal human being has hearing ability which depends on the frequency of the sound. Figure 11.1 shows the gain plot for this hearing sense. The curve is determined by measuring, at distinct frequencies, the weakest signal the man can hear. His hearing is most sensitive about 4000 Hz, and actually is within 10 dB over the range from 700 to 6500 Hz.

Such a "hearing-gain" curve shows immediately which noises are likely to be most annoying or injurious to the human being (if we are anxious to control the noise environment). If similar curves are measured regularly for an individual during his life, we can obtain a picture of the aging and deterioration of hearing, often an anticipatory warning of more serious medical problems. The gain characteristic (gain versus frequency) is an important measure of the performance of the hearing system.

Figure 11.2 shows the experimental measurement of the gain-versus-frequency characteristics of all components except the radar in a system for tracking of aircraft near an airport. The complete system contains the components indicated in Fig. 11.3. The gain characteristic refers to the open-loop system from the radar output to the antenna position; each point on the curve is measured by applying a sinusoidal signal

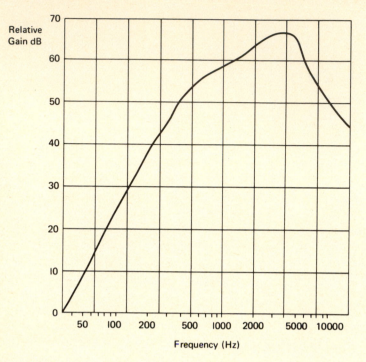

FIG. 11.1 "Gain" of the hearing sense as a function of sound frequency.*

*Throughout this chapter, we shall measure gain in deciBels (dB). The deciBel is defined in the Appendix to the chapter, where we also explain the conversion between the magnitude of the transfer function and the dB gain.

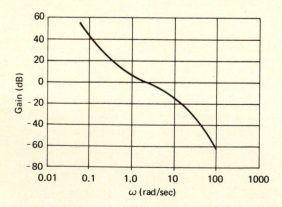

FIG. 11.2 Open-loop gain characteristic for a tracking-radar system.

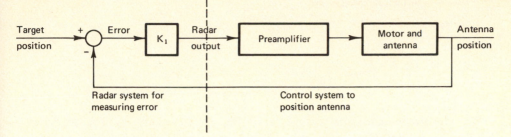

FIG. 11.3 Radar tracking system.

at the preamplifier input and determining the amplitude of the resulting sinusoidal variation in antenna position.

The design engineer must often use the motor, antenna, and radar that have been selected on the basis of considerations of performance, cost, space, and the like. He is free, however, to add components to the preamplifier to change the gain characteristic of Fig. 11.2. The design task is to select the preamplifier networks.

In both the above examples, we describe the system by its *gain characteristic*, the plot of gain versus frequency. Throughout electrical engineering, we find that this gain characteristic is a fundamental tool for showing quantitatively how a system behaves, for guiding the design of the system, and for studying the interaction of two or more systems.

This chapter is devoted to techniques for plotting the gain characteristic corresponding to a known transfer function, for interpreting a measured gain characteristic, and finally for plotting the phase characteristic—the phase shift as a function of frequency.

11.1 FORM OF THE GAIN PLOT

In the following paragraphs, a simple, rapid technique is presented for plotting the gain of a system versus ω from the transfer function in factored form. The gain-frequency plots which result are also called Bode plots, after Hendrik Bode who developed the technique in the 1930s at the Bell Telephone Laboratories.

Statement of the problem
Given the transfer function in the form

$$T(s) = K \frac{s(as + b)}{(cs + d)(es^2 + fs + g)} \qquad (11\text{-}1)$$

we desire to plot $|T(j\omega)|$ versus ω—the gain versus frequency, from values of ω near zero through to very large values.*

---

*In some texts, the authors insist that each term be written in the form $(s + \alpha)$—i.e., with the coefficient of the highest s power equal to unity. In other texts, the term is always written in the form $(Ts + 1)$—the constant term unity. Discussion in the following pages applies to either of these forms or to the more general $(as + b)$.

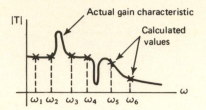

FIG. 11.4 Example of a gain characteristic.

We can always substitute $s = j\omega_1$ into Eq. (11-1), where ω_1 is a particular frequency, evaluate $|T(j\omega_1)|$ by the ordinary algebra of complex numbers, and then repeat this calculation for each different value of ω_1 required to cover the frequency range of interest. This brute-force approach has two shortcomings: (1) a great amount of tedious algebra is required; and (2) we are apt to miss entirely sharp peaks or nulls in the gain characteristic (as in Fig. 11.4, which shows a common gain characteristic and the specific values of ω we might choose). Hence, we want a simpler method which will reveal any sharp minima and maxima; indeed, for these advantages, we are willing to sacrifice great accuracy.

Separation of factors of $T(s)$

If we continue to use the $T(s)$ of Eq. (11-1) as an example, the gain can be written

$$|T(j\omega)| = |K| \frac{|j\omega||aj\omega + b|}{|cj\omega + d||-e\omega^2 + fj\omega + g|} \tag{11-2}$$

In other words, the gain involves the multiplication (or division) of five factors. We could simplify the gain calculation if these factors were added rather than multiplied.

In order to convert a product to a sum, we want to take the logarithm. What logarithm should be used? Since engineers often measure gain in decibels, we might as well go directly to that form and consider the gain G in dB, defined by

$$G = 20 \log |T(j\omega)| \quad \text{dB} \tag{11-3}$$

(Here the logarithm is to the base 10, not the base e.) Thus, instead of plotting $|T(j\omega)|$ versus ω, we want a plot of G versus ω.

Substitution of Eq. (11-2) into (11-3) reveals the way in which the various factors of $|T|$ are added in the calculation of G:

$$G = 20 \log |K| + 20 \log |j\omega| + 20 \log |aj\omega + b| - 20 \log |cj\omega + d|$$
$$-20 \log |-e\omega^2 + fj\omega + g| \tag{11-4}$$

In other words, if the transfer function is

$$T = \frac{ABC}{DEF} \tag{11-5}$$

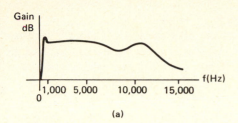

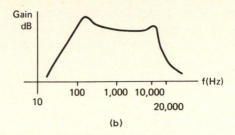

FIG. 11.5 Amplifier gain characteristic. (a) Gain on linear f scale; (b) gain on logarithmic f scale.

the dB gain is

$$G = G_A + G_B + G_C - G_D - G_E - G_F \qquad (11\text{-}6)$$

where G_A is the dB gain for A, and so on. When we use dB, the total dB gain is the sum of all numerator factors minus the sum of all denominator factors.

Frequency scale

The gain plot is much simpler to construct if we plot G versus log ω rather than ω. Again it "happens" that this is usually the way gain characteristics are plotted. For example, in describing a hi-fi amplifier, we are interested in the frequency range from 50 Hz to 15 kHz, and particularly in the lows (50-200 Hz) and the highs (6 kHz to 15 kHz). Figure 11.5 shows a typical gain characteristic plotted against both ω and log ω. Clearly, the logarithmic frequency scale permits us to portray meaningfully both the low-frequency and the high-frequency behaviors on the same plot.

Thus, our plots of the gain characteristic involve dB gain plotted against ω (or f) on a logarithmic scale. The easiest way to plot is to use semilog graph paper (ω along the logarithmic scale) as indicated in Fig. 11.6(a). If we have only linear graph paper

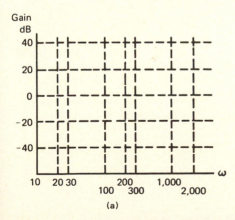

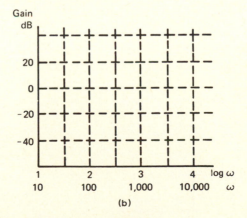

FIG. 11.6 Form of the gain plots. (a) On semilog paper; (b) on linear graph paper.

available, essentially we construct semilog paper as shown in part (b) of the figure—that is, we plot gain versus $\log \omega$ or $\log f$.

Thus, we wish to plot dB gain (the sum of the dB gains of the various T factors) versus a logarithmic frequency scale. Now we consider how the curve looks for a single factor of $T(s)$. First we consider the linear factor $(as + b)$; after that, we turn to a quadratic factor.

11.2 dB GAIN FOR A LINEAR FACTOR

A general linear factor of $T(s)$ has the form

$$as + b$$

We often write the factor as $(s + \alpha)$ or $(\beta s + 1)$, but the form $(as + b)$ includes both special cases. For this factor, the contribution to the dB gain is

$$G_1 = 20 \log |aj\omega + b| \tag{11-7}$$

One way to plot this G_1 is to consider the behavior at both very low and very high frequencies. When ω is very small (Fig. 11.7),

$$G_{1_{LF}} = 20 \log b \tag{11-8}*$$

When ω is very large,

$$G_{1_{HF}} = 20 \log a\omega = 20 \log a + 20 \log \omega \tag{11-9}$$

If G_1 is plotted versus $\log \omega$, we obtain a straight line (since G_1 is a constant plus 20 times $\log \omega$). The slope of this straight line is

$$20 \text{ dB/decade}$$

In other words, when ω increases by one decade (from α to 10α or, for example, from 10 to 100), $\log \omega$ changes by +1, and G_1 increases by 20 dB. *Each decade increase in ω*

---

*We assume here, in Eq. (11-9), and thereafter that a and b are both positive. If either is negative, the system is unstable, but the gain can still be evaluated; we need to change Eq. (11-8) or (11-9) then to take $|a|$ or $|b|$.

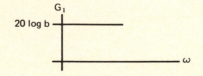

FIG. 11.7 Low-frequency behavior of G_1.

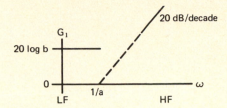

FIG. 11.8 Low-frequency and high-frequency behaviors of G_1.

corresponds to 20 dB increase in G_1. Thus, the straight line describing G_1 at high frequencies is entirely located (now that we know its slope) by any one point. Equation (11-9) reveals that $G_{1_{HF}} = 0$ dB when $\omega = 1/a$. Hence, Fig. 11.8 can be drawn showing the high-frequency behavior of G_1.

The two straight lines we have constructed are called the *low-frequency asymptote* and the *high-frequency asymptote* for G_1. As ω becomes smaller and smaller, G_1 approaches the low-frequency asymptote; at higher and higher ω, G_1 tends toward the high-frequency asymptote. The two asymptotes accurately describe the behavior of G_1 in the extreme frequency ranges (low and high).

Actually, we are now lucky: it turns out that simply extending the straight lines to their point of intersection (Fig. 11.9) gives a reasonably good piecewise-linear approximation to the actual G_1 curve.

The two asymptotes meet when

$$20 \log b = 20 \log a\omega \tag{11-10}$$

or

$$b = a\omega \tag{11-11}$$

In other words, this frequency, called the *break frequency* for the factor $(aj\omega + b)$, occurs when the low-frequency and high-frequency approximations (b and $a\omega$, respectively) are equal (Fig. 11.10).

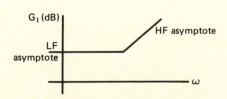

FIG. 11.9 Piecewise-linear approximation for G_1. (In all these plots, the horizontal plot is logarithmic.)

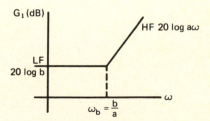

FIG. 11.10 Break frequency is where *LF* value, b, equals *HG* value, $a\omega$.

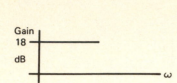

FIG. 11.11 *LF* asymptote of $(2s + 8)$.

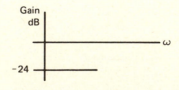

FIG. 11.12 *LF* and *HF* asymptotes of $(2s + 8)$.

Summary—Asymptotic plot for a single factor

What have we accomplished at this point? We can now see how to construct the *asymptotic gain plot* for a single factor (for example, $3s + 4$, or as another example, $s + 6$). Since the method is fundamental to the next few sections, we give two more examples:

(1) First example:

$$T(s) = 2s + 8 \qquad\qquad (11\text{-}12)$$

We proceed with the following steps:

(a) , The *LF* behavior (s small) is 8, or 18 dB.

(b) The *HF* behavior (s large) is 2s, or 2ω. Hence, the slope is +20 dB/decade, and the straight line hits 0 dB at $2\omega = 1$ or $\omega = 1/2$.

(c) If we now extend the two asymptotes until they meet, we obtain the approximation to the gain of $(2s +8)$—Fig. 11.13. We have a check since the break frequency can be read from the figure or calculated by equating the *LF* and *HF* approximations:

$$8 = 2\omega \qquad \text{or} \qquad \omega = 4 \qquad\qquad (11\text{-}13)$$

(2) Second example:

$$T(s) = \frac{1}{4s + 16} \qquad\qquad (11\text{-}14)$$

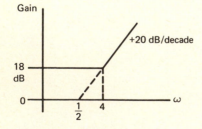

FIG. 11.13 Asymptotic gain characteristic for $(2s + 8)$.

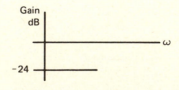

FIG. 11.14 *LF* asymptote for $\dfrac{1}{4s + 16}$.

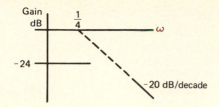

FIG. 11.15 *LF* and *HF* asymptotes for $\dfrac{1}{4s + 16}$.

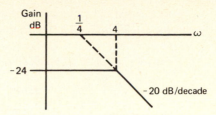

FIG. 11.16 Asymptotic gain characteristic for $\dfrac{1}{4s + 16}$.

In this case, the factor $(4s + 16)$ is in the denominator; the logarithmic gain is then just the negative of what it would be if the factor were in the numerator.

(a) *LF*: Factor is 16 or 24 dB

   Hence gain is −24 dB

(b) *HF*: Factor is $4s$ or 4ω. Hence slope is −20 dB/decade (minus since we have a denominator term) and the line hits 0 dB at $\omega = 1/4$.

(c) Finally, the complete asymptotic gain is drawn. The break frequency is

$$4\omega_b = 16 \quad \text{or} \quad \omega_b = 4 \tag{11-15}$$

11.3 ASYMPTOTIC PLOTS FOR MORE COMPLEX TRANSFER FUNCTIONS

In the preceding section, we saw how to construct the asymptotic gain plot for a single linear factor

$$T(s) = as + b \quad \text{or} \quad T(s) = \frac{1}{cs + d} \tag{11-16}$$

With this background, we can now construct a piecewise-linear (asymptotic) characteristic for *any* transfer function which contains only real poles and zeros—that is, linear factors in both the numerator and denominator. This construction is based on our observation that:

   At a numerator break frequency, the slope increases by 20 dB/decade.
   At a denominator break frequency, the slope decreases by 20 dB/decade.

Each break frequency is found by equating the *LF* and *HF* approximations for the factor.

   To illustrate the general procedure, we choose a specific example. The given transfer function is

$$T(s) = \frac{192 (s + 2)(3s + 30)}{(s + 1)(s + 5)(s + 8)(6s + 72)} \tag{11-17}$$

We want to construct a piecewise-linear approximation to the gain characteristic (in dB versus log ω). We proceed in the following steps.

(1) First we calculate all break frequencies.

	Factor	Equation (HF = LF)	Break ω
For the numerator:	$(s + 2)$	$\omega = 2$	2
	$3s + 30$	$3\omega = 30$	10
For the denominator:	$s + 1$	$\omega = 1$	1
	$s + 5$	$\omega = 5$	5
	$s + 8$	$\omega = 8$	8
	$6s + 72$	$6\omega = 72$	12

(2) The frequency range of interest is determined (so we can scale the graph paper along the horizontal axis). The break frequencies, which range from 1 to 12, define the frequency range within which interesting variations occur. Usually, we want to cover from one decade below the smallest break frequency (here from 0.1 rad/sec) to a decade above the largest break (here to 120 rad/sec or about 100 rad/sec). Hence, our ω scale should run from 0.1 to 100 rad/sec. If we use semilog paper, we need three cycles (Fig. 11.17).

(3) Next we find the LF and HF approximations for the entire transfer function,

$$T(s) = 192 \frac{(s + 2)(3s + 30)}{(s + 1)(s + 5)(s + 8)(6s + 72)} \tag{11-18}$$

At low frequencies (ω much less than the smallest break frequency), each factor is replaced by its LF equivalent:

$$T_{LF} = 192 \frac{(2)(30)}{(1)(5)(8)(72)} = 4 \tag{11-19}$$

Hence, the low-frequency asymptote is at a gain of 4 or + 12 dB. Since the LF behavior of T is a constant, the LF asymptote is horizontal (of zero slope).

At high frequencies (well above the maximum break), each factor of T is replaced by its HF equivalent:

$$T_{HF} = \frac{192 \times s \times 3s}{s \times s \times s \times 6s} = \frac{96}{s^2}$$

$$|T_{HF}(j\omega)| = \frac{96}{\omega^2} \tag{11-20}$$

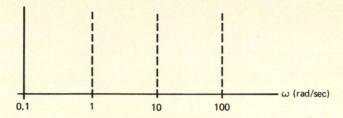

FIG. 11.17 Labeling the horizontal scale or frequency axis.

Since the *HF* gain falls as a constant over ω^2, the dB gain falls with a slope of -40 dB/decade (-20 dB/decade for each $1/\omega$ factor). At the highest ω on our graph paper ($\omega = 100$ in Fig. 11.17), the gain is

$$|T_{HF}(j100)| = \frac{96}{10^4} \tag{11-21}$$

or approximately

$$G(\omega = 100) = -40 \text{ dB} \tag{11-22}$$

The *HF* asymptote is then a straight line defined by the two conditions:

Through $G = -40$ dB at $\omega = 100$,   Slope $= -40$ dB/decade

(4) Next, we must label the vertical axis of our graph paper. We certainly need to cover the range from -40 dB to $+12$ dB (the values at $\omega = 100$ and 0.1); the gain may exceed this range at intermediate frequencies, so we might mark our vertical scale from -60 dB to $+20$ dB (Fig. 11.18).

(5) On the graph paper, we draw the *LF* and *HF* asymptotes. (Fig. 11.19).

(6) We now construct the total asymptotic gain characteristic, starting from the *LF* asymptote and extending this to the smallest break frequency ($\omega = 1$). This is a *denominator* factor, so at $\omega = 1$ the characteristic breaks *downward* (with the slope changing by -20 dB/decade). Figure 11.20 shows the construction thus far.

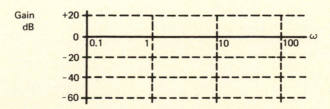

FIG. 11.18 Horizontal and vertical scales selected.

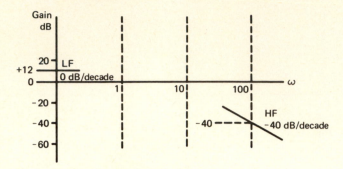

FIG. 11.19 *LF* and *HF* asymptotes drawn for $T(s)$.

This straight line is extended to the next break frequency ($\omega = 2$), which occurs in the *numerator*, so the slope *increases* by 20 dB/decade (to zero). This horizontal line is extended to $\omega = 5$, the next break frequency, where the slope breaks downward to -20 dB/decade. Continuing, we find a slope of -40 dB/decade starting at $\omega = 8$, then -20 dB/decade at $\omega = 10$, and finally -40 dB/decade from $\omega = 12$ on (Fig. 11.21).

(7) A check on our work is provided by the earlier plotting of the *HF* asymptote of $T(s)$. As we work our way up in frequency, the last slope change (at the largest break frequency) should leave us on the HF asymptote.*

*Actually, we can also check at any selected ω. At a particular ω, each factor is represented by either its *LF* or its *HF* approximation, depending on whether we are below or above the break frequency for that factor. For example, at $\omega = 4$, the asymptotic curve should give

$$T(s) = 192 \frac{(s + 2)(3s + 30)}{(s + 1)(s + 5)(s + 8)(6s + 72)} = 192 \frac{(s)(30)}{(s)(5)(8)(72)} = 2 \text{ or } 6 \text{ dB}.$$

This is not the actual gain at $s = j4$ (which we would have to find by substituting $s = j4$ into T and going through complex algebra). Instead, it is the asymptotic gain at $\omega = 4$—the gain with each term replaced by its appropriate asymptote. The fact that this value is close to the accurate gain (which is 1.46 rather than 2) indicates that we have an easy method for calculating approximately the magnitude of a transfer function at a particular frequency.

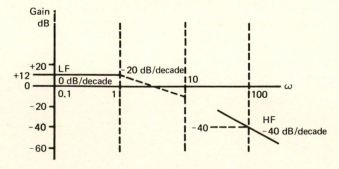

FIG. 11.20 First break included at $\omega = 1$.

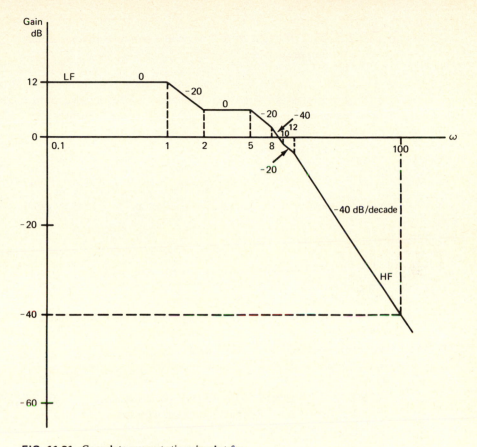

FIG. 11.21 Complete asymptotic gain plot for

$$T(s) = 192 \frac{(s + 2)(3s + 30)}{(s + 1)(s + 5)(s + 8)(6s + 72)} .$$

11.4 SECOND EXAMPLE OF ASYMPTOTIC PLOTS

One additional example illustrates the method for constructing the asymptotic gain plot (we still are considering only transfer functions with real poles and zeros). The transfer function is

$$H(s) = \frac{160(s + 10)}{s(s + 2)^2(s + 20)} \tag{11-23}$$

If we follow the same steps as those in the preceding example, the asymptotic plot is constructed as:

(1) The break frequencies are:

Numerator   10

Denominator   2, 2, 20   (we might write 2 twice as a reminder there is a double break here)

(2) ω range of interest: 0.2 to 200, so we scale horizontally from 0.1 to 1000. Then the lowest frequency of interest is 0.1 rad/sec, the highest is 1000 rad/sec. In each case we are approximately one decade or more from the nearest break frequency.

(3) *LF* asymptote:*

$$\frac{160\,(10)}{s\,(4)\,(20)} = \frac{20}{s} \quad \text{or} \quad \frac{20}{\omega}$$

Slope = -20 db/decade; at $\omega = 0.1$ [lowest ω of interest as found in (2)], gain is 200 or 46 dB

HF asymptote:

$$\frac{160\,(s)}{s\,(s)^2\,(s)} = \frac{160}{s^3} \quad \text{or} \quad \frac{160}{\omega^3}$$

Slope = -60 dB/decade; at $\omega = 1000$ (highest ω of interest), gain is 160×10^{-9} or -136 dB

(4) Gain range of interest: $+60$ dB to -140 dB

(5) The asymptotic gain characteristic is now drawn by

     (a) constructing *LF* and *HF* asymptotes;

     (b) working from *LF* asymptote upward in ω. The initial slope of -20 dB/decade changes to -60 dB/decade at $\omega = 2$ (a double, denominator break), to -40 dB/decade at $\omega = 10$, and finally to -60 dB/decade at $\omega = 20$.

Figure 11.22 shows the result. The example illustrates the tremendous change in gain which sometimes occurs over the frequency range of interest. Here we go from $+46$ dB to -136 dB (i.e., from a gain of 200 to a gain of 1.6×10^{-7}).

Finally, the example illustrates the ease with which the asymptotic gain plots can be constructed. Fortunately, these asymptotic plots are good approximations to the actual gain curves; indeed, in most cases, we can use the asymptotic plots as the system gain. If we want a more accurate plot, however, appropriate corrections can be added to the asymptotic characteristic.

11.5 CORRECTIONS FOR ASYMPTOTIC GAIN CHARACTERISTIC

The logarithmic or dB gain is the *sum* of the gains for each factor in the transfer function. Therefore, the total correction converting from the asymptotic to the actual plot is simply the sum of the corrections for each separate factor.

---

*The $1/s$ factor has no break frequency (or we might say the break frequency is at $\omega = 0$, which is off our logarithmic horizontal scale). Any power of s in the numerator or denominator is treated in this way: the *LF* asymptote is affected, but there is no associated break frequency.

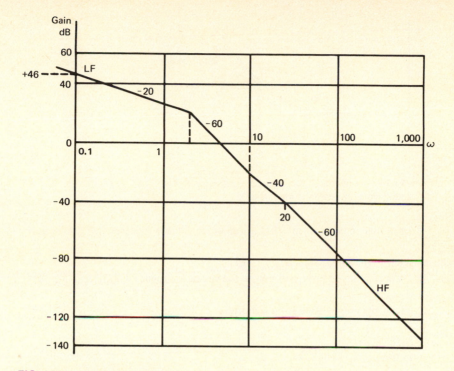

FIG. 11.22 Asymptotic gain for

$$H(s) = \frac{160(s + 10)}{s(s + 2)^2(s + 20)}.$$

A single, numerator factor

$$(as + b)$$

has the asymptotic plot shown in Fig. 11.23. At the break frequency ($\omega_b = b/a$), the actual gain is $|aj(b/a) + b|$ or $b\sqrt{2}$ or 3 dB greater than the *LF* asymptote. Thus, the actual curve lies 3 dB above the asymptotic characteristic at the break frequency.

Calculation of the error between the actual and asymptotic characteristics reveals that the deviation is a maximum at ω_b, and that the error is 1 dB at one-half and at twice the break frequency. The actual gain curve is shown in Fig. 11.24.

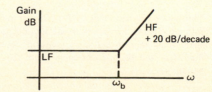

FIG. 11.23 Asymptotic plot for a single numerator factor.

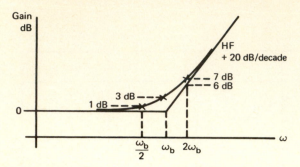

FIG. 11.24 Actual curve for a single numerator factor (*LF* asymptote arbitrarily shown at 0 dB).

To summarize, we can say that a numerator factor requires this correction:

+ 3 dB at the break frequency (ω_b)
+ 1 dB an octave above and below the break frequency ($2\omega_b$ and $\omega_b/2$)*

Similarly, the corrections for a denominator factor are:

−3 dB at ω_b
−1 dB at $2\omega_b$ and $\dfrac{\omega_b}{2}$

More than an octave from the break frequency, the corrections are less than 1 dB; the asymptotic curve is an excellent approximation to the actual gain.

When the above relationships are known, we can convert the asymptotic plot for $T(s)$ to an actual gain plot by inserting the corrections for each break frequency, one at a time. For instance, in the second example above, the corrections are inserted as shown in Fig. 11.25. (Here the double break at $\omega = 2$ can be handled by taking each $s + 2$ factor separately, or simply by doubling the corrections for a single denominator factor.)

It should be emphasized that many applications do not require the accuracy associated with the actual rather than the asymptotic gain characteristic. We can often make adequate correction merely by drawing a smooth curve following the asymptotic plot. For example, Fig. 11.26 shows an asymptotic plot and, as a dashed line, a rough approximation to the actual gain characteristic. This dashed curve can often be drawn without calculation and still be correct within a dB.

---

*An octave change refers to doubling or halving the frequency. The term is from music where the eighth note above middle *A* is the second harmonic at twice the frequency—i.e., the *A* one octave higher. Some engineering texts measure slopes of asymptotes in terms of changes over an octave, rather than the decade we have used. 20 dB/decade corresponds to very nearly 6 dB/octave (actually 6.02).

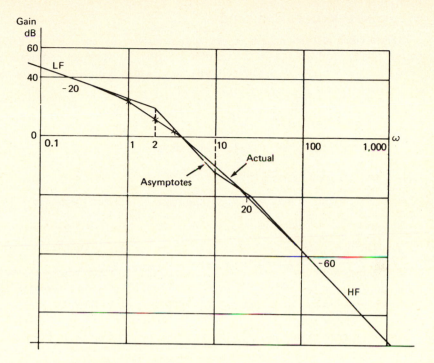

FIG. 11.25 Corrections inserted for

$$T(s) = \frac{160(s + 10)}{s(s + 2)^2(s + 20)}$$

from the asymptotic plot of Fig. 11.22.

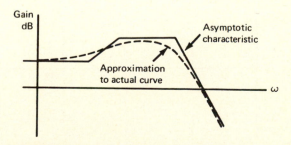

FIG. 11.26 Rough approximation to actual gain.

11.6 COMPLEX POLES AND ZEROS

The preceding sections have been devoted exclusively to real poles and zeros (linear factors of the form $as + b$ in the transfer function). If there are conjugate complex poles or zeros, we have to modify our approach, which is illustrated by the example:

$$T(s) = \frac{12(s + 16)}{(3s^2 + 6s + 48)} \qquad (11\text{-}24)$$

The gain plot is constructed in the following steps.

(1) The break frequencies are:

	Factor	*Equation*	*Break frequency*
Numerator:	$s + 16$	$\omega = 16$	$\omega = 16$
Denominator:	$3s^2 + 6s + 48$	$3\omega^2 = 48$	$\omega = 4$ (quadratic)

The quadratic factor $(3s^2 + 6s + 48)$ cannot be split into real factors; hence, we treat the entire factor at once. The *HF* behavior (s large) is $3s^2$ or, in magnitude, $3\omega^2$. The *LF* behavior (s small) is 48. The break frequency is determined by equating the *LF* and *HF* approximations.

(2) The ω range of interest is then 0.4 to 160; we scale our paper from 0.1 to 1000 rad/sec.

(3) The *LF* and *HF* asymptotes for the entire $T(s)$ are found.

$$LF: \qquad T(s) = \frac{12(16)}{(48)} = 4 \quad \text{or} \quad 12 \text{ dB}$$

Slope = 0; gain at 0.1 rad/sec is 12 dB

$$HF: \qquad T(s) = \frac{12(s)}{(3s^2)} = \frac{4}{s} \quad \text{or} \quad \frac{4}{\omega}$$

Slope = -20 dB/decade; gain at 1000 rad/sec is 4/1000 or -48 dB

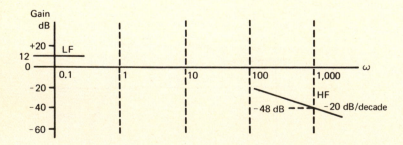

FIG. 11.27 Graph paper prepared for gain plot; *LF* and *HF* behavior shown.

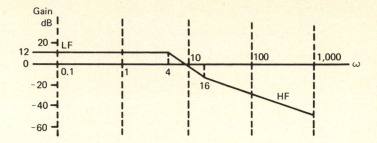

FIG. 11.28 Complete asymptotic plot for

$$T(s) = \frac{12(s + 16)}{3s^2 + 6s + 48} \cdot$$

(4) The gain scale is selected to run from +20 dB to −60 dB (Fig. 11.27).

(5) The asymptotic plot is constructed. The *LF* asymptote of +12 dB continues to $\omega = 4$; at this frequency, the plot breaks downward with a slope of −40 dB/decade (the *LF* approximation 48 is replaced by the *HF* value $3s^2$ for the quadratic factor, so the slope change is twice that for a linear factor). This −40 dB/decade slope continues until $\omega = 16$, beyond which the slope is −20 dB/decade—and we are on the *HF* asymptote for $T(s)$. The complete asymptotic plot is shown in Fig. 11.28.

(6) We next insert the corrections for the linear factors. For our $T(s)$, there is only one factor; it appears in the numerator with a break frequency of 16 rad/sec. Hence, the appropriate corrections are +3 dB at $\omega = 16$, +1 dB at $\omega = 8$ and 32, and the new curve is shown in Fig. 11.29.

(7) Finally, we insert the correction for each quadratic factor, one at a time. In our case, the single quadratic factor is

$$3s^2 + 6s + 48 \qquad \text{or} \qquad 3(s^2 + 2s + 16)$$

Here the undamped natural frequency or break frequency ω_b is 4 rad/sec (the square root of the constant term after the coefficient of s^2 is made unity). The coefficient of

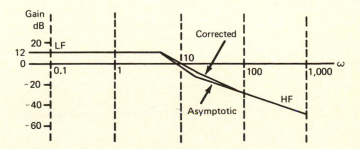

FIG. 11.29 Gain with corrections included for linear factor.

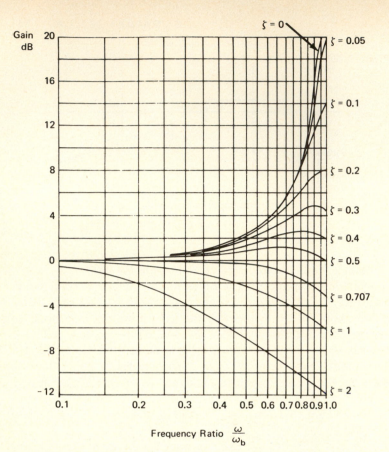

Frequency Ratio $\dfrac{\omega}{\omega_b}$

FIG. 11.30 Corrections for a quadratic factor $(s^2 + 2\zeta\omega_b s + \omega_b^2)$ (shown for a denominator factor; for a numerator factor, change sign of correction). Correction is the same for $\alpha\omega_b$ as for $(1/\alpha)\omega_b$; in other words correction shown at $0.8\omega_b$ also occurs at $\omega_b/0.8$ or $1.25\omega_b$.

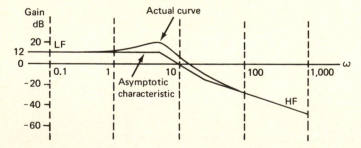

FIG. 11.31 Gain characteristic for

$$T(s) = \frac{12(s + 16)}{3s^2 + 6s + 48}.$$

362

s is $2\zeta\omega_b$ (again with s^2 coefficient unity). Hence, the relative damping ratio ζ is

$$2\zeta\omega_b = 2 \quad \text{or} \quad \zeta = \frac{1}{4} \tag{11-25}$$

The correction for $\zeta = 1/4$ can be read from Fig. 11.30, where we find

Frequency	Correction (with $\zeta = 1/4$)
ω_b	+6 dB
$0.9\,\omega_b$ or $1.11\,\omega_b$	+6 dB
$0.8\,\omega_b$ or $1.25\,\omega_b$	+5 dB
$0.7\,\omega_b$ or $1.42\,\omega_b$	+4 dB
$0.5\,\omega_b$ or $2\,\omega_b$	+2 dB

Insertion of these corrections yields the total gain characteristic shown in Fig. 11.31.

It is revealed in Fig. 11.30 that the asymptotic characterisitc may be a *very poor* approximation to the actual gain characteristic when the transfer function has quadratic factors. When there are only linear factors (i.e., real poles and zeros), we often do not bother with corrections; when one or more quadratic factors are present, corrections must be included unless all relative damping ratios (ζ's) are between 0.5 and 1. The appropriate corrections depend on ζ and can be found by referring to Fig. 11.30.

11.7 FREQUENCY MEASUREMENT OF TRANSFER FUNCTION

The simplicity of the Bode or asymptotic gain plots means that we can often work in the reverse direction: we measure the gain characteristic experimentally, then determine an appropriate $T(s)$ to fit this characteristic.

Because of the form of the asymptotic gain plots, first we fit the measured characteristic by straight-line segments, each with a slope of $20n$ dB/decade, where n is a positive or negative integer or zero. Sharp peaks or minima are realized by appropriate quadratic factors in the denominator or numerator of $T(s)$. Once the various break frequencies are selected (i.e., the poles and zeros of T), we can plot the gain of the estimated $T(s)$ and then adjust the constants slightly to improve the approximation.

We describe the procedure by a specific example.

Statement of the problem

We are given a process to be controlled—a process too complex to be analyzed by writing the differential equations based upon physical laws. For example, we might have the steering mechanism of an automobile on a wet road, or a vibrating missile in which liquid fuel is sloshing about in the tanks. In order to control the process, we must determine its transfer function.

As a first step, we apply sinusoidal signals at angular frequencies varying from 0.1 rad/sec to 200 rad/sec. At each value of ω, we measure output and input amplitudes and calculate the ratio

$$| T(j\omega) | = \frac{\text{Output amplitude}}{\text{Input amplitude}} \tag{11-26}$$

We convert this ratio to decibels and plot the characteristic shown in Fig. 11.32. From these measured data, $T(s)$ is to be determined.

A note on accuracy

Before starting the solution, we should note that ordinarily accuracy is not of great importance in problems of this sort. In many cases, the measured gain characteristic varies somewhat as the input signal amplitude is changed (in other words, the process is slightly nonlinear). Furthermore, the measured gain is seldom accurate to better than 1 dB (instruments always have some error, any measurement is corrupted by noise and other interferences). Consequently, we are looking for a $T(s)$ which has a gain which *approximates* the measured characteristic of Fig. 11.32.

This emphasis on an *approximate* answer is fortunate, since two transfer functions with markedly different poles and zeros may have very similar gain characteristics. A simple example is provided by the three transfer functions

$$T_1(s) = \frac{10}{s + 10} \tag{11-27}$$

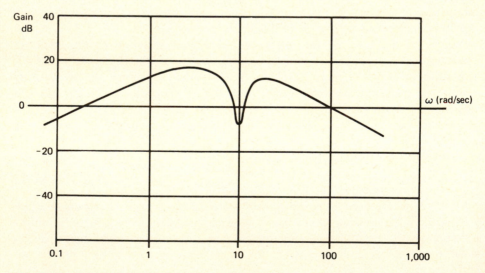

FIG. 11.32 Gain characteristic determined experimentally.

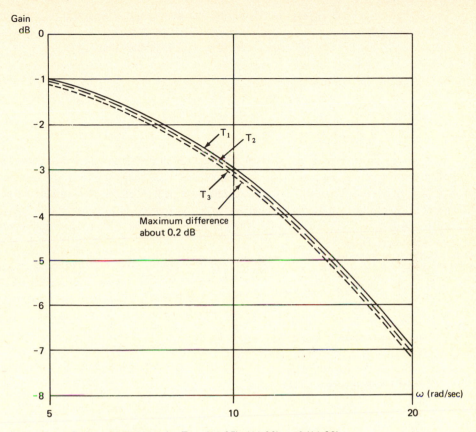

FIG. 11.33 Gain characteristics for Eqs. (11.27), (11.28), and (11.29).

$$T_2(s) = \frac{10(s + 10)}{(s + 9.1)(s + 11)} \qquad\qquad (11\text{-}28)$$

$$T_3(s) = \frac{10(s + 10)}{(s + 8)(s + 12.5)} \qquad\qquad (11\text{-}29)$$

The corresponding gain characteristics are shown in Fig. 11.33. Obviously, only slight changes in the gain measurement are required to switch from one characteristic to another. The examples illustrate that often we cannot even determine the order of the system from the measured gain characteristic.

First trial on $T(s)$

Returning to our example of Fig. 11.32, we note first that the measured LF and HF behaviors must be matched by slopes of $20n$ dB/decade. Inspection of the figure reveals that the LF slope is +20 dB/decade; hence the $T(s)$ has a factor s^1.

The *HF* slope is −20 dB/decade. Therefore, the denominator degree is one greater than the numerator.

Inspection of the *LF* and *HF* asymptotes (Fig. 11.34) reveals that:

$$LF \quad T(s) = 5s \quad \text{since } LF \text{ gain} = 0 \text{ dB or } 1 \text{ at } \omega = \frac{1}{5} \tag{11-30}$$

$$HF \quad T(s) = \frac{100}{s} \quad \text{since } HF \text{ gain} = 0 \text{ dB or } 1 \text{ at } \omega = 100 \tag{11-31}$$

We wish to approximate the rest of the measured characteristic by a series of straight-line segments, each of slope $20n$ dB/decade. The sharp dip in the vicinity of $\omega = 10$ is neglected, since this can be realized by a pair of conjugate complex zeros. Figure 11.35 shows a first approximation, which corresponds to denominator break frequencies at $\omega = 1.42$ and $\omega = 14.8$. The numerator must be $100s$ to give the correct behavior at *HF*. Then

$$T(s) = \frac{100s}{(s + 1.42)(s + 14.8)} \tag{11-32}$$

The resulting *LF* behavior should be $5s$. Equation (11-32) gives $100s/(1.42)(14.8)$ or $4.8s$. The error results from an inaccurate reading of the break frequencies in Fig. 11.35. The error is small enough to ignore.

Next, we must realize the sharp minimum at $\omega = 10$. This is achieved by multiplying the $T(s)$ of Eq. 11-32 by a factor

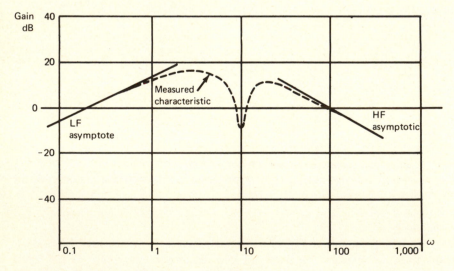

FIG. 11.34 Establishing *LF* and *HF* asymptotes.

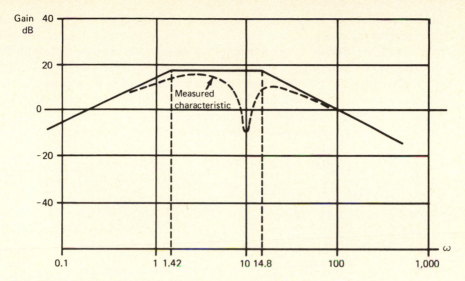

FIG. 11.35 First total approximation to measured gain characteristic.

$$\frac{s^2 + \alpha s + 100}{(s + 10)^2}$$

where α is to be adjusted to yield the correct gain for $T(s)$ at $\omega = 10$. The denominator $(s + 10)^2$ is used, so that multiplication of the transfer function by this factor will not change the gain at either LF or HF. (The factor is unity when $|s|$ is large or very small.)

The total transfer function is now

$$T(s) = \frac{100s(s^2 + \alpha s + 100)}{(s + 1.42)(s + 10)^2(s + 14.8)} \tag{11-33}$$

To select α, we evaluate the gain at $\omega = 10$:

$$|T(j10)| = \frac{(100)(10)(\alpha 10)}{(10.1)(200)(17.9)} = 0.276\alpha \tag{11-34}$$

The measured gain at $\omega = 10$ is -6 dB or 0.5. Therefore

$$0.276\alpha = 0.5 \quad \text{or} \quad \alpha = 1.8 \tag{11-35}$$

and

$$T(s) = \frac{100s(s^2 + 1.8s + 100)}{(s + 1.42)(s + 10)^2(s + 14.8)} \tag{11-36}$$

Check on this $T(s)$

In order to determine if we need further readjustment of the various break frequencies (the poles and zeros selected for T), we now plot the dB gain for this $T(s)$ of Eq. (11-36) and compare this plot with the measured characteristic (Fig. 11.36).

In most cases, the near-agreement indicated is adequate. If we wished greater accuracy, we could now adjust slightly the break frequencies (or add additional break frequencies close together to give a slight increase or decrease in the gain).

Actually, the $T(s)$ for the system shown as the measured characteristic was

$$T_{actual}(s) = \frac{100s(s^2 + 2s + 100)}{(s + 2)(s + 5)(s + 10)(s + 20)} \tag{11-37}$$

compared to our approximation

$$T_{approx}(s) = \frac{100s(s^2 + 1.8s + 100)}{(s + 1.42)(s + 10)^2(s + 14.8)} \tag{11-38}$$

Final comment

The preceding example emphasizes that the procedure involves appreciable trial and error. We can find a rough approximation easily, simply by fitting with asymptotes and inserting resonances and antiresonances to realize the sharp minima and maxima. Finding a very accurate realization may be a tedious chore. Fortunately, most engineering applications do not require great accuracy.

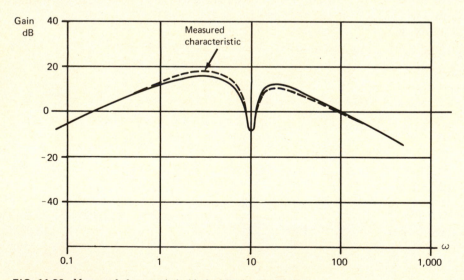

FIG. 11.36 Measured characteristic (dashed line) and gain for
$$T(s) = \frac{100s(s^2 + 1.8s + 100)}{(s + 1.42)(s + 10)^2(s + 14.8)}.$$

11.8 PHASE CHARACTERISTIC

The complete description of a system in terms of frequency response requires not only the gain characteristic, but also the *phase characteristic*—the phase shift through the system as a function of frequency. Figure 11.37 shows a complete frequency-domain description of the system. At each frequency, both gain and phase are given; both characteristics are plotted versus a logarithmic frequency scale. The gain is measured in decibels, the phase in degrees.*

The gain and phase plots show directly the sinusoidal, steady state response of the system for any input signal frequency. For example, at $\omega = 600$, the curves of Fig. 11.37 give

$$\text{Gain} = 17 \text{ dB} \qquad \text{Phase} = -85° \tag{11-39}$$

The gain of 17 dB corresponds to a $|T|$ gain of 7.1:

$$|T(j600)| = 7.1 \qquad \angle T(j600) = -85° \tag{11-40}$$

If the input signal is

$$x(t) = 2 \sin(600t + 30°) \tag{11-41}$$

we can immediately write the steady state output

$$y_{ss}(t) = 14.2 \sin(600t - 55°) \tag{11-42}$$

In passing through the system, the sinusoid has its amplitude multiplied by $|T|$ and its phase shifted by $\angle T$. The transfer function is just the complex "gain" of the system.

Distortionless transmission

From this interpretation of $T(j\omega)$ as the complex gain, the conditions for "distortionless" transmission are apparent. Distortionless means that the output waveform should be the same as the input. If the input is a triangular pulse (Fig. 11.38), of duration t_1, the undistorted output is a pulse of the same shape, which may be delayed in time (by t_2) and changed in amplitude. The system introduces no waveform distortion.

The Fourier theorem states that the input x can be decomposed into a sum of sinusoids. The frequencies of these components span the region over which the spectrum or transform $X(\omega)$ is not zero (or essentially zero). In other words, if the

---

*Phase shift really should be measured in radians. For the signal $\sin(\omega_1 t + \theta)$, $\omega_1 t$ is in radians and, for addition, θ must be also. Engineers so often think and work in terms of degrees that we follow that practice here.

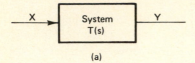

(a)

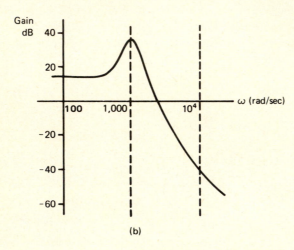

(b)

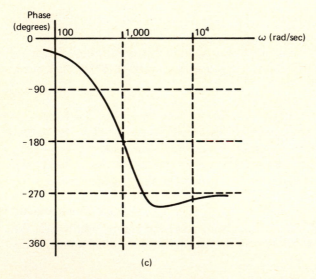

(c)

FIG. 11.37 System characterization in the frequency domain.
(a) System with an input X and one output Y; (b) gain
characteristic; (c) phase characteristic.

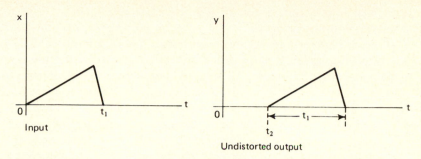

FIG. 11.38 Distortionless transmission.

spectrum of x is significant from 0 to 100 rad/sec, the Fourier components are sinusoids with frequencies in this range.

When these sinusoids pass through the system, each is amplified by $|T(j\omega)|$ and shifted in phase by $\underline{/T(j\omega)}$. If, at the output, they are to add up to the same waveform, the gain must be the same for all components. In other words, over the frequency range of the input signal spectrum, the gain must be constant. Furthermore, the time delay of each component must be the same (and equal to t_2 in Fig. 11.38), if the output components are to add up properly. For a sinusoid at an angular frequency ω_1, the time delay is

$$t_d = \frac{\theta}{\omega_1} \tag{11-43}$$

where θ is the phase lag (negative phase shift) in radians. Thus, the phase lag θ must be $\omega_1 t_d$—the phase lag must be proportional to the frequency ω_1. If a component at 20 rad/sec is shifted by $-40°$, a component at 60 rad/sec must be shifted $-120°$ for the same time delay (because at the higher frequency each degree corresponds to less time).*

Thus, if the spectrum of the input signal spans the frequency range from ω_1 to ω_2, the conditions for transmission with no waveform distortion are

(1) Gain equal to a constant K, ω_1 to ω_2
(2) Phase equal to $-k\omega$, ω_1 to ω_2

Then the output is a replica of the input but amplified by K and delayed by k seconds.

Failure to satisfy either condition results in an output waveform which differs from the input. Hence, the phase characteristic is fully as important as the gain in

---

*We should note particularly that distortionless transmission does *not* mean a system phase shift constant over the frequency band of interest. The time delay is the quantity which should be constant.

determining waveform distortion. The complete description of a system requires display of both characteristics.

There are important applications where phase distortion is of secondary importance. Voice communication is an example. The human hearing system can not distinguish rather large phase distortions. Consequently, in a system for transmitting voice signals, if we can keep the gain essentially constant over the frequency range of importance, the communication fidelity is good, even though, because of phase distortion, the waveform of the received signal may differ markedly from that transmitted.

In the broad variety of applications in which information is carried by pulses, however, it can be argued that phase distortion is even more important than gain distortion. Preservation of the pulse shape certainly requires satisfaction of both the above conditions.

11.9 PHASE FOR REAL POLES AND ZEROS

Just as the logarithmic gain in dB is the sum of the gain of individual factors, the total phase shift of $T(s)$ is the sum of the contributions of individual terms. Unfortunately, in the phase case, the straight-line asymptotes are not particularly useful. Instead, we have to plot the phase of one term, then add to this curve the phase of a second term, and continue adding the effects of one term at a time until we have the complete phase characteristic.

Phase of a zero or pole at $s = 0$

A factor s^m contributes a phase angle of $m\pi/2$ radians or $m(90°)$ at all frequencies. Here m can be any integer; a negative value corresponds to a pole of $T(s)$ at the origin.

Phase of a linear factor

First we consider a simple linear factor

$$(as + b)$$

in the numerator of $T(s)$. The corresponding break frequency (in the gain plot) is

$$\omega_b = \frac{b}{a} \tag{11-44}$$

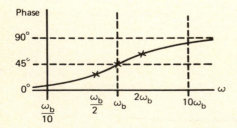

FIG. 11.39 Phase shift for $(as + b)$.

The phase shift contributed by such a factor is described by the following values:

ω	Factor	Phase
Very small	b	$0°$
$\omega_b/10$	$b\left(1 + j\frac{1}{10}\right)$	$5.7°$
$\omega_b/2$	$b\left(1 + j\frac{1}{2}\right)$	$26.5°$
ω_b	$b(1 + j)$	$45°$
$2\omega_b$	$b(1 + j2)$	$63.5°$ (equal to 90 – 26.5)
$10\omega_b$	$b(1 + j10)$	$84.3°$ (equal to 90 – 5.7)
Very large	$aj\omega$	$90°$

Thus, the phase varies as shown in Fig. 11.39 where we use a logarithmic ω scale just as we did for the gain plots. The characteristic can be sketched with sufficient accuracy for most purposes from the fact that:

at break frequency ω_b, phase = $45°$
at half ω_b, phase is $26.5°$
at twice ω_b, $26.5°$ from $90°$ or $63.5°$
at $\omega_b/10$, 1/10 radian or $5.7°$
at $10\omega_b$, $5.7°$ from $90°$ or $84.3°$

In other words, we only need to remember that the tangent of $26.5°$ is 1/2, and that $\tan\theta = \theta$ when θ is small.

Figure 11.39 shows that the phase characteristic changes over a much wider frequency range around the break frequency than the gain. In the gain for a linear factor, the deviation from *LF* or *HF* behavior is restricted to an octave either side of the break frequency. The phase characteristic requires two decades around the break to change from *LF* to *HF* behavior.

It is suggested in some texts that the phase characteristic can be adequately approximated by the piecewise-linear approximation shown in Fig. 11.40: $0°$ to a

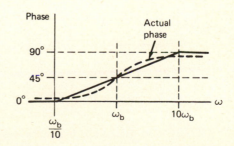

FIG. 11.40 Straight-line approximation for phase of $as + b$.

decade below the break frequency, a line of slope 45°/decade to a decade above the break, and then a constant 90°. The figure indicates that the approximation is reasonably good (the maximum error is 5.7°).

When the factor $(as + b)$ appears in the denominator of $T(s)$, the phase contribution is simply changed in sign.

When the factor $(as + b)$ is raised to a power higher than one, each $(as + b)$ contributes the characteristic shown in Fig. 11.39.

Phase for a $T(s)$ with real poles and zeros

As an example, we now plot the phase characteristic for

$$T(s) \doteq \frac{30s(s + 10)}{(2s + 4)(s + 20)} \tag{11-45}$$

We use the following steps:

(1) The break frequencies are

$\omega = 2$ Denominator

$\omega = 10$ Numerator

$\omega = 20$ Denominator

(2) Hence, the frequency range of interest is 0.2 to 200 (a decade from the nearest break frequency). We scale the ω axis from 0.1 to 1000 (four cycles).

(3) The phase at very low and high frequencies is:

$$LF \quad T(s) \rightarrow \frac{300s}{80} \qquad \angle T = 90°$$

$$HF \quad T(s) \rightarrow \frac{30s^2}{2s^2} = 15 \quad \angle T = 0°$$

Therefore, we mark the vertical scale from 180° down to −90°. Figure 11.41 shows the graph paper prepared for the plotting, also the *LF* and *HF* values.

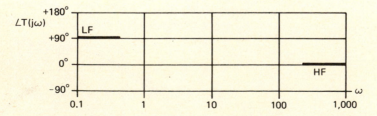

FIG. 11.41 Horizontal and vertical scales constructed.

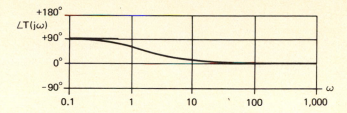

FIG. 11.42 Phase for $\dfrac{30s}{2s+4}$ portion of $T(s)$.

(4) The first factor is $1/(2s+4)$ with a break frequency of 2 rad/sec. Hence, the phase contribution is

$-5.7°$ at 0.2 rad/sec $-63.5°$ at 4
$-26.5°$ at 1 $-84.3°$ at 20
$-45°$ at 2

By using these five points, we sketch the characteristic of Fig. 11.42. (Each value is added from the *LF* value of $+90°$.)

(5) The next break frequency (as we move up in frequency) is 10 rad/sec in the numerator. Therefore, at 1, 5, 10, 20, and 100 rad/sec, we move *up* from the curve of Fig. 11.42 by the amounts: $5.7°$, $26.5°$, $45°$, $63.5°$, and $84.3°$, respectively. Figure 11.43 results.

(6) Finally, we add to Fig. 11.43 the contributions from $1/(s+20)$ to obtain the total phase characteristic for $T(s)$ (Fig. 11.44).

We have constructed the total phase characteristic by adding one-by-one the contributions from each factor of $T(s)$. Each factor is described in terms of five values: at the break, an octave below and above the break, and a decade below and above the break.

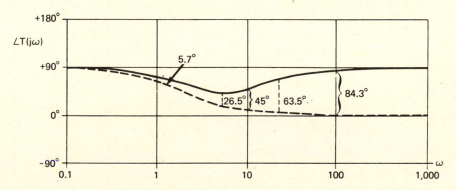

FIG. 11.43 Phase for $\dfrac{30s(s+10)}{2s+4}$ portion of $T(s)$.

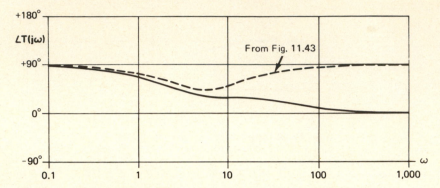

FIG. 11.44 Phase characteristic for

$$T(s) = \frac{30s(s + 10)}{(2s + 4)(s + 20)} \cdot$$

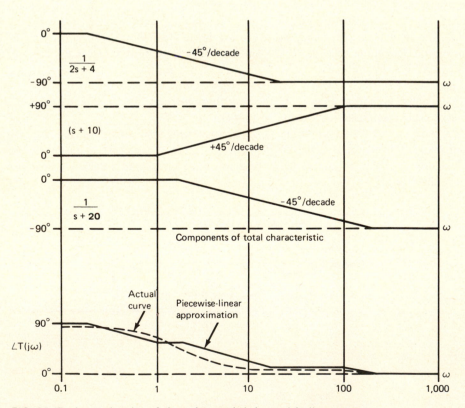

FIG. 11.45 Approximation of phase characteristic by straight-line segments.

StraInght-line approximation for $\angle T(j\omega)$

It was noted above that when $T(s)$ has only real poles and zeros, we can obtain a reasonable approximation to $\angle T(j\omega)$ by piecewise-linear characteristics. Each approximation involves a straight-line change over two decades around the break. If the same $T(s)$ is considered, Fig. 11.45 shows the piecewise-linear approximation.

Even this simple example illustrates that the piecewise-linear approximation often is not sufficiently accurate. Furthermore, addition of corrections is not a simple matter, and we usually prefer to use the addition of contributions from single factors.

11.10 PHASE FOR COMPLEX POLES AND ZEROS

When the transfer function contains a quadratic factor

$$as^2 + bs + c$$

we factor out the coefficient of s^2 so that the factor appears as

$$s^2 + 2\zeta\omega_n s + \omega_n^{\,2}$$

Here ω_n is the break frequency and ζ the damping ratio.

The phase contribution of such a factor depends on ζ, as shown in Fig. 11.46 (which is drawn for the factor in the denominator). Regardless of ζ, the phase is $0°$ at very low frequencies, $-90°$ at ω_n, and $-180°$ at very high frequencies. The smaller ζ, the more rapidly the phase changes in the vicinity of ω_n—indeed, if $\zeta = 0$ (the poles on the $j\omega$ axis), the phase is $0°$ below ω_n, $-180°$ above ω_n.

A final example

The circuit of Fig. 11.47 is described by the transfer function

$$\frac{E_2}{E_1} = \frac{s(s + 10^4)}{s^2 + 10^4 s + 10^{10}} \tag{11-46}$$

We want to plot the phase versus ω.

(1) The break frequencies are 10^4 (numerator) and 10^5 (denominator); hence we scale the frequency axis from 10^3 rad/sec to 10^6 rad/sec.

(2) The phase at the extreme frequencies is:

$$LF: \quad \frac{E_2}{E_1} \to 10^{-6}\,s \qquad \text{Phase} = +90°$$

$$HF: \quad \frac{E_2}{E_1} \to 1 \qquad \text{Phase} = 0°$$

Therefore, we scale the vertical axis from $-90°$ to $+180°$.

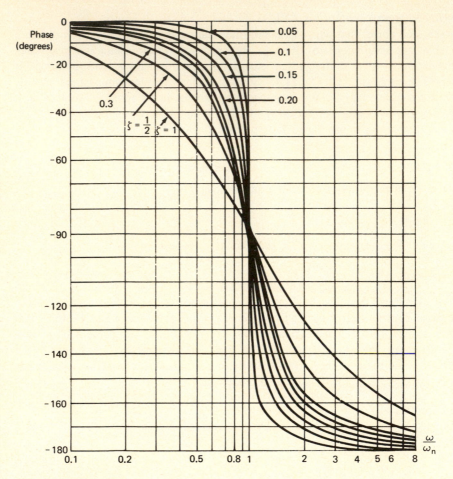

FIG. 11.46 Phase for

$$\frac{1}{s^2 + 2\zeta\omega_n s + \omega_n^2} .$$

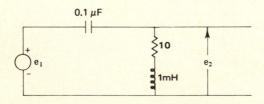

FIG. 11.47 System for final example.

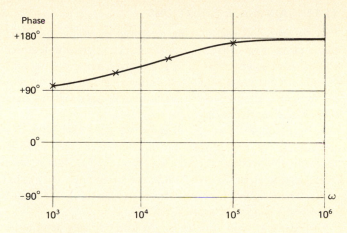

FIG. 11.48 Phase of $s(s + 10^4)$.

(3) The effects of the $(s + 10^4)$ factor are plotted in Fig. 11.48 by using the deviations from the *LF* behavior of

$+45°$ at $\omega = 10^4$

$+26.5°$ at $\frac{1}{2} \times 10^4$ $+63.5°$ at 2×10^4

$+5.7°$ at 10^3 $+84.3°$ at 10^5

(4) Finally, the phase contributed by the complex poles is added ($\zeta = 0.05$ and $\omega_n = 10^5$ in Fig. 11.46) to give the total phase characteristic shown in Fig. 11.49.

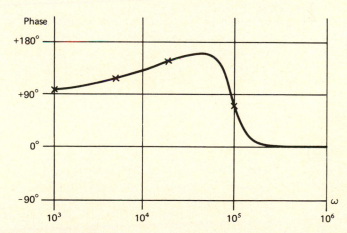

FIG. 11.49 Phase of

$$\frac{E_2}{E_1} = \frac{s(s + 10)^4}{s^2 + 10^4 s + 10^{10}}.$$

11.11 GAIN-PHASE RELATIONS

In the preceding sections, we discussed techniques for rapid plotting of the gain and phase characteristics from a transfer function given in factored form. If $T(s)$ is known as a ratio of polynomials in s, it may be simpler just to substitute $s = j\omega_1$, with ω_1 varied over the range of interest, rather than to factor the numerator and denominator polynomials. Alternatively, standard computer programs are available for plotting gain and phase characteristics.

As we look back at the chapter, we note that both characteristics are uniquely determined by the various break frequencies and relative damping ratios. *Either* characteristic portrays these parameters of $T(s)$. Hence, we might suspect that either characteristic alone is really sufficient to describe the system. Indeed, given the gain characteristic only (not the transfer function), we should be able to find the corresponding phase characteristic.

This idea is valid in most cases: the gain uniquely determines the phase. The gain and phase are the real and imaginary parts of $\log T(j\omega)$:

$$\log T(j\omega) = \underbrace{\log|T(j\omega)|}_{\substack{\text{Proportional to} \\ \text{gain in dB}}} + \underbrace{j\,\underline{/T(j\omega)}}_{\text{Phase}} \tag{11-47}$$

In mathematics, the Hilbert transforms relate the real and imaginary parts of a function of a complex variable. There are two important situations in which the phase is not determined by the gain.

Nonminimum-phase systems

If the system function $T(s)$ possesses one or more zeros in the right half of the s plane, the system is called *nonminimum-phase*. An example is

$$T(s) = 10\,\frac{(s-2)(s+3)}{(s+1)(s^2+s+1)} \tag{11-48}$$

where there is a zero at $s = +2$. This transfer function can be rewritten in the form

$$T(s) = 10\,\underbrace{\frac{(s+2)(s+3)}{(s+1)(s^2+s+1)}}_{T_{mp}(s)}\underbrace{\left(\frac{s-2}{s+2}\right)}_{T_{ap}(s)} \tag{11-49}$$

$T(s)$ is the product of two transfer functions: the first $T_{mp}(s)$ with all poles and zeros in the left half plane; the second $T_{ap}(s)$ with zeros in the right half plane at the images of the poles in the left half plane.

The form of Eq. (11-49) is revealing. The gain of $T_{ap}(s)$ is unity at all angular frequencies. If $s = j\omega$ is substituted, the numerator is just the conjugate of the denominator. In the more general case of Fig. 11.50 with three poles and three zeros, the gain at ω_1 depends on the magnitudes of the six vectors

$$\frac{abc}{ABC}$$

But since $a = A$, $b = B$, and $c = C$, this gain is unity for any frequency. For this reason, $T_{ap}(s)$ is called an *all-pass system function*—all frequencies are passed or transmitted with unity gain. (The subscript *ap* was originally selected to denote *all-pass*.)

In Eq. (11-49), since the gain of $T_{ap}(j\omega)$ is unity, the gain of the factor $T_{mp}(s)$ is identical with the gain of $T(s)$. In other words, the gain of a transfer function is unchanged if any right-half-plane zeros are moved to their images in the left half plane. When this change is made, we have a system function with all zeros and poles in the left half plane—the $T_{mp}(s)$ in Eq. (11-49), which is called a *minimum-phase transfer function*.

Equation (11-49) is an example of the theorem:

> Any nonminimum-phase transfer function can be viewed as the product of a minimum-phase transfer function and an all-pass transfer function.

Phase characteristic for nonminimum-phase systems
The phase of $T(s)$,

$$T(s) = 10 \underbrace{\frac{(s+2)(s+3)}{(s+1)(s^2+s+1)}}_{T_{mp}(s)} \underbrace{\frac{s-2}{s+2}}_{T_{ap}(s)} \tag{11-50}$$

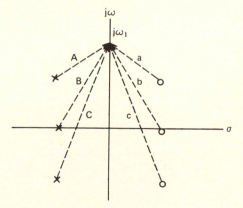

FIG. 11.50 Calculation of gain at ω_1 when poles and zeros are symmetrical about $j\omega$ axis.

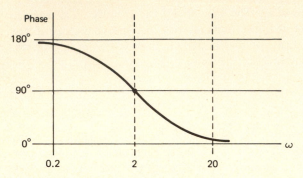

FIG. 11.51 Phase characteristic for $(s - 2)/(s + 2)$.

is the phase of the "minimum-phase system," T_{mp}, augmented by the phase of the all-pass system T_{ap}. The $\underline{/T_{ap}}(j\omega)$ is shown in Fig. 11.51. As the frequency increases, the lag continually increases. If, in Eq. (11-50), the all-pass function were replaced by its $s = 0$ value (-1 or $+180°$), the phase of $T(j\omega)$ would be less negative at all frequencies; the phase lag would be decreased. This is the origin of the term minimum-phase; this system has the minimum phase *lag* for a given gain characteristic. Adding an all-pass term always increases the phase lag.

Now we can return to the original problem of this section: can the phase characteristic be uniquely determined from the gain characteristic? If the system function is known to be a ratio of polynomials in s, the answer is:

> The gain characteristic uniquely determines the phase characteristic for the minimum-phase system.

If the system is nonminimum-phase, the gain characteristic gives no indication of which all-pass function should be included.

Application of an all-pass network

Figure 11.52 shows a pulse train which we want to transmit by radio. If we send this signal directly, the transmitter must be able to deliver a large peak power for the full

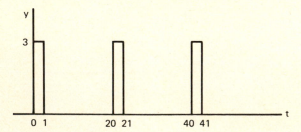

FIG. 11.52 Pulse train.

second during the pulse; it then sits idle for the next 19 seconds. Can we pass the signal through a network which spreads out this power requirement, but does not discard any of the important frequency components of the signal? In Fig. 11.53, can we predistort the signal purposely to decrease power required, then restore the pulse waveform after reception?

Since we do not want to discard any frequency components of the signal, the predistortion network should be an all-pass system. If we arbitrarily select the transfer function

$$T(s) = \frac{s - 1.15}{s + 1.15} \tag{11-51}$$

the output of the predistortion network is as shown in Fig. 11.54.*

Transportation lag

There is one additional, important class of problems in which the gain characteristic does not determine the phase characteristic. In this entire chapter, so far, we have assumed that the transfer functions are ratios of polynomials in s. There are many important problems, particularly in control engineering, where the system involves *transportation lags*: components for which the output starts only after a time interval has passed after the application of the input. The corresponding component in the transfer function is the exponential e^{-Ts}, where T is the transportation lag.†

When the e^{-Ts} simply appears as a multiplier in the transfer function,

$$T(s) = T_0(s) e^{-Ts} \tag{11-52}$$

the term contributes nothing to the gain characteristic (since $|e^{-j\omega T}|$ is unity) and merely adds a phase lag proportional to frequency:

$$\angle e^{-j\omega T} = -\omega T \tag{11-53}$$

Clearly, the measured gain characteristic can give no indication of the transportation lag.

11.12 FINAL COMMENT

The gain and phase characteristics are graphical portrayal of the frequency behavior of the system—essentially the generalization of the a-c or sinusoidal circuit analysis

---

*We would have to design the restorer network to return the waveform to close to its original pulse shape. We want the two tandem networks (predistorter and restorer) together to satisfy the conditions for distortionless transmission.

†Transportation lags are discussed in Section 3.7.

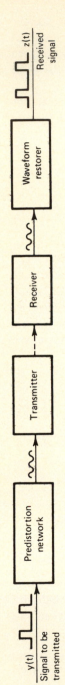

FIG. 11.53 Structure of the communication system.

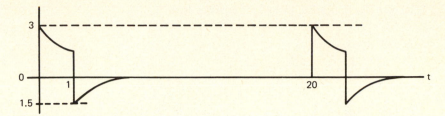

FIG. 11.54 Predistorted signal, with reduced power requirements during a longer pulse.

which is fundamental in electrical engineering. We shall see in subsequent chapters that the gain and phase characteristics are also important in the design of feedback systems, and in the study of system stability.

APPENDIX—dB

The gain of an amplifier or two-port system is the output-signal amplitude divided by the input-signal amplitude. Thus, a gain of 100 signifies that the system magnifies the signal by this factor; a gain of 0.4 means that the output signal is 4/10 as large as the input.

For several reasons, engineers dislike such gain measurements and prefer instead to use the logarithm of the signal ratio. Mathematically, we define the gain of the system of Fig. A11.1 as

$$G = 20 \log \frac{Y}{X} \tag{A11-1}$$

The gain G is 20 times the logarithm to the base 10 of the ratio: Y/X, or (amplitude of output)/(amplitude of input). When gain is measured in this way, G has the units of dB (or deciBels). In other words, if $Y/X = 10$, the system has a gain of 20 dB.

Origin of Eq. (A11-1)

This measurement of gain in terms of the logarithm of the output/input ratio was originally made in terms of power rather than signal amplitudes.

$$G = \log \frac{P_{out}}{P_{in}} \text{ Bels} \tag{A11-2}$$

and given the unit name Bels in honor of Alexander Graham Bell. A deciBel is simply

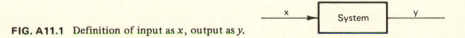

FIG. A11.1 Definition of input as x, output as y.

1/10 of a Bel; hence Eq. (A11-2) is equivalent to

$$G = 10 \log \frac{P_{\text{out}}}{P_{\text{in}}} \text{ dB} \qquad (A11-3)$$

The emphasis on power (and energy) measurements in the early years of this century gradually gave way to describing signals in terms of voltage (or current) amplitude. If x and y in Fig. A11.1 are sinusoidal voltages across resistances R_x and R_y, respectively, the power ratio of Eq. (A11-3) is

$$\frac{P_{\text{out}}}{P_{\text{in}}} = \frac{Y^2/R_y}{X^2/R_x} \qquad (A11-4)$$

In the special case when $R_x = R_y$,

$$G = 10 \log \frac{Y^2}{X^2} \text{ dB} \qquad (A11-5)$$

$$\boxed{G = 20 \log \frac{Y}{X} \text{ dB}} \qquad (A11-6)$$

In an actual system, it is unusual for $R_x = R_y$. If we are interested in the signal gain, Y/X, however, we can arbitrarily define this gain G in dB by the formula of Eq. (A11-6). In other words, today the engineer defines the dB signal gain of a system as 20 times the logarithm of Y/X. Thus, Eq. (A11-6) is a fundamental definition used widely in many different fields.*

Rapid calculation of dB *gain*
The gain is so commonly measured in dB that the engineer learns to convert back and forth, approximately and rapidly, between dB and signal ratios. This mental conversion is simplified by a few relationships apparent from the values of the logarithms of common numbers:

---

*Indeed, control engineers even use Eq. (A11-6) as a definition of dB gain when y and x are dimensionally different (y might be a mechanical displacement, x a voltage). Clearly such a definition makes no sense on a power basis (or any logical basis), but it provides a convenient tool for system analysis.

$\dfrac{Y}{X}$	dB	Basis
1	0	$\log 1 = 0$
10	20	$\log 10 = 1$
100	40	$\log 100 = 2$
10^n	$20n$	$\log 10^n = n \log 10 = n$
2	6	$\log 2 = 0.301$
3	9.5	$\log 3 = 0.477$
4	12	$\log 4 = 2 \log 2$
6	15.5	$\log 6 = \log 2 + \log 3$
8	18	$\log 8 = 3 \log 2$
20	26	$\log 20 = \log 2 + \log 10$

Thus, the above table is constructed entirely from knowledge of log 1, log 2, and log 3. Furthermore, the table can be extended by including any product of the numbers in the Y/X column; the corresponding dB-column entry is the sum of the original two values. Finally, the reciprocal of a Y/X value yields a negative dB gain (e.g., if $Y/X = 1/2$, the gain is -6 dB).

Examples. (1) Signal gain of 40,000:

$$\frac{Y}{X} = 4 \times 10^4 \qquad G = 12 + 80 = 92 \text{ dB}$$

(2) Y/X given as 120

$$\frac{Y}{X} = 2 \times 6 \times 10 \qquad G = 6 + 15.5 + 20 = 41.5 \text{ dB}$$

(3) $G = 86$ dB

$$G = 80 + 6 \qquad \frac{Y}{X} = 10^4 \times 2 = 20,000$$

(4) $G = -14$ dB

$$G = -20 + 6 \qquad \frac{Y}{X} = \frac{1}{10} \times 2 = \frac{1}{5}$$

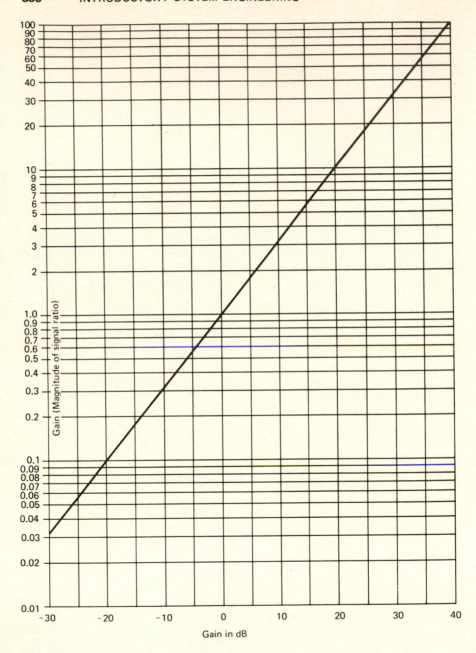

FIG. A11.2 Relation between dB and signal ratio.

(5) $G = 25$ dB

Here we can estimate the Y/X value:

$$G = 26 \text{ dB corresponds to } \frac{Y}{X} = 10 \times 2 = 20$$

$$G = 24.5 \text{ dB} = 40 - 15.5 \text{ corresponds to } \frac{Y}{X} = 100 \times \frac{1}{6} = 16.7$$

Hence $G = 25$ dB is approximately $\dfrac{Y}{X} = 18$

Accurate determination. Because Y/X can always be written as $A \times 10^n$, where A lies between 1 and 10, we can find the dB gain for any given Y/X value from Fig. A11.2. For example,

$$\frac{Y}{X} = 1.125 \qquad\qquad \text{corresponds to } G = 1 \text{ dB}$$

$$G = 53.4 \text{ dB } (= 40 + 13.4) \text{ corresponds to } \frac{Y}{X} = 4.7 \times 10^2 = 470$$

Advantages of logarithmic measure of gain. The popularity of dB descriptions of gain is based on the following properties.

(1) When two systems are in tandem (Fig. A11.3), the total dB gain is merely the *sum* of the separate, subsystem gains:

$$G = G_A + G_B \qquad\qquad\qquad\qquad\qquad\qquad \text{(A11-7)}$$

For example, if we need a total amplifier gain of 90 dB, this can be achieved by three tandem amplifiers, each with a gain of 30 dB.

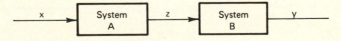

FIG. A11.3 Two systems in tandem.

(2) The numbers involved in describing the gain of a system are often more convenient. For example, 216 dB is equivalent to 2,000,000.

(3) Certain output receptors respond approximately to the logarithm of the signal amplitude. For example, the human ear and brain estimate roughly the same increase in a sound intensity for each doubling of the amplitude (i.e., each 6 dB increase is given the same interpretation by the human listener).*

Of these three reasons, the first is by far the most important. The many problems in which systems occur in tandem fully justify the use of a logarithmic gain measurement.

PROBLEMS

11.1   (a) Determine the dB gain corresponding to the transfer function magnitudes:

4,000,000   1/20   30

(b) The noise level in New York City is rising 1 dB per year. What is the annual percentage increase in pressure level? In how many years does the noise pressure double?

11.2   An experimental measurement of the frequency response of a linear system led to a gain plot shown. Suggest a transfer function consistent with this plot.

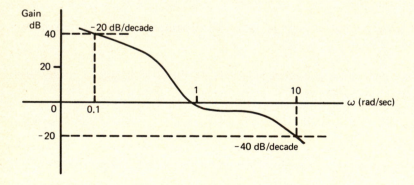

11.3   The transfer function of interest is

$$T(s) = \frac{1}{(s + 1)(s^2 + s + 1)}$$

*This statement is only very approximate, as substantiated by recent studies of hearing.

(a) Show the location of the poles in the s plane.

(b) Construct the asymptotic gain or Bode plot.

(c) Add the corrections to obtain the correct plot. Indicate clearly how you are determining corrections, even if you need a series of plots.

11.4 The transfer function of a piezoelectric crystal is

$$T(s) = \frac{10^{16}}{s^2 + 10^5 s + 10^{16}}$$

(a) Determine the undamped natural frequency ω_n, the resonant frequency ω_r, the bandwidth in rad/sec, the relative damping ratio ζ, and Q.

(b) Plot the gain in dB versus a logarithmic frequency scale; show the gain at the half-power frequencies and a few frequencies off resonance.

11.5 We consider the transfer function

$$T(s) = \frac{250 s (s + 20)}{(s + 1)(s + 10)(s^2 + 20s + 500)}$$

(a) Draw the asymptotic gain characteristic.

(b) From $T(s)$, what is the asymptotic gain at 5 rad/sec? Does this agree with the answer in (a)?

(c) Insert the corrections to obtain a smooth gain characteristic.

(d) Sketch the phase characteristic approximately.

11.6 Sketch the phase characteristic versus frequency for

$$G(s) = \frac{s - 2}{s + 2} e^{-0.2s}$$

11.7 (a) Construct the asymptotic gain characteristic for the transfer function

$$G(s) = \frac{16(s + 20)^2}{s(s + 2)^2(s + 8)}$$

(b) Sketch the phase characteristic for the same transfer function.

(c) If the input is $4 \cos(8t + 170°)$, what is the corresponding, steady state component of the output?

11.8 Determine the transfer function determined by the measurements shown. A sinusoid of adjustable frequency and constant amplitude was applied to the input of the unknown system, and the amplitude of the s-s output was measured.

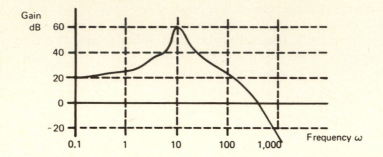

11.9 A transfer function has a magnitude which varies with frequency as

$$|G(\omega)| = \frac{1}{\sqrt{1 + \omega^6}}$$

(a) Sketch the plot of the gain as the magnitude of a voltage ratio (not in dB) versus a linear frequency scale.

(b) Plot the gain in dB versus a logarithmic frequency scale.

(c) At what frequency is the gain −52 dB?

(d) What is the slope of the high-frequency asymptote in dB/decade? In dB/octave?

11.10 What phase characteristics are realizable by a stable, all-pass network with two poles and two zeros? Sketch the range.

11.11 A linear system transmits an input without distortion if its phase response is linear in frequency over the frequency band of interest (and the gain is constant). Suppose you have to design such a system for the band from 1 to 10 rad/sec. Suppose further that the phase at 1 rad/sec is required to be −180°, at 10 rad/sec it is to be 180°.

(a) What would the relation for $\phi(\omega)$, the phase shift, have to be?

(b) Draw a phase plot of ϕ versus $\log \omega$.

(c) Suppose now you have to design a system with the phase characteristic you have obtained, but you are restricted to systems with transfer functions of the form $(s + a)^n / s^m$. How would you choose n and m? Why?

(d) What could be your choice of a? Plot the resulting phase characteristic. Is the approximation very good?

11.12 The gain and phase characteristics of a system are measured experimentally as shown in the figure. We apply a signal to this system which has no significant frequency components above 1.0 rad/sec. We wish to add a network $C(s)$, which compensates so that the output waveform is a replica of the input, although possibly delayed in time. What do you recommend for the gain and phase of $C(s)$?

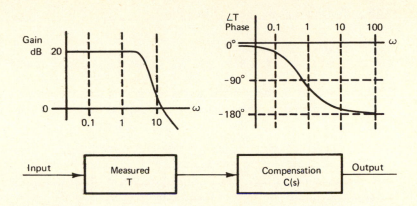

11.13 For a seismographic instrument to measure earthquakes, we want a transfer function $H(s)$ which has a gain which is constant (say zero dB) from zero frequency to 2 rad/sec, then above 2 rad/sec falls off with frequency at a rate of -60 dB/decade. For simplicity, we decide to use an $H(s)$ with no finite zero. Select appropriate poles for $H(s)$ and explain your choice.

11.14 Plot the gain characteristic for

$$\frac{Y(s)}{X(s)} = T(s) = \frac{400s^2}{(s + 100)(s + 200)(s + 400)}$$

(a) Determine the Bode plot.

(b) Show that your piecewise-linear characteristic agrees with the given $T(s)$ at 300 rad/sec.

(c) Add the appropriate corrections to approximate the actual curve.

(d) A signal

$$x(t) = 10 \sin 40t + 2 \sin 400t$$

is applied. Calculate the amplitudes of the two components of the steady state response.

(e) At low frequencies, the system yields an output which is proportional to the second derivative of the input. Below what frequency is this statement true within one dB?

11.15 Sketch the gain characteristic for the network shown, with values given in ohms, henrys, and farads.

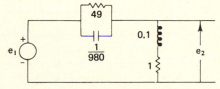

11.16 Sketch the phase characteristic for

$$T(s) = 320 \frac{s}{(s + 4)^2} e^{-s/4}$$

11.17 A portion of the hospital diagnostic system tests the patient's visual acuity and perception. These characteristics depend on the functional relationships

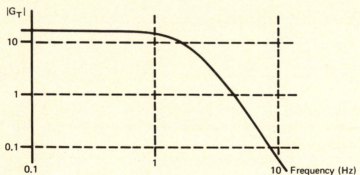

(a)

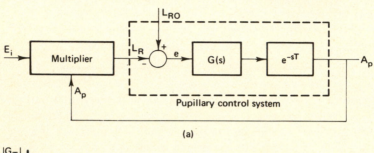

Measured, open-loop frequency response of Ge^{-sT}

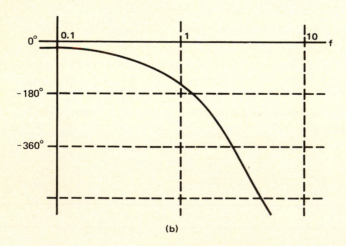

(b)

of the pupillary control system of the eye. A possible block diagram of this system is shown in the figure. The pupil aperture responds to increases or decreases of incident light intensity so as to regulate the total incident light (lumens) on the retinal light receptors (the rods and cones).

The excitation light falling on the retina is the product of the instantaneous pupil aperture area in meters squared and incident light intensity (lumens/m^2), thus introducing a nonlinear aspect to the overall system. The error signal is the difference between the "set point" value of incident light L_{R0} and the true incident light L_R. The set point is an internally set, quiescent operating point for the system, and is arrived at by a still-unknown internal mechanism.

The portion of the figure enclosed by dotted lines is the pupillary light reflex system. An optical excitation and monitoring technique has been developed which permits light excitation of the retina, and hence change in pupil aperture, without allowing this changing aperture to affect the total excitation (i.e., allows an open-loop study between L_R and A_p). In this process, a small-signal sinusoidally varying light intensity is introduced (varying about L_{R0}), and the open-loop response curves of the figure are obtained.

Three suggestions have been made for $G(s)$:

$$\frac{K_a(s + a_1)}{(s + a_2)(s + a_3)} \qquad \frac{K_b}{(s + b_1)(s + b_2)} \qquad \frac{K_c}{(s + c)^n}$$

In view of the experimental evidence, determine a form for $G(s)$, and find the appropriate parameter values. Also determine the value of the transportation lag for the measured system.

12 FEEDBACK

Feedback is a fundamental engineering concept which we have been using throughout this book. In this chapter, we define feedback then enumerate the various system characteristics that can be realized when using feedback intentionally and purposefully.

12.1 WHAT IS FEEDBACK?

In the usual engineering system, signals tend to travel from input to output. There may be parallel paths, but in general, the cause-effect relationships are sequential (e.g., as we move from left to right in the signal flow diagram of Fig. 12.1).

Often the system designer finds it useful to make the signal at a certain point dependent not only on the preceding signals (to the left), but also on the subsequent signals (to the right). For example, in Fig. 12.2 the signal z depends on y as well as on x; y in turn depends on z. There is a closed cycle or *closed loop* of cause-effect relationships: z determines y at least in part; y determines z in part. When this closed-loop situation exists, we have a *feedback system*.

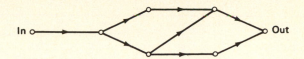

FIG. 12.1 A simple system without feedback.

Feedback is really just a viewpoint

In most cases, there is really no way to decide whether a system has feedback or not; the answer depends on your viewpoint. This is apparent from the signal flow diagram: we can always manipulate to find an equivalent diagram with no feedback loops.

Figure 12.3 shows an electrical circuit with e_1 the input and e_2 the output. The engineer would ordinarily write

$$e_2 = \frac{R_2}{R_1 + R_2} e_1 \tag{12-1}$$

The output is just a constant times the input; assuredly, there is no feedback in such a trivial system. If we forget the voltage-divider relationship, however, we might write the two simultaneous equations

$$i = \frac{1}{R_1}(e_1 - e_2) \tag{12-2}$$

$$e_2 = R_2 i \tag{12-3}$$

which are represented by the signal flow diagram of Fig. 12.4, where there is feedback.

Thus, any system can be considered as with or without feedback. In practice, feedback is a useful viewpoint when we purposefully insert a feedback path to achieve some desired system characteristic, or when the model of the system is simplified by using feedback. In the following paragraphs, we consider various situations in which the feedback viewpoint is useful.

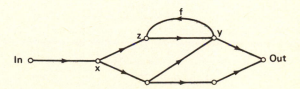

FIG. 12.2 System with single feedback path *f*.

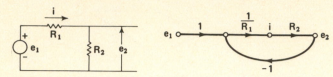

FIG. 12.3 Voltage-divider circuit.

FIG. 12.4 A possible signal flow diagram for Fig. 12.3.

Simple examples of feedback.   According to the definition given above, a feedback system includes at least the elements depicted in Fig. 12.5, which also serves to introduce our notation. The output signal c is measured by the sensor to yield the feedback signal b. The comparator generates an error e which depends upon a comparison of the input or desired output r and the measured output b; in the simplest case, e is simply given by the difference

$$e = r - b \tag{12-4}$$

as depicted by the + and − signs in the figure. More generally, the comparator might measure any function of r and b—e.g., $(r - b)^3$—but here we consider only comparators described by Eq. 12-4. Finally, the error signal e drives (or modifies) the output c through the *process*.

Illustrations of the basic system of Fig. 12.5 are plentiful throughout our daily life. In a broad class of problems, the human being is an important element of the feedback system. For example, if a man reaches forward to pick up a pencil from a desk, a feedback system drives the grasping hand. In this case, the desired output (the desired position of the hand) is at the pencil; the actual output is the instantaneous position of the hand. The sensor is now the eye visualizing the present position of the hand; this position b is compared (in the eye and brain) with the pencil position (also visualized by the eye), and an error signal actuates the muscles that move the hand (Fig. 12.6).

The *feedback* aspect of the operation arises because of the continual measurement of the hand position and comparison with the measured pencil location; actually the eye functions to measure directly the error, which the brain then converts to the appropriate drive signal for the muscles. Improvement in performance which results

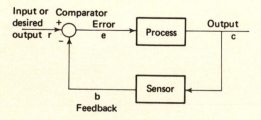

FIG. 12.5 Simplest feedback system.

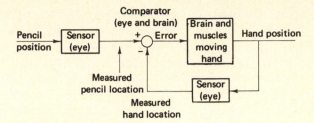

FIG. 12.6 A human feedback system.

from the feedback can be demonstrated, if the same task is attempted with the eyes closed, after only an initial measurement of the pencil position relative to the initial hand position.

The fact that we can pick up the pencil with closed eyes indicates that Fig. 12.6 is only a very rough model. In addition to the visual feedback loop shown, there are additional feedback paths. For example, as I move my hand toward the pencil, I sense that motion and corresponding signals are sent to the brain. Furthermore, as my hand touches the pencil (or the table where the pencil lies), a tactile signal also travels to the brain. These different feedback paths mean that the system can be represented by a multiloop structure, much more detailed than that shown in Fig. 12.6 (cf. Problem 12.1).

The feedback concept has been important in engineering from the earliest times. The Romans, during the days of the Empire, used the same feedback system still employed today to maintain control over the water level in the reservoir of the common bathroom toilet. After flushing is completed, the outlet valve is closed, the water rushes into the reservoir; as the water rises, a ball float senses the water level, closing the supply valve at the predetermined level. This is shown in the block diagram in Fig. 12.7. The float position measures the actual water level; when the error exceeds a predetermined magnitude, the valve is open and water flows in to increase the actual water level. As soon as the error falls below the specified value, the valve is shut off, and the system remains in a steady condition. When the water level is abruptly lowered by a separate system, the cycle of operation starts again.*

*This water-level control system is often cited as the first example of a man-made system intentionally designed as a feedback system.

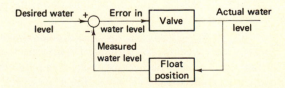

FIG. 12.7 The plumbing feedback system.

Both the above systems demonstrate not only the phenomenon defined as feedback, but also several remarkable characteristics achievable with feedback. For example, each system possesses a reliability rarely surpassed in engineering design; in each system, the performance is practically independent of system parameters. For example, in the case of the human pencil-grasper, almost everyone achieves success, whether an Olympic weight lifter, a ballet dancer, or a scholarly, bespectacled engineer. Similarly, the plumbing fixture operates almost regardless of the value of the water supply pressure; the only common difficulty occurs when the auxiliary system emptying the water fails to shut off, so that the water level is never able to rise. In view of the remarkable reliability of both systems, it is not surprising that both have been in use without modification for many centuries.

In spite of the obvious merits of such feedback structures, only very recently did engineers recognize the importance of feedback as a design tool. Feedback theory, as we know it today, came into being in the 1920s with work at the Bell Telephone Laboratories. It received its greatest impetus during the Second World War with extensive work on automatic control systems for such military purposes as fire control (the positioning of gun turrets), radar antenna control (the positioning and measurement of the antenna), and aircraft control (for stabilization as well as navigation). During the last decade, major advances in feedback theory have been associated with missile guidance and control, industrial automation (for example, the automatic production line or the automatic adjustment of steam boiler operating conditions to maximize the economy of operation), and the application of automation to social and urban problems (e.g., the control of city traffic).

12.2 WHY USE FEEDBACK?

The system with feedback is certainly more complex than the system without it. What do we gain by introducing feedback? Quantitatively, what does feedback accomplish in systems similar to those described above? In what ways can we measure the effectiveness of the feedback?

In the majority of systems encountered in engineering, the feedback is used to control the degree to which the overall transmission characteristics of the system depend upon variation of one or more of the system parameters. For example, in the plumbing system, the operation of the water level control is independent (at least over a wide range) of the pressure of the in-flowing water. As a result of this independence, performance of the system (the success of the system in controlling the water level at the time the valve is shut off) is the same in the many homes and buildings where the system is used.

Early work on feedback theory by engineers at Bell Telephone Laboratories was largely motivated by the advent of transcontinental telephone transmission systems, which used a large number of repeater amplifiers. In order to maintain the overall system characteristics constant as the gains of the individual amplifiers varied because of tube aging, temperature and voltage variations, and the like, design engineers turned to utilization of large amounts of feedback around each of the amplifiers. The reliability requirements are illustrated by the present-day demand that amplifiers in the transatlantic telephone cable should operate for 20 years without maintenance (because of the expense and difficulty of pulling up cable from the bottom of the ocean to make repairs and adjustments).

Thus, the critical characteristic of feedback systems most often is the tendency of the system to operate satisfactorily even when the specific parameters of the various components are changing radically. In order to place this discussion on a more quantitative basis, it is helpful to consider the characteristics of the two systems of Fig. 12.8—the first involving no feedback, and the second utilizing feedback embodied in the sensor and comparator elements. (Conventionally, we call the former system *open-loop*, the latter *closed-loop*—to emphasize that there is a closed signal-transmission loop around which signals may flow.)

In order to consider a specific example, we specify that the gain (i.e., the ratio of the output to the input signals) of each of the two systems should be 10. In the open-loop system, this specification may be realized by a single amplifier with a gain of 10; however, when the amplifier gain changes by 2 percent, the overall transmission of the system obviously changes by the same 2 percent.

The closed-loop system is quite different in its behavior. In this case, the overall transmission can be found by solving simultaneously the three equations that describe the performance of each of the elements of the system:

$$E = R - B \qquad\qquad (12\text{-}5)$$

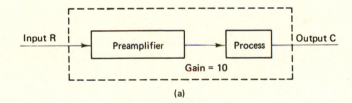

(a)

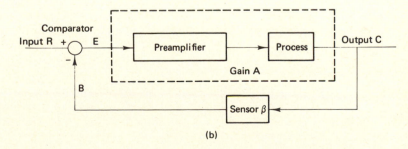

(b)

FIG. 12.8 Evaluation of feedback. (a) Open-loop system; (b) Feedback system.

$$B = \beta C \tag{12-6}$$

$$C = AE \tag{12-7}$$

When the variables E and B are eliminated from these three equations, the overall transmission is

$$\frac{C}{R} = \frac{A}{1 + \beta A} \tag{12-8}$$

If β is selected (arbitrarily for the moment) as 99/1000, the realization of an overall gain of 10 requires that the quantity A be 1000; hence we have the feedback configuration shown in Fig. 12.9.

In order to realize our overall gain of 10, we require an amplifier with a gain of 1000 (or three tandem amplifiers, each with a gain of 10). The introduction of feedback complicates the system design—not only is the number of required amplifiers raised from one to three, but we now need the sensor or β circuit and the comparator. What have we gained from this extra complexity?

The advantage of feedback is demonstrated if we consider the effects of variation in the gain of the amplifiers. If the total amplifier gain of 1000 changes by 2 percent (e.g., if A drops to 980), the overall transmission of our complete system decreases to the value

$$T = \frac{980}{1 + (99/1000)\,980} = 9.998 \tag{12-9}$$

As a result of the feedback, the *overall gain drops only 0.02 percent*, even though the gain of the amplifier falls 2 percent. We have achieved an improvement in performance by a factor of 100, in terms of making the system insensitive to the variations of system parameters.

It is natural to object that insertion of the feedback actually requires more amplifiers, hence the above improvement is deceptive. Even if all three amplifiers change in the same direction by the full 2 percent in our example, the total amplifier gain changes by approximately 6 percent. The gain of our overall system changes by only 0.06 percent—thus we still achieve a marked improvement in comparison to the system with no feedback.

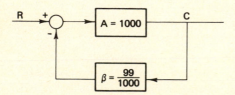

FIG. 12.9 System for $C/R = 10$.

12.3 SENSITIVITY

Feedback is used to reduce the sensitivity of the system characteristics to a specific parameter. We now wish to put this concept of sensitivity control in quantitative terms.

The sensitivity of the system transfer function $T(s)$ to changes in a parameter x is defined as

$$S_x^{\ T} = \frac{dT/T}{dx/x} \tag{12-10}$$

In other words,

$$\text{Sensitivity of } T \text{ to } x \; = \; \frac{\% \text{ change in } T}{\% \text{ change in } x} \tag{12-11}$$

when both percentage changes are small.*

In terms of this definition, the open-loop system of Fig. 12.10(a) possesses a sensitivity of T to G_2:

$$S_{G_2}^{\ T} = 1 \tag{12-12}$$

A 1 percent change in G_2 causes the same change in T. In contrast, the feedback system of Fig. 12.10(b) can give sensitivity control. Straightforward differentiation of T with respect to G_b and use of the definition of Eq. (12-10) give†

---

*Equation (12-10) is sometimes written in the equivalent form

$$S_x^{\ T} = \frac{d \ln T}{d \ln x}$$

but the form given for Eq. (12-10) is usually the one we use for calculating the sensitivity.

†In a single-loop system, the sensitivity of T to a *forward* transfer function (G_a or G_b) comes out to be just $1/(1\text{-loop gain})$.

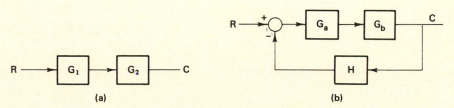

FIG. 12.10 System without and with feedback, $T = C/R$. (a) Open-loop system; (b) closed-loop system.

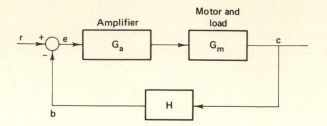

FIG. 12.11 Positioning control system.

$$S_{G_b}^{\ T} = \frac{1}{1 + G_a G_b H} \qquad (12\text{-}13)$$

Thus, the sensitivity can be made any desired small value by choosing G_a, G_b, and H appropriately. *Feedback permits the simultaneous realization of a desired overall transfer function $T(s)$ and a required sensitivity S.*

Figure 12.11 illustrates a simple design problem. The system is used for positioning material in a production line. The output c is the position which is to be controlled; the input r is the reference signal or the desired value of the output. The comparator measures the error e which is then amplified by G_a and used to drive the motor G_m which is controlling c.

In the design of such a system, first we select a motor which can provide the power and force required to position the output c. Once this G_m is chosen, we need to select G_a and H to achieve the desired transfer function and sensitivity.

The overall transfer function $T(s)$ is (from Mason's theorem or simply the solution of the equations for Fig. 12.11)

$$T = \frac{C}{R} = \frac{G_a G_m}{1 + G_a G_m H} \qquad (12\text{-}14)$$

The sensitivity of T to changes in G_m is

$$S_{G_m}^{\ T} = \frac{1}{1 + G_a G_m H} \qquad (12\text{-}15)$$

If the sensitivity is to be appreciably less than unity, $G_a G_m H$ must be much greater than unity, and the two equations reduce to

$$T = \frac{1}{H} \qquad (12\text{-}16)$$

$$S_{G_m}^{\ T} = \frac{1}{G_a G_m H} \qquad (12\text{-}17)$$

Now the design is straightforward. For example, if T is to be unity, we must select

$$H = 1 \qquad\qquad (12\text{-}18)$$

Then the sensitivity is just $1/G_a G_m$ where G_m is known and G_a is to be chosen. For example, if $G_m = 2$ and we desire a sensitivity of 0.01,

$$G_a = 50 \qquad\qquad (12\text{-}19)$$

The above "design" is actually deceptive, since in practice G_m varies with frequency (G_m is a transfer function, the ratio of polynomials in s). Under these circumstances, we have to define a frequency range of interest (the frequency band within which r has significant components). The minimum value of $|G_m|$ over this band is calculated; if this is 2, then we can select a $|G_a|$ which is at least 50 over the band and be confident that $|S|$ will be less than 0.01 at all frequencies of interest.

In actual design problems, the above procedure is a common engineering approach. Having found the minimum G_a to ensure the desired sensitivity, we next can select $G_a(s)$ as a transfer function varying with s. Ordinarily, $G_a(s)$ must be chosen in such a way that the system is stable.* We shall defer consideration of the stability problem until Chapter 15; here we only want to emphasize that feedback system design to realize specified $T(s)$ and S is conceptually rather trivial (as indicated above). It is only when we consider the stability problem that we find design by no means a trivial problem. In any problem of interesting complexity, it turns out to be very difficult to find a $G_a(s)$ which gives a suitably large $|G_a|$ over the frequency band of interest and, at the same time, results in a stable system.

12.4 TECHNIQUES OF SENSITIVITY CONTROL

The control of sensitivity is a primary motivation for the use of feedback. Even in the very simple single-loop system (Fig. 12.12), we can obtain some independence in the specification of the overall system function and the sensitivity. This same system points out, however, the design problem.

*Usually, the larger $|G_a|$—i.e., the smaller the desired sensitivity—the more the tendency for the feedback system to be unstable.

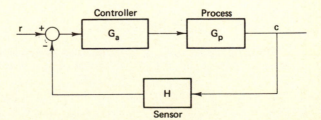

FIG. 12.12 Single-loop system.

For example, for this system

$$T(s) = \frac{C}{R} = \frac{G_a G_p}{1 + G_a G_p H} \qquad (12\text{-}20)$$

$$S_{G_p}{}^T = \frac{1}{1 + G_a G_p H} \qquad (12\text{-}21)$$

In a typical design problem, G_p is given. We might arbitrarily select a $T(s)$ and a $S_{G_p}{}^T$ which would meet the specifications, then solve these two equations for the required $G_a(s)$ and $H(s)$. This approach would appear to constitute totally logical design.

We now try this procedure in an example. We are given

$$G_p(s) = \frac{10}{s(s + 10)} \qquad (12\text{-}22)$$

Perhaps a suitable $T(s)$ is*

$$T(s) = \frac{400}{s^2 + 20s + 400} \qquad (12\text{-}23)$$

We want a sensitivity of $1/100$.

If we solve algebraically Eqs. (12-20) and (12-21) for G_a and H, we find

$$G_a = \frac{1}{G_p} \frac{T}{S_{G_p}{}^T} \qquad (12\text{-}24)$$

$$H = \frac{1 - S_{G_p}{}^T}{T} \qquad (12\text{-}25)$$

Substitution of the given G_p and the desired T and $S_{G_p}{}^T$ yields

$$G_a = 4000 \frac{s(s + 10)}{s^2 + 20s + 400} \qquad (12\text{-}26)$$

$$H = \frac{99}{100}\left(1 + \frac{1}{20} s + \frac{1}{400} s^2\right) \qquad (12\text{-}27)$$

The resulting H is difficult to build accurately; we need to feed back a signal which includes the output and its first *two* derivatives. If there is any high-frequency noise or

---

*Since the process has a double zero at ∞, $T(s)$ should also—otherwise G_a will have to have a pole at ∞. In Eq. (12-20) as $s \to \infty$, the denominator tends to 1, and T is just $G_a G_p$.

random motion of the output, this signal will be accentuated by the second derivative measurement.*

This very simple example illustrates the problems that can arise in feedback-system design when we try to control (or specify) both T and $S_{G_p}{}^T$. In more complex cases, H or G_a are apt to be totally unrealizable by feasible networks (e.g., G_a might turn out to be unstable).

It is precisely because of such difficulties that feedback-system design has become an interesting and challenging field. If the engineer encounters trouble with the single-loop system of Fig. 12.12, one of the first things he does is to consider whether he can simplify his job by going to other system configurations. In the remainder of this section, we look at a very few of the possibilities.

A two-loop system

We might try feedback around the process as well as around the total system. Figure 12.13 shows the two possible systems for obtaining a desired $T(s)$ when one is given the process G_p. In (a), the desired transfer function is achieved by utilizing a controller placed in tandem with the process; in (b) the controller function is supplied by a local feedback path encircling the plant and the tandem element is merely an amplifier. Our interest is centered in comparing the two systems from the point of view of how the overall transfer function is affected by variations in the plant parameters K and a, over

---

*Since Eq. (12-27) calls for only 1/400 of the second derivative of c, we might simply neglect the $s^2/400$ term (or assume it is negligible). Unfortunately, we then find that the sensitivity is only 1/10 (instead of 1/100) at low frequencies, and $T(s)$ is no longer underdamped—both $S_{G_p}{}^T$ and T are changed radically.

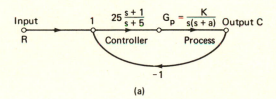

(a)

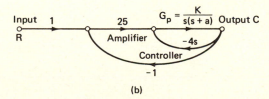

(b)

FIG. 12.13 Control systems with normal process parameter values $K = 1$; $a = 1$. (a) Tandem controller; (b) minor-loop controller.

which we presumably have no direct design control. In other words, we want to investigate the sensitivity of $T(s)$ to both K and a for each of the two cases.

For the single-loop system of (a), the overall system function is

$$T(s) = \frac{C}{R} = \frac{25K(s + 1)}{s(s + 5)(s + a) + 25K(s + 1)} \tag{12-28}$$

If this is placed in the form

$$T(s) = \frac{1}{1 + s(s + 5)(s + a)/[25K(s + 1)]} \tag{12-29}$$

the sensitivity $S_K{}^T$ can be evaluated by direct differentiation:

$$S_K{}^T = \frac{s(s + 5)(s + a)}{25K(s + 1) + s(s + a)(s + 5)} \tag{12-30}$$

When $a = 1$ and $K = 1$, this reduces to

$$S_K{}^T = \frac{s(s + 5)}{s^2 + 5s + 25} \tag{12-31}$$

Similarly, the sensitivity $S_K{}^T$ for the two-loop system of Fig. 12.13(b) is carried out by operating with the transmittance

$$T(s) = \frac{25K/[s(s + a)]}{1 + 4Ks/[s(s + a)] + 25K/[s(s + a)]} \tag{12-32}$$

$$= \frac{25K}{s(s + a) + K(25 + 4s)} \tag{12-33}$$

Computing the sensitivity $S_K{}^T$ we obtain

$$S_K{}^T = \frac{s(s + a)1/K}{(1/K)s(s + a) + 25 + 4s} \tag{12-34}$$

Again, inserting the normal values of the process parameters, we find

$$S_K{}^T = \frac{s(s + 1)}{s^2 + 5s + 25} \tag{12-35}$$

Thus, even though the two system functions are identical, the two sensitivity functions differ. The significance of the difference is somewhat difficult to interpret, but one way of achieving a relative evaluation might be to plot $|S_K{}^T|$ versus the angular frequency (i.e., with $s = j\omega$ so that we are focusing attention on the system performance under sinusoidal excitation). The results shown in Fig. 12.14 demonstrate the superiority of the two-loop system (b) (which utilizes local feedback around the process) when the angular frequencies are low. Thus, through the definition of the sensitivity function we have expressed in quantitative terms the degree to which the feedback accomplishes its appointed task, and we have a firm basis for the comparison of alternate designs.

Evaluation of the competitive designs of Fig. 12.13 might also require determination of the sensitivities with respect to the parameter a. By direct differentiation, we learn that for case (a)

$$S_a{}^T = \frac{-s(s + 5)}{(s + 1)(s^2 + 5s + 25)} \tag{12-36}$$

and for case (b)

$$S_a{}^T = \frac{-s}{s^2 + 5s + 25} \tag{12-37}$$

Plotting graphs of $|S_a{}^T|$ versus ω again reveals the superiority of the minor-loop compensation [case (b)] at low frequencies.

In this section, our objective is not, however, to consider a specific example in great detail, but rather to indicate the general way in which the feedback engineer attempts to make a quantitative assessment of the role of feedback in a system design—an assessment based on the sensitivity function.

The above discussion demonstrates that:

(1) Feedback permits the purposeful control of system sensitivity to particular parameters.

(2) The sensitivity function allows a quantitative comparison of different feedback structures for achieving the same transfer function.

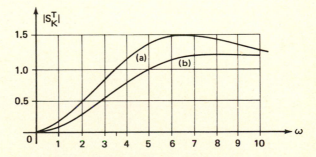

FIG. 12.14 Sensitivities of the systems of Fig. 12.13.

Zero-sensitivity systems

As we consider systems with more than one feedback loop, we find that there are systems with a sensitivity function equal to *zero*. In other words, for small changes, the system function is independent of the parameter.

Figure 12.15 shows a two-loop system, including a process G_p and now three controller transfer functions: G_a, G_b, and H. For this system,

$$T = \frac{C}{R} = \frac{G_a G_p}{1 + G_a G_p H - G_a G_b H} \tag{12-38}$$

The sensitivity of T with respect to G_p is just $1/(1 - \text{loop gain through } G_p)$, or

$$S_{G_p}{}^T = \frac{1}{1 + G_p G_a H/(1 - G_a G_b H)} = \frac{1 - G_a G_b H}{1 + G_a H(G_p - G_b)} \tag{12-39}$$

Inspection of Eq. (12-39) reveals that $S_{G_p}{}^T$ can be made equal to zero if we choose

$$G_a G_b H = 1 \tag{12-40}$$

Over the frequency band where this equation is approximately satisfied, the sensitivity of T to changes in G_p is negligible. In other words, by selecting the three controller transfer functions (G_a, G_b, and H) to satisfy Eq. (12-40), we can build a total system in which the process can change significantly without affecting T. If Eq. (12-40) is satisfied, the T of Eq. (12-38) becomes

$$T = \frac{G_a G_p}{G_a G_p H + \underbrace{(1 - G_a G_b H)}_{\text{Near zero}}} = \frac{1}{H} \tag{12-41}$$

In very simple terms, the reason the sensitivity of T to G_p is zero in Fig. 12.15 is because G_p is in tandem with essentially an infinite gain. With $G_a G_b H = 1$, there is a feedback loop with a gain of $+1$.

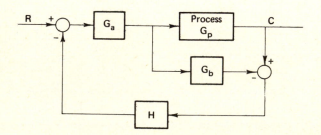

FIG. 12.15 Two-loop system.

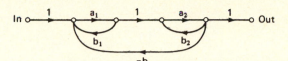

FIG. 12.16   System with three loops .

Figure 12.16 shows a similar system, but with two amplifiers (a_1 and a_2) and two positive feedback loops ($a_1 b_1$ and $a_2 b_2$). In this system, if we choose b_1 and b_2 as potentiometer gains which satisfy

$$b_1 = \frac{1}{a_1} \qquad b_2 = \frac{1}{a_2} \tag{12-42}$$

we find that the total system is described by

$$\left.\begin{aligned}
T &= \frac{a_1 a_2}{(1 - a_1 b_1)(1 - a_2 b_2) + a_1 a_2 b} = \frac{1}{b} \\[2ex]
S_{a_2}{}^T &= \frac{1 - a_1 b_1}{(1 - a_1 b_1)(1 - a_2 b_2) + a_1 a_2 b} = 0 \\[2ex]
S_{a_1}{}^T &= \frac{1 - a_2 b_2}{(1 - a_1 b_1)(1 - a_2 b_2) + a_1 a_2 b} = 0
\end{aligned}\right\} \tag{12-43}$$

This is a remarkable system. If each amplifier (a_1 and a_2) possesses a gain of 10 and $b = 1/1000$, we find

$$\left\{\begin{aligned}
&\text{The total system has a transfer function, } T = 1000 \\
&\text{If } a_2 \text{ only changes from 10 to any nonzero value, } T \text{ stays at 1000} \\
&\text{If } a_1 \text{ only changes from 10 to any nonzero value, } T \text{ stays at 1000}
\end{aligned}\right.$$

If a_1 and a_2 both change, the overall gain T does actually change. But if either amplifier deteriorates in performance, the response is not affected.

The multiloop systems of Figs. 12.15 and 12.16 are merely examples of the way the system designer can attempt to find a feedback configuration which allows him to meet performance specifications simply, reliably, and economically. Fortunately, for the fun of engineering, there is no straightforward theory which leads from specifications to the choice of configuration. This step is left largely to the engineer's creative imagination, built upon his familiarity with configurations used in the past.

12.5 SENSITIVITY IN ACTIVE NETWORKS

The concept of sensitivity is extended beyond our use of $S_x{}^T$, the sensitivity of the overall transfer function T to changes in a parameter x. The basic definition

$$S_x^T = \frac{d \ln T}{d \ln x} = \frac{dT/T}{dx/x} \tag{12-44}$$

measures how sensitive T is to x changes.

For example, active networks (Rs, Cs, and voltage amplifiers) can be used to realize any desired transfer function. When hybrid integrated circuits are so common, we often wish to build such a circuit to realize the resonant characteristics familiar in an RLC network: the sharp gain peak at the angular resonant frequency ω_0 and with the sharpness described by a bandwidth B (in rad/sec) or the Q ($= \omega_0/B$). The corresponding transfer function is

$$T(s) = -\frac{As}{s^2 + Bs + \omega_0^2} = -\frac{As}{s^2 + (\omega_0/Q)s + \omega_0^2} \tag{12-45}$$

When we design such a network for a specified ω_0 and Q, the "quality" of the network is described by the sensitivity of ω_0 or the sensitivity of Q with respect to the circuit parameters R, C, and amplifier gain K. Since we can anticipate variations in R, C, and K because of manufacturing tolerances and changes with time, we want the sensitivities to be as small as possible. Alternatively, of the many different active networks realizing $T(s)$ of Eq. (12-45), we prefer that system with the minimum sensitivities.

As a specific example, the circuit of Fig. 12.17 has been suggested for the realization of the simple resonance transfer function* A corresponding signal flow is shown in Fig. 12.18. From this, we can calculate the overall transfer function and the various sensitivities. In particular,

$$T = \frac{-K_1 K_2 R_2 C_2 s}{(1 + R_1 C_1 s)(1 + R_2 C_2 s) + K_1 K_2 R_1 R_2 C_1 C_2 s^2} \tag{12-46}$$

Before analyzing further, we can simplify somewhat by assuming that $R_1 = R_2$,

*M. A. Soderstrand and S. K. Mitra, Extremely Low Sensitivity Active RC Filter, *Proc. IEEE*, Dec. 1969, pp. 2175-6.

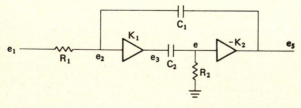

FIG. 12.17 Active circuit for realizing resonance (K_1 and K_2 both positive).

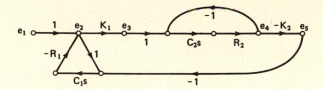

FIG. 12.18 Possible signal flow diagram for active network of Fig. 12.17.

$C_1 = C_2$, and $K_1 = K_2$ (a common situation). Then the circuit and flow diagram are as shown in Fig. 12.19, and by direct application of Mason's theorem, we have

$$T = \frac{-K^2 RCs}{(1 + RCs)^2 + K^2 R^2 C^2 s^2} \tag{12-47}$$

If we divide numerator and denominator by the coefficient of s^2 in the denominator, we obtain

$$T(s) = \frac{-[K^2/(1 + K^2)](1/RC)s}{s^2 + 2s/[RC(1 + K^2)] + 1/[R^2 C^2 (1 + K^2)]} \tag{12-48}$$

We now want to select RC and K to give the transfer function of a resonant system

$$T(s) = \frac{-As}{s^2 + (\omega_0/Q)s + \omega_0^2} \tag{12-49}$$

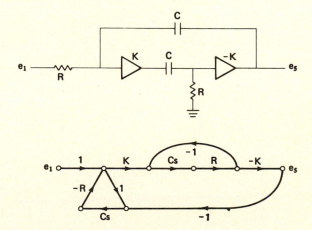

FIG. 12.19 Circuit simplified with equal resistances, capacitances, and voltage gains.

Algebraic manipulation gives

$$\left.\begin{array}{l} K = \sqrt{4Q^2 - 1} \\[2ex] RC = \dfrac{1}{2\omega_0 Q} \end{array}\right\} \tag{12-50}$$

As a specific example, if we specify

$$\omega_0 = 1000 \text{ rad/sec} \qquad Q = 50$$

we obtain

$$K = 100 \qquad RC = 10^{-5} \quad \left\{ \text{e.g.,} \quad \begin{array}{l} C = 10^{-11} \\[1ex] R = 10^6 \end{array} \right. \tag{12-51}$$

and the circuit is shown in Fig. 12.20.

The extent to which the circuit is satisfactory and economical depends on the sensitivity of ω_0, Q, and B to changes in any resistance, capacitance, or amplifier gain. Equation (12-46) reveals that, if we divide the denominator by the coefficient of s^2, the new denominator constant term ($\omega_0{}^2$) is

$$\omega_0{}^2 = \frac{1}{R_1 R_2 C_1 C_2 (1 + K_1 K_2)} \tag{12-52}$$

Now

$$2 \ln \omega_0 = - \ln R_1 - \ln R_2 - \ln C_1 - \ln C_2 - \ln(1 + K_1 K_2) \quad . \tag{12-53}$$

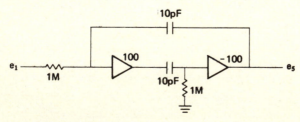

FIG. 12.20 Circuit to realize
$$\frac{E_5}{E_1} = \frac{-10^5 s}{s^2 + 20s + 10^6} \, .$$

By the logarithmic definition of sensitivity,

$$S_{R_1}^{\omega_0} = \frac{d \ln \omega_0}{d \ln R_1} \tag{12-54}$$

and recognizing that $K_1 K_2 \gg 1$, we have

$$S_{R_1}^{\omega_0} = S_{R_2}^{\omega_0} = S_{C_1}^{\omega_0} = S_{C_2}^{\omega_0} = S_{K_1}^{\omega_0} = S_{K_2}^{\omega_0} = -1/2 \tag{12-55}$$

For each parameter, a 1 percent increase in the parameter represents a 1/2 percent decrease in ω_0.

Similarly, the bandwidth B is the coefficient of s in the denominator (when the coefficient of s^2 is unity), and we find that if normal values are $R_1 = R_2, C_1 = C_2$, and $K_1 = K_2$

$$S_{R_1}^{B} = S_{R_2}^{B} = S_{C_1}^{B} = S_{C_2}^{B} = -\frac{1}{2} \qquad S_{K_1}^{R} = S_{K_2}^{B} = -1 \quad (12\text{-}56)$$

Finally,

$$Q = \frac{\omega_0}{B}$$

Hence

$$S_x^{Q} = S_x^{\omega_0/B} = \frac{d \ln \omega_0/B}{d \ln x} = \frac{d \ln \omega_0}{d \ln x} - \frac{d \ln B}{d \ln x} = S_x^{\omega_0} - S_x^{B}$$

Therefore,

$$S_{R_1}^{Q} = S_{R_2}^{Q} = S_{C_1}^{Q} = S_{C_2}^{Q} = 0 \qquad S_{K_1}^{Q} = S_{K_2}^{Q} = \frac{1}{2} \tag{12-57}$$

and we have described all sensitivities.

The circuit of Fig. 12.20 (or more generally Fig. 12.19) is really rather remarkable. Using two capacitors, two resistors, and two amplifiers, we realize a resonant circuit or band-pass filter. The Q is insensitive to the passive element values (for small changes, of course) and the sensitivity of Q to the gain of each amplifier is only 1/2. Problem 12.6 shows that these sensitivities are consistent with or better than the values obtained with an RLC network realizing the same resonance characteristic.

From the standpoint of electrical engineering, the important point of this example is that we can realize any desired transfer function without using any inductors, and our circuit characteristics need not be excessively sensitive to parameter variations.

12.6 DISTURBANCE CONTROL

Feedback is also a useful design tool when there is a problem of an unwanted disturbance signal that enters at some point within the system. Figure 12.21 shows an open-loop system in which the input signal r passes through two tandem subsystems, G_1 and G_2. Between these two parts, a noise or corruption signal u is added to the desired signal m. Then the response c is the sum of two components:

$$C = \underbrace{G_1 G_2 R}_{\substack{\text{desired} \\ \text{output}}} + \underbrace{G_2 U}_{\substack{\text{unwanted} \\ \text{output}}}$$

(12-58)

The difficulty revealed in Fig. 12.21 arises in a wide variety of control and communication examples. In attempting to control the elevator surface of an airplane to change altitude, the pilot moves the stick back and forth; the elevator should respond to this input signal. Unfortunately, other forces also act on the elevator due to air pressures, wind gusts, and the like. These forces constitute an undesired or disturbance signal which contributes to the system output and results in an altitude change not desired by the pilot and not under his direct control.

Similarly, Fig. 12.21 might represent a complex radio communication system. r is the input signal, G_1 the transmitter, G_2 the receiver, and c the output. Between the transmitter and the receiver, noise is added to the signal in the form of static, interfering signals, and so forth. The received signal c is then the desired signal corrupted by the noise.

Feedback allows control of the effect of u if we can add components to the system of Fig. 12.21. Figure 12.22 is one possible form for such an augmented feedback system; here G_a, H, and the comparator have been added. We wish to select G_a and H so that the transfer function from r to c is unchanged, but the effect of u is reduced by a specified amount. In other words, comparison with Eq. (12-58) shows that we would like to choose G_a and H so that

$$C = G_1 G_2 R + \alpha G_2 U \tag{12-59}$$

where α is small (perhaps 0.01).

Direct application of Mason's reduction theorem to Fig. 12.22 gives

$$\frac{C}{R} = \frac{G_a G_1 G_2}{1 + G_a G_1 G_2 H} \tag{12-60}$$

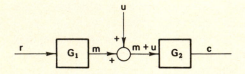

FIG. 12.21 System with disturbance input u.

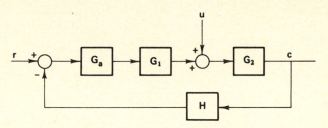

FIG. 12.22 Feedback system with disturbance input u.

$$\frac{C}{U} = \frac{G_2}{1 + G_a G_1 G_2 H} \tag{12-61}$$

Hence, Eq. (12-59) states that we want

$$\frac{1}{1 + G_a G_1 G_2 H} = \alpha$$

$$\alpha G_a = 1$$

Algebraical solution of these two relations for G_a and H yields

$$G_a = \frac{1}{\alpha} \tag{12-62}$$

$$H = \frac{G_a - 1}{G_a G_1 G_2} \simeq \frac{1}{G_1 G_2} \quad \text{if} \quad \alpha \ll 1 \tag{12-63}$$

Thus, over the frequency band of interest, we must select G_a to be $1/\alpha$ and H to be $(1/G_1 G_2)$; if these two choices are realized, insertion of the feedback leaves the C/R transmission unchanged, but reduces the effect of the disturbance signal to α of its original value.

Feedback can be used to control the effect of a disturbance signal entering the system at a point different from the primary input.

12.7 IMPORTANCE OF RETURN DIFFERENCE

In the preceding sections, we have found that feedback is useful to

(1) control the sensitivity
(2) control the effects of disturbances

In our analysis, we have compared systems with and without feedback in order to determine quantitatively what is to be gained by adding feedback.

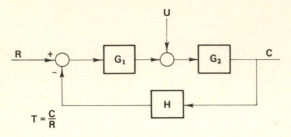

FIG. 12.23 Feedback system with disturbance input U.

In both discussions, we studied the system redrawn in Fig. 12.23. We found that

$$S_{G_2}{}^T = \frac{1}{1 + G_1 G_2 H} \tag{12-64}$$

and

$$\frac{C}{U} = \frac{G_2}{1 + G_1 G_2 H} \tag{12-65}$$

In the former equation, feedback reduces the sensitivity by division by $(1 + G_1 G_2 H)$. In Eq. (12-65), the effect of disturbances is the direct U-to-C transmittance, again reduced by division by $(1 + G_1 G_2 H)$. At least in this simple single-loop system, the effect of feedback is measured by

$$1 + G_1 G_2 H$$

What is this quantity? Since the sensitivity is with respect to G_2 and the disturbance enters the system at the input of G_2, we are interested in the feedback around G_2. If we open the system *at the input of* G_2 (Fig. 12.24) and send out a unit test signal from the forward side of the break, the signal returned at the left side due to that test signal is

$$-G_2 H G_1$$

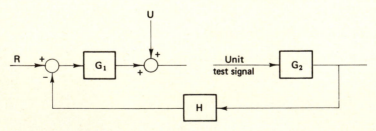

FIG. 12.24 Measuring feedback around G_2.

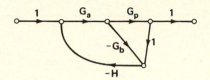

FIG. 12.25 A two-loop system.

Hence, $(1 + G_1 G_2 H)$ is the difference between the test signal and the returned signal. $(1 + G_1 G_2 H)$ is called the *return difference with respect to* G_2. Indeed, this return difference is a quantitative measure of the amount of *feedback* around G_2, and is denoted by the symbol F_{G_2}.

Thus, the return difference with respect to any element x or the feedback around x, F_x, is determined by breaking the system at the input of x and then finding 1 minus the gain from the output of the break back to the input.

In the system of Fig. 12.25 (which we studied earlier as Fig. 12.15), we calculate the feedback around G_p as unity minus the transmission from A to B (Fig. 12.26). The feedback around G_a is unity minus the gain from A' to B' (Fig. 12.27).

$$\left.\begin{aligned} F_{G_p} &= 1 + \frac{G_p H G_a}{1 - G_a G_b H} \\ F_{G_a} &= 1 + G_a(G_p - G_b)H \end{aligned}\right\} \tag{12-66}$$

One of the important features of F is that it can often be measured experimentally (as a function of frequency since it is unity minus a transfer function). If we want gain and phase curves, we apply a sinusoid at the input of the reference element and measure the returned signal.

Sensitivity in terms of F

In the single-loop example of the beginning of this section (Fig. 12.23),

$$S_{G_2}{}^T = \frac{1}{F_{G_2}} \tag{12-67}$$

The sensitivity is just one over the feedback. The more feedback, the smaller the sensitivity.

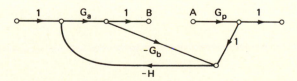

FIG. 12.26 Calculation of F_{G_p}.

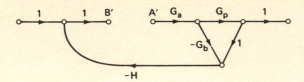

FIG. 12.27   Calculation of F_{G_a}.

Actually, Eq. (12-67) is valid only for elements (here G_2) in which, when we set $G_2 = 0$, the overall transfer function goes to zero. In other words

$$S_x^T = \frac{1}{F_x} \quad \text{only if} \quad x = 0 \quad \text{means} \quad T = 0 \tag{12-68}$$

In Fig. 12.23, this condition is satisfied for G_1 or G_2, but not for the feedback sensor element H. More generally, we can show (Prob. 12.12) that

$$S_x^T = \frac{1}{F_x}\left(1 - \frac{T_0}{T}\right) \quad \begin{array}{l} \text{where } T_0 \text{ is the value of } T \\ \text{when } x = 0. \end{array} \tag{12-69}$$

For example, For Fig. 12.28,

$$T = \frac{G}{1 + GH} \qquad T\,(H = 0) = G$$

$$F_H = 1 + GH$$

and

$$S_H^T = \frac{1}{1 + GH}\left[1 - \frac{G}{G/(1 + GH)}\right] \tag{12-70}$$

or

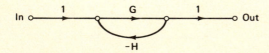

FIG. 12.28   Single-loop system.

$$S_H{}^T = \frac{-GH}{1 + GH} \qquad\qquad (12\text{-}71)^*$$

Thus, the feedback F around an element affects the sensitivity, but $S_x{}^T$ equals $1/F_x$ only if x is an element with no leakage around it (no signal reaches the output when $x = 0$).

Positive and negative feedback

In many texts we find the terms, "positive feedback," and "negative feedback," often used rather loosely. Originally the terms were used to differentiate between the two configurations of Fig. 12.29: the comparator system called negative feedback, the adder system called positive feedback. The assumption is made that G and H are both positive constants.

Actually, in essentially all of the feedback systems of interest, G at least is a transfer function in s rather than a constant. Consequently, the product GH possesses a phase angle which varies with frequency. In either system, there are likely to be frequencies at which $\underline{/GH}$ is $0°$, other frequencies where it is $180°$. When system (a) has $\underline{/GH} = 180°$, it behaves like system (b) with $\underline{/GH} = 0°$.

In other words, whether a comparator or an adder is used (or whether H is positive or negative) does not really distinguish the two systems. In an attempt to preserve the concept of positive and negative feedback, other authors have distinguished on the basis of the magnitude of the return difference. If $|1 + GH|$ in (a) or $|1 - GH|$ in (b) is less than unity, the total gain is greater than $|G|$ and the system has positive feedback. If $|F|$ is greater than unity, the feedback is called negative. In an actual system, this definition is likely to lead to frequency bands where feedback is alternately negative and positive.

The principal purpose of this brief discussion is to emphasize that the terms positive and negative feedback are of very limited value, or, at least, they must be interpreted

*With a large amount of feedback ($1 + GH$ or GH very large), the sensitivity of T with respect to H is nearly -1. The feedback really transfers the system sensitivity from G to H. But this is exactly what we usually want. G is an element which must provide the power and force to move the output; hence its parameters are likely to vary widely. H (the sensor) need drive only the comparator and can usually be a passive device with carefully controlled parameters.

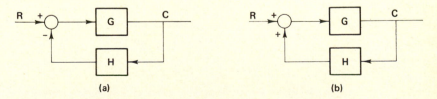

(a)   (b)

FIG. 12.29 Two systems, one with a comparator and the second with an adder.

(a) $T = \dfrac{G}{1 + GH}$, (b) $T = \dfrac{G}{1 - GH}$.

with care. In simple systems (such as the single-loop systems above), the important quantity is the feedback or return difference (unity minus the loop gain). This quantity determines sensitivity, disturbance rejection, and the change in gain—all resulting from the use of feedback. In more complex systems, the overall system function, sensitivity, and disturbance rejection have to be evaluated for the particular system configuration.

12.8 DYNAMICS CONTROL

Feedback can also be used to realize a desired dynamic characteristic while using a device with entirely different dynamics. A simple example is the use of feedback to change the time constant of an instrument.

The field of instrumentation provides many important examples of the use of feedback. Here we consider only one: the measurement of temperature, involving the generation of an electric voltage proportional to the instantaneous temperature within an enclosure.

The basic nonfeedback system is depicted in Fig. 12.30. The temperature-sensitive element is a thermistor, with the model, shown in part (b) of the figure, including a voltage source proportional to the change in temperature from a quiescent or normal operating value. In series with this voltage source there is a second voltage source dependent on the thermistor current, since an increase of the current causes an increase in temperature as a result of the power dissipated in this resistive element. The

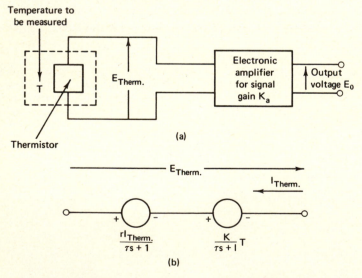

FIG. 12.30 Temperature instrument. (a) Measuring system; (b) model for the thermistor (T temperature; r, K positive constants).

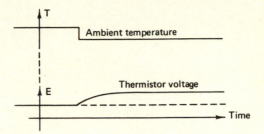

FIG. 12.31 Change of thermistor voltage with temperature.

voltage source is related to the temperature by the transfer function $-K/(\tau s + 1)$, where τ is the thermal time constant. In other words, when the ambient temperature changes abruptly in the oven or container within which the thermistor is located, the terminal voltage approaches exponentially the new value as depicted in Fig. 12.31.

If the amplifier of Fig. 12.30(a) draws no input current, the input voltage equals the voltage of the source in the thermistor model; consequently the output E_0 is related to the temperature by the overall transfer function

$$\frac{E_0}{T} = -\frac{K_a K}{\tau s + 1} \tag{12-72}$$

Since the thermal time constant is typically of the order of magnitude of two seconds (or greater), the full reading of the output voltage is not approached until about eight seconds (four time constants) after a change in termperature. In an alternative viewpoint, we can say that the system does not respond to temperature variations at frequencies greater than 1/2 radian/sec (the reciprocal of the time constant), or about 1/12 Hz.

In order to speed up the response of our instrument, we can pass the output voltage E_0 of Fig. 12.30 through an electrical network, as depicted in Fig. 12.32. If this compensation network possesses the transfer function $(\tau s + 1)/(\tau_c s + 1)$, the overall instrument transfer function is

$$\frac{E_{\text{out}}}{T} = -\frac{KK_a}{\tau s + 1}\frac{\tau s + 1}{\tau_c s + 1} = \frac{KK_a}{\tau_c s + 1} \tag{12-73}$$

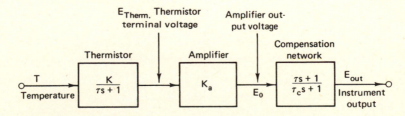

FIG. 12.32 Improved instrument.

In other words, the thermal time constant τ is replaced by the electrical time constant τ_c. If τ_c is made 1/100th of τ, we speed up the instrument response by the factor of 100; in other words, the instrument responds to signals at frequencies 100 times greater.

We seldom achieve such significant results without an associated penalty; in the present case, the penalty is apparent if we consider the realization of the compensation network required in Fig. 12.32. Search for a suitable network yields the system drawn in Fig. 12.33; in order to effect a 100:1 decrease in the instrument time constant, it is necessary to include an amplifier of gain 100. The greater the improvement demanded in dynamic response, the larger the gain required.

The instrument of Fig. 12.32 is quite satisfactory for many purposes, but the modern systems engineer (armed with the tools of feedback theory) naturally asks, if, since an additional gain of τ/τ_c is required, better results might not be achieved through the use of feedback. A principal disadvantage of the above instrument is that any changes in the gain of either amplifier result in corresponding (and equal) changes in the overall instrument transfer function. Hence, as amplifiers age or environmental conditions vary, the instrument requires continual recalibration. If feedback is used, not only can the speed of response be controlled—but also the recalibration can be made automatic and self-contained.

An appropriate feedback configuration is shown in Fig. 12.34, where a current proportional to the output voltage is passed through the thermistor. Under these conditions, the terminal voltage of the thermistor is determined from Fig. 12.30(b) as

$$E_{\text{therm}} = \frac{-K}{\tau s + 1} T - \frac{r\beta}{\tau s + 1} E_0 \tag{12-74}$$

and the system is described by the block diagrams of Fig. 12.35 [the second diagram is derived from the first if we recognize that addition after multiplication of each variable by $1/(\tau s + 1)$ is equivalent to addition followed by multiplication of the sum].

The dynamic characteristics of the feedback instrument are given by the overall transmittance:

$$\frac{E_0}{T} = (-K) \frac{K_a/(\tau s + 1)}{1 + r\beta K_a/(\tau s + 1)} \tag{12-75}$$

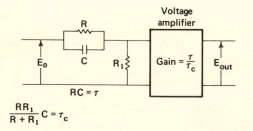

FIG. 12.33

$$\frac{E_{\text{out}}}{E_0} = \frac{\tau s + 1}{\tau_c s + 1} \cdot$$

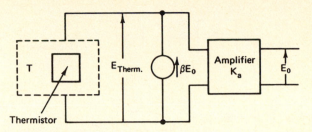

FIG. 12.34 Instrument with feedback.

or, if we clear of fractions,

$$\frac{E_0}{T} = \frac{-KK_a}{\tau s + 1 + r\beta K_a} \qquad (12\text{-}76)$$

This equation can be placed in somewhat more convenient form if we divide numerator and denominator by $(1 + r\beta K_a)$

$$\frac{E_0}{T} = \frac{-KK_a/(1 + r\beta K_a)}{[\tau/(1 + \beta K_a)]\,s + 1} \qquad (12\text{-}77)$$

The above equation reveals two important facts:

(1) The time constant of the feedback system is $\tau/(1 + r\beta K_a)$. In other words, if we wish to decrease the time constant by a factor of 100, we need only select β so that

$$1 + r\beta K_a = 100 \qquad (12\text{-}78)$$

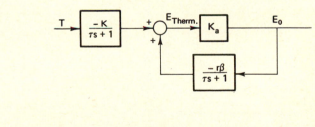

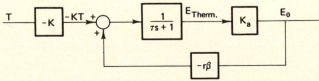

FIG. 12.35 Equivalent block diagrams for the feedback instrument.

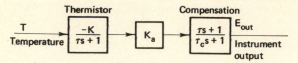

Tandem compensation

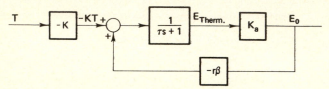

Feedback compensation

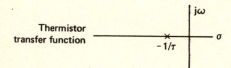

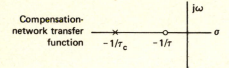

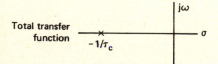

Behavior of tandem compensation

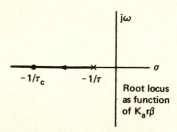

Behavior of feedback compensation

FIG. 12.36 Alternative ways of compensating the thermistor instrument.

Thus, feedback can be used to control the time constant, just as we did with the tandem compensation network.

(2) As a result of the feedback, the gain is reduced by division by $(1 + r\beta K_a)$; in other words, the gain is reduced in exactly the same ratio as the time constant. This result is rather startling, since it is identical with our earlier conclusions about the tandem compensation scheme. There, too, if we wished to decrease the time constant by a factor of 100, we had to insert an amplifer with a gain of 100 to restore the level of the output signal.

The surprising similarity between the two instruments ends at this point, however. In the tandem scheme, a 1 percent change in amplifier gain results in 1 percent change in output: that is, $S_{K_a}{}^{E_0/T} = 1$. In the feedback configuration of Fig. 12.34, however,

$$S_{K_a}{}^{E_0/T} = \frac{1}{1 + r\beta K_a/(\tau s + 1)} = \frac{\tau s + 1}{\tau s + 1 + r\beta K_a} \tag{12-79}$$

At zero frequency (that is, in steady state conditions),

$$S_{K_a}{}^{E_0/T} = \frac{1}{1 + r\beta K_a} \left(= \frac{1}{100} \text{ for our numerical example} \right) \tag{12-80}$$

The sensitivity of the feedback system is reduced by the *same* factor as the time constant and gain! The feedback instrument contains automatic self-calibration!

Thus, the feedback system possesses all of the engineering-design flexibility of the tandem compensation scheme, as well as the possibility of design control over the sensitivity of the instrument transfer function to changes in system parameters. Because of this added significant advantage, feedback systems are used extensively in instrumentation. The only penalty we must pay for this sensitivity control is the extra equipment (often trivial) to realize the feedback path—in Fig. 12.34, the current generator with the value of current proportional to the output voltage.

If we refer to the two schemes for reducing the instrument time constant, we can summarize the possibilities shown in Fig. 12.36. With tandem compensation, the designer just cancels the long time-constant pole and reinserts a desired pole at $-1/\tau_c$. With feedback, the root locus plot shows that the pole is drawn out along the negative real axis to its desired location. Conceptually, the two types of compensation are quite different.

The figure suggests an important advantage of feedback compensation: we do not need to know precisely the position of the original pole; indeed, this pole $(-1/\tau)$ can change significantly without any deterioration of system performance when feedback is used.

A simple calculation illustrates this advantage of the feedback approach. In designing, we believe that the transistor time constant is 1 sec and we want an improvement of 100. Then the compensation-network transfer function is, with the amplifier gain

$$100 \, \frac{s + 1}{s + 100}$$

If the actual thermal time constant is 2 sec, the total instrument transfer function is proportional to

$$100 \, \frac{s + 1}{(s + 1/2)(s + 100)}$$

In the feedback approach to the same compensation problem, $1 + r\beta K_a$ is 100 and the total system is described by a single pole at $(-1/2)$ 100 or -50:

$$100 \, \frac{1}{s + 50}$$

The step response in the two cases is shown in Fig. 12.37.

This example certainly represents an extreme deviation of the thermal time constant from its expected value, but the response curves illustrate that the cancellation approach places a high premium on the precision with which we know r. If we miss cancellation, there is a component of the response which is a simple exponential of time constant r.

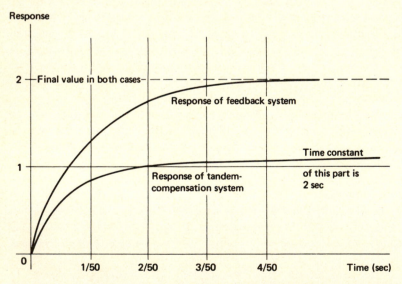

FIG. 12.37 Response of each compensation scheme when thermal time constant is 2 sec rather than the design value of 1 sec.

Thus, feedback provides a mechanism for controlling system dynamics even when we know only approximately the dynamic characteristics of the components to be used.

12.9   RADAR TRACKING SYSTEM

In spite of emphasis in the preceding sections on relatively simple system configurations (one loop or at most two), many real-life, feedback systems involve a more complex interconnection of components. The detailed study of multiloop systems and of systems with more than one input and output must be left to texts devoted solely to feedback. In this section and the next, however, we want to conclude this chapter with brief descriptions of two systems which display some of the potentialities of feedback.

Figure 12.38 is the basic configuration for antenna position control in a radar tracking system. The system can be broadly divided into two parts: the radar equipment and the servo elements.* The communications engineer designs the radar (and the antenna); a description of the control engineer's task follows. The radar is trying to track or follow a particular target. The beam radiated from the antenna generates an echo with characteristics which depend on the error between the antenna direction (the center of the beam) and the target direction.† The "radar output" is a signal

$$K_r(\theta_t - \theta_a)$$

which measures the extent to which the antenna direction is in error. Actually there are usually two systems, one for azimuth angle and one for elevation angle, so that the antenna is driven in both azimuth (angle horizontally or along the horizon) and elevation (angle up and down from the horizontal or horizon). Here we consider only the azimuth system.

This radar output is amplified and modified by the controller transfer function $G_c(s)$. The signal then drives the motor which, through the gear train, moves the antenna—hopefully in a direction to reduce the error. In many installations, the antenna may be an enormous "dish," for example 80 feet in diameter. There are tremendous horsepower requirements on the motors, and the antenna must track the target within an accuracy of a small fraction of a degree. (An error of $1°$ at 5.7 miles corresponds to a distance of 1/10 mile—insufficient accuracy for air traffic control around a busy airport.)

How does the control engineer design such a system (that is, select the servo elements in Fig. 12.38)? Normally the antenna has already been selected when the

---

*A feedback system to control position is a servomechanism, often abbreviated servo.

†The antenna may, for example, be vibrated rapidly back and forth with the echo strength measured under various antenna directions. Alternatively, the radar can use the changes in the radiated beam as the echo is received from off-center directions. In any case, the radar output is proportional to the error between target and antenna directions.

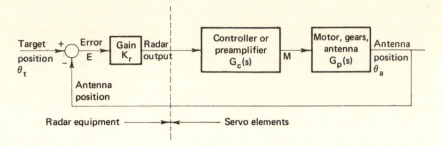

FIG. 12.38 Simplest radar tracking configuration (azimuth system only is shown).

control engineer enters the picture. His first task is to select motor and gears that can drive the antenna of known inertia at the speeds required for system performance. (These depend on anticipated target speeds and maneuvers, as well as desired tracking accuracy.) Choice of this high-power portion of the system is also strongly influenced by cost, size, reliability, expected life, and maintainability.

Once the power elements are chosen, all of Fig. 12.38 is specified except the preamplifier or controller $G_c(s)$. In a sense, the engineer must select this controller to compensate for all the limitations and shortcomings of the other system components. This block, operating at a low-power level, is the heart of the feedback system; in this block the signal processing can be done to shape the dynamic characteristics of the total system.

Before launching the design of $G_c(s)$, the engineer should consider those configuration changes that are possible and might be desirable. For example, the power elements $G_p(s)$ are often nonlinear and certainly the parameters can be expected to vary significantly over the range of operating conditions. Such considerations (and our earlier discussions of the advantages of feedback) suggest

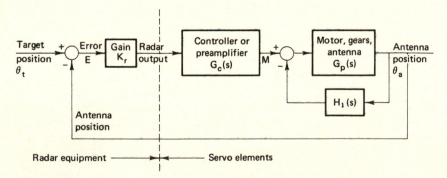

FIG. 12.39 Feedback used around $G_p(s)$ to improve the performance of the high-power portion of the system.

adding local feedback around just the power elements (Fig. 12.39), in order to improve the performance of the θ_a/M portion of the system before adding the controller G_c.

As soon as $H_1(s)$ is added to make up the system of Fig. 12.39, we can proceed to design $G_c(s)$ to provide a suitable overall system. Exactly this was done for many years. Then in the late 1950s, several imaginative engineers looked at this configuration and asked: What do we really want the system to do?

There are two answers:

(1) The system must drive the antenna to follow the target with sufficient accuracy to be sure we always receive an echo.

(2) The system must have an output which is a measure of target position (for example, to provide data to air traffic controllers, or to furnish data to a computer which predicts future target position).

In the early systems, these two functions were combined: the antenna position θ_a was used as the measurement of target position.

If the two functions are separated, we can drive the antenna with only enough tracking accuracy to ensure that the echo always appears. It is not necessary to make the antenna position a precise replica of target position. Instead, looking at the block diagram of Fig. 12.40, we ask: Where in the system to the right of the dashed line can we measure target position θ_t?

There are two possibilities. First, the signal $K_r E$ can be measured (at the input of the controller G_c). If we know K_r, we can then generate the signal E. When E is added to θ_a, the target position θ_t is found (Fig. 12.41). The accuracy of the measured θ_t depends primarily on the accuracy with which we can estimate K_r. (Unfortunately, K_r tends to vary somewhat as the echo strength changes.)

A second scheme for accomplishing essentially the same measurement of θ_t is shown in Fig. 12.42. To see how this system works, we consider the feedback from θ_a to the signal N. There are two feedback paths: one through the input comparator and K_r, the other through H_1 and H_2. The two transfer functions from θ_a to N are

$$-K_r \quad \text{and} \quad +H_1 H_2$$

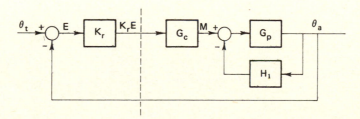

FIG. 12.40   System of Fig. 12.39 redrawn.

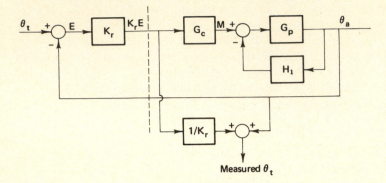

FIG. 12.41   Feedback control system supplemented by a system to measure θ_t.

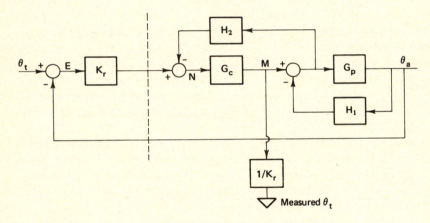

FIG. 12.42   A three-loop feedback configuration for radar tracking.

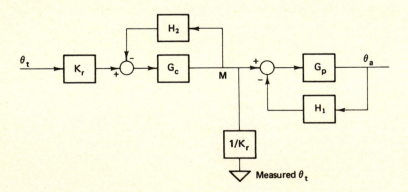

FIG. 12.43   System equivalent to Fig. 12.42 when

$$H_2 = \frac{K_r}{H_1}.$$

If H_2 is chosen so that

$$H_2 = \frac{K_r}{H_1} \tag{12-81}$$

the two paths cancel; N is independent of θ_a and depends only on θ_t.

In order to study this operation in more detail, we can use Mason's theorem to find both M/θ_t and θ_a/θ_t, subject to the choice of H_2 given above. Then

$$\frac{M}{\theta_t} = \frac{K_r G_c}{1 + G_c H_2} \tag{12-82}$$

$$\frac{\theta_a}{\theta_t} = \frac{K_r G_c}{1 + G_c H_2} \frac{G_p}{1 + G_p H_1} \tag{12-83}$$

In other words, the three-loop system of Fig. 12.42 really behaves as the two simpler tandem systems of Fig. 12.43—as long as $H_2 = K_r/H_1$. M can be used as a measure of θ_t, since the $G_c - H_2$ loop is a low-power system which can be built with large bandwidth (to give an output M which is an accurate replica of the input $K_r \theta_t$). The high-power, low-accuracy part of the system is the $G_p - H_1$ loop.

The three-loop system of Fig. 12.42 has one important advantage. By introducing the H_2 block, we have split the one complex system into the tandem combination of two simpler systems. Each loop of Fig. 12.43 can now be designed without regard for the other.*

The example of this section, hopefully, indicates some of the flexibility which the control system designer enjoys as he develops the system configuration.

12.10 HOMEOSTASIS†

Most of the feedback systems discussed in the preceding sections are strikingly simple. The actual system output is compared with the desired value of the output; the error is used to drive the system in a way that changes the output toward its desired value.

There are several difficulties inherent in such a simple system: (1) If the output is to be changed rapidly, large error signals are required. (2) Of more importance, if any component fails, the entire system becomes inoperative. (3) Furthermore, each part of the system is used for only one function. If we are considering a very complex control problem with many different objectives, we would like each part to contribute to a variety of different functions.

---

*Ultimately, we have to worry about what happens when K_r changes and $H_2 \neq K_r/H_1$. Then the total system of Fig. 12.42 must be considered: a system so complex that usually simulation studies are necessary, rather than theoretical analysis.

†This section was originally written by the author for "The Man Made World," preliminary edition, vol. 1, McGraw-Hill Book Company, 1969.

These points can be illustrated by the design of a modern airplane, in which difficulty (3) is at least partially avoided. Here there are many different functions for the electronic equipment: navigation or determination of present location; control of the rudder, ailerons, and elevator which are used to adjust the position and velocity of the plane; control of the engines; control of the temperature and humidity within the cabin; and so forth. In a military plane, other functions are added, such as aiming and firing of guns and missiles; detection of enemy aircraft; control of cameras used to photograph enemy territory; and release of bombs.

In many of these functions, we would like to use a digital computer to carry out the required calculations. It is impractical, however, for the plane to include a separate computer for each task. Consequently, a single computer is used which automatically switches from one job to the next. For example, in a small fraction of a second, calculations are made for navigation; then the computer calculates the signals which should be applied to the rudder; and so on. One part of the system performs many different functions.

Such a system becomes more interesting than the simple examples we have considered in this chapter. Even in a more complex system such as this, however, each function is clearly defined, and we are simply dividing up the time of the computer. Consequently, difficulty (2) above still exists; if the computer becomes inoperative, all systems fail.* Much greater complexity and elegance are manifest if we consider some of the feedback systems that have evolved in the human body for the control of important signals, such as blood pressure or temperature. While scientists are just beginning to understand the ways these systems operate, it is apparent that many of the physiological feedback systems possess performance capabilities which are awesome when compared with the control systems man builds for automobiles, airplanes, and industrial plants.

Homeostasis defined

One of the intriguing characteristics of several feedback systems in the human body is *homeostasis*. This term refers to the ability of the overall feedback system to maintain one or more important signals very close to the desired values.

For example, the system for the control of body core temperature is a homeostatic system. Several different subsystems (shivering, metabolism, blood flow to the skin, and so forth) combine in their effects to keep the core temperature within $1°$ to $2°$ of the normal value of $98.6°$F, even though the surrounding temperature may change by $100°$ and the human being may engage in vigorous exercise. In such a homeostatic system, subsidiary signals (for example, the signals driving the muscles into shivering

---

*To minimize the chance of such a catastrophe, the designer may use two computers. When one becomes inoperative, the other is automatically switched into the system. In designing a military plane, there is an interesting question: Is it better to use such extra equipment for improved reliability, or to use only one set of equipment and thereby achieve greater speed and maneuverability through reduced weight? The U.S. policy has often been toward reliability, while the British have favored emphasis on performance in combat.

action) may change by large amounts; the signal which is the primary concern of the control remains remarkably constant.

Such homeostatic behavior is found in a wide variety of the feedback systems within the human body.* In these cases, the impressive system performance is achieved by interplay of several different subsystems—there are several alternate or parallel means in which control is achieved. Furthermore, the typical component or part of any one subsystem is used simultaneously for many different functions. For example, the muscles used for shivering are also used by the human being for motion and work. Thus, entirely different control systems of the human body are interrelated by the sharing of particular components; in this way, the body realizes its remarkable efficiency and compactness.

Effects of hemorrhage. Significant characteristics of a homeostatic system can be illustrated by consideration of some of the reactions of the human body to a sudden loss of blood. After a hemorrhage and the abrupt drop in blood volume, the blood pressure falls. Very rapidly, parts of the circulatory system constrict in most of the body in a response designed to accommodate the volume of the circulatory system to the available volume of blood (and hence to maintain pressure). The heart and brain vessels are not constricted, so that flow is maintained there, even when blood flow through the kidneys or the extremities may be reduced to almost zero. Concurrently (within a minute after the hemorrhage), the heart rate rises rapidly in order to maintain flow—for example, from the normal 70 beats per minute to as high as 200.

Other physical changes occur somewhat more slowly. In the hours following the hemorrhage, after contraction of blood vessels, large quantities of fluid are absorbed from the intestines and the body as a whole in order to restore blood volume toward normal (even though the chemical constitution of "blood" may temporarily be very abnormal). The patient's thirst and appetite for salt rise significantly in order to augment the body's fluids.

These changes are paralleled by activity of the pituitary and adrenal glands. The pituitary gland† secretes vasopressin, a hormone which limits loss of fluid from the body and allows the kidneys to retain water and salt, and the hormone ACTH, which stimulates the adrenal gland‡ to secrete cortisol. Cortisol simplifies the flow of blood into the veins from the capillaries and restricts the blood into the capillaries from the arteries; it also increases the protein concentration in the blood. As a result of the increased level of cortisol in the blood plasma, blood is formed to restore the volume toward normal level.

*Indeed, the term homeostasis was first used by the physiologist, Walter Cannon, in his 1929 study of control within the human body.

†The pituitary gland, about one centimeter in diameter and located at the base of the brain, secretes hormones influencing such factors as growth, metabolic rate, and activity of the thyroid gland, which in turn controls the rate of chemical reactions throughout the body.

‡The adrenal glands, located at the kidneys, secrete hormones which control the amounts of sodium, chloride, and potassium in the body fluids and important chemical properties of the blood.

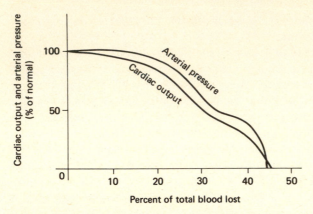

FIG. 12.44 Effect of hemorrhage (A. C. Guyton, "Textbook of Medical Physiology," W. B. Saunders Company, 4th ed., 1971, p. 236).

Thus, the response of the human body to hemorrhage results from a wide variety of different subsystems, all operating in parallel and combining their effects to maintain blood pressure and adequate flow to the heart and brain. As a result of these combined reactions, the human body is able to recover from major losses of blood. Figure 12.44 shows that the loss of 10 percent of the blood yields no particular effect on pressure or cardiac output (the flow of blood from the heart). As the size of the hemorrhage is increased, flow starts to fall, but pressure is maintained. When approximately 40 percent of the blood is lost, both flow and pressure drop abruptly to zero: the system tends to become unstable (or lose control) if more than about one-third of the blood volume is lost.

Control of cortisol secretion. The complexity of the system is illustrated by consideration of just a portion—in particular, the elements that control secretion of cortisol by the adrenal gland. The elements of this system, which are of primary concern, are shown in Fig. 12.45. The construction of a total model requires determination of the various ways in which the hemorrhage signal influences the pituitary gland, which in turn stimulates cortisol secretion by the adrenal glands.

The basic feedback system is shown in Fig. 12.46. The stimuli from the nervous system travel to the recticular activity system of the brain (the portion that controls

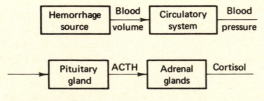

FIG. 12.45 Start of model.

sleep, coma, and consciousness). Here, a desired value is determined for the level of cortisol (the desired value rises sharply after hemorrhage). The actual level is compared with this desired value in the hypothalamus portion of the brain, and a corresponding error signal is sent to the pituitary gland. In response, this gland secretes ACTH, which stimulates the adrenal gland to secrete cortisol. The nervous system responds in fractions of a second, but the pituitary and adrenal glands respond more slowly—in a matter of several minutes. Until the early 1960s, the model of Fig. 12.46 was used to describe the process.

A series of experiments in the last six years have indicated the inadequacy of this model. The model indicates, for example, that the action of the control system can be blocked by administering drugs which essentially saturate the feedback element measuring cortisol level, and by deactivating the hypothalamus control. When experiments were performed this way, medical investigators found an alternate path through the kidney, included in Fig. 12.47. The kidney releases the chemical, renin, which stimulates the pituitary gland. Furthermore, experiments demonstrated that the kidney responded both to blood pressure directly, and to the measurement of blood pressure by the primary pressure sensors in the chest and head—indeed, the latter signal is the dominant input to the kidney.

The model of Fig. 12.47 suggested further experiments on dogs, where the feedback was blocked and the kidneys removed. These experiments showed that there is a path directly into the hypothalamus from the pressure sensors—the high threshold path shown in Fig. 12.48. Indeed, this path can be blocked by administering the drug Iproniazid. If it is not blocked, there is a direct effect of low pressure on the pituitary, which is not related to the feedback by measured cortisol level. This model gives results which correspond with a wide variety of experiments performed during the last few years.*

Comments on the model. The model of Fig. 12.48 shows vividly the way in which homeostasis is realized by a variety of feedback subsystems. The various paths from the hemorrhage input signal to the pituitary gland permit the system to respond rapidly and easily to small changes in blood volume (the low threshold path), and to utilize additional channels (the moderate and high threshold paths) in the case of large losses of blood. The model shown is, as we noted earlier, only a portion of the entire system that represents the response of the body to hemorrhage.

Comments on the modeling procedure. The procedure by which the model has been derived over the past few years is, in itself, significant. From the early scientific

---

*Donald S. Gann, Systems Analysis in the Study of Homeostasis, with Special Reference to Cortisol Secretion, *Amer. J. Surg.*, vol. 114, no. 1, pp. 95-102, July, 1967. Gann also reports that, when all links shown between the hemorrhage and the pituitary gland are blocked chemically and surgically, there is no response for 15 minutes after hemorrhage. After 30 minutes, there is a slow increase in cortisol secretion—an observation which would indicate an additional path in the block diagram with long time delay.

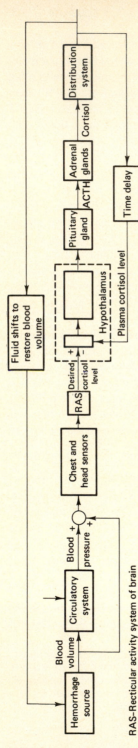

RAS–Recticular activity system of brain

FIG. 12.46 Model showing feedback comparison of desired and measured values of cortisol level.

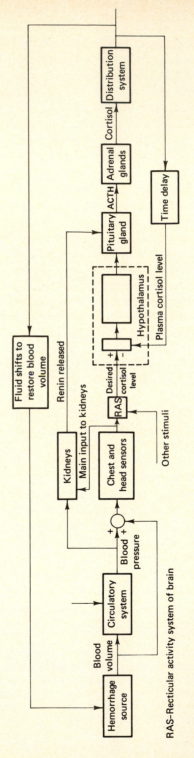

RAS–Recticular activity system of brain

FIG. 12.47 Model including path through kidney.

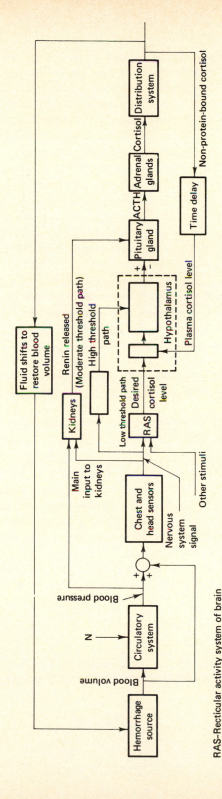

RAS—Recticular activity system of brain

N—norepinephrine (from adrenal gland to hold blood pressure up when volume drops)

FIG. 12.48 Complete model for control of cortisol secretion after hemorrhage.

knowledge, a crude model (Fig. 12.46) was developed. This model suggested experiments; the result of the experiments demanded modification of the model. This evolutionary process, a feedback mechanism in its own right, led ultimately to the model of Fig. 12.48, representing the current level of understanding of the processes through which cortisol secretion is controlled.

Thus, the model is used not only as a lucid display of knowledge, but also as a basis for the formulation of additional experiments designed to augment that knowledge. The construction of the model is an integral part of the overall search for understanding of the way the system operates.

Use of the completed model. When a reasonably complete model is developed, the physician can use it to help determine which of the subsystems or particular parts is malfunctioning when the overall homeostatic system fails to perform. Medical diagnosis and treatment are severely limited by the doctor's difficulty in determining the exact location or source of trouble. Even the relatively simple block diagram of Fig. 12.48 illustrates the complexity of this problem, particularly when we recognize that each component has many different functions within the body,* and that improper functioning of any portion is usually partially compensated by an automatic change in the operation of other components.

Finally, fuller understanding of such models and of the remarkable performance capabilities of human feedback systems provides the scientist and engineer with a forecast of the potentialities of man-made feedback systems.

PROBLEMS

12.1   In Sec. 12.1, we discuss briefly the feedback system utilized when a man picks up a pencil lying on a table top. Construct the complete block diagram for this system—the diagram including not only the visual feedback, but also the feedback through the neuromuscular system which indicates to the man where his hand is and also the tactile feedback. Write a brief description of system operation to explain the sequence of events when the visual feedback is interrupted for long periods, when the pencil is moving in a regular or a random way, and when the man himself is in motion with regard to the table.

12.2   In the feedback system shown, the sensitivity of $T = Y/X$ to changes in G is

$$S_G{}^T = \frac{1}{1 + GH}$$

(a) What is the sensitivity of T to changes in K?

(b) What is the sensitivity of T to changes in H?

*The same system may be used for different functions; the system here, for example, is also used to control the level of blood sugar.

(c) If $|GH| = 75$ and $|G|$ decreases by 2 percent, what is the corresponding change in $|T|$?

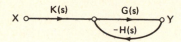

12.3 In the feedback system shown, determine the sensitivity of T with respect to K_1 when $K_2 = 6$ and $K_3 = 1$. Repeat with respect to K_2 when $K_1 = 1$ and $K_3 = 1$. Why do we often want to keep the sensitivity with respect to K_3 small, while we are much less concerned about the sensitivity with respect to K_2?

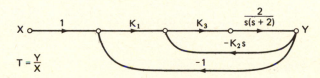

12.4 For the system shown, determine S_K^T, where $T = Y/X$ and

$$G(s) = \frac{3}{s(s + 1)^2}$$

Find the resulting sensitivity function as the ratio of polynomials in s. Sketch the magnitude of this sensitivity (in dB) as a function of frequency (i.e., in the form of a Bode plot), with K equal to 1/8.

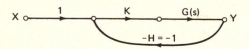

12.5 Repeat Prob. 12.4 for the sensitivity of T with respect to H, and with $K = 1/8$.

12.6 For the series-resonant RLC circuit, determine the sensitivities of the resonant frequency, Q and the bandwidth with respect to R, L, and C (i.e., nine sensitivity functions in all). The interesting case is when the Q is high, so that the resonance is pronounced.

How would the corresponding sensitivities for the parallel resonant circuit differ from these answers?

12.7 (a) Design a feedback structure which realizes (using amplifiers, resistors, and capacitors) a transfer function which exhibits resonance at 200 rad/sec with a bandwidth of 10 rad/sec.

(b) Sketch the gain and phase characteristics of your system from 180 to 220 rad/sec.

(c) An alternate realization of the same resonance phenomenon is possible with a single-loop feedback structure as shown. Here K is a constant gain, and the transfer function $H(s)$ is to be realizable by an RC network (this means H can have poles only on the negative real axis and the poles must be simple; $s = 0$ or infinity is not permitted). Determine an appropriate $H(s)$.

Any comparison of the two schemes for realizing resonance with an active RC system must take into account the sensitivity of the system bandwidth and resonant frequency to parameter values.

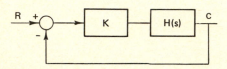

12.8   In the single-loop system shown,

$$G(s) = \frac{Ks}{(s + 1)(s + 2)}$$

(a) Sketch the way the poles of $T = Y/X$ vary as K varies from 0 to very large values and also as K varies from 0 to very large negative values.

(b) Select a K which yields a resonant circuit with a Q of 100. What is the resonant frequency?

(c) Find a $G(s)$ which yields the same Q but a resonant frequency of 40 rad/sec.

(d) Realize the system of (c) with op-amp circuits.

(e) Repeat the design with

$$G(s) = \frac{K}{(s + 1)^3}$$

Which would be the preferable system to build?

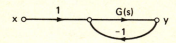

12.9   A simple feedback system with two inputs, x and u, and one output, y, is shown.

(a) Draw a signal flow diagram for the block diagram.

(b) Find the transfer functions H_1 and H_2 in the relation

$$Y = H_1 X + H_2 U$$

(c) If the loop gain, BC, is much greater than unity, show that H_1 is independent of changes in the plant B. .

(d) Is H_2 also independent of B under the circumstances of (c)? Discuss.

(e) Is H_1 independent of A if the loop gain is large? Is H_2?

(f) Is H_1 independent of C if the loop gain is large?

(g) The claim is made that feedback causes the system performance to be independent of plant variations and load disturbances. Discuss.

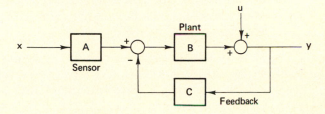

12.10 A notch filter has the transfer function

$$H(s) = \frac{s^2 + 100}{(s + 5)(s + 20)}$$

This system is used in the feedback configuration shown. Sketch the gain characteristic for H, then the same characteristic for Y/X.

The notch filter can be realized by an RC network. The circuit without feedback is used to delete one unwanted frequency component (e.g., the line frequency or the wobble frequency of the antenna in a radar system). The circuit with feedback is used to accentuate or pull out a single frequency component. For example, in locating leaks in an underground gas line, a microphone is pulled through the line and the noise signal spectrum is measured by varying continuously the notch frequency. This variation can be achieved by making the 100 factor in the numerator of H proportional to the gain of a variable-gain amplifier.

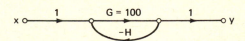

12.11 The basic mechanical accelerometer is shown in Fig. 4.9 and described in Sec. 4.2. As mentioned there, practical accelerometers for reasonable accuracy require certain refinements (and use of feedback).

(a) When the device is used in a vertical position (to measure vertical acceleration), the acceleration of gravity gives an output with no other input signal. What compromise is necessary in the design as a result?

(b) In order to overcome this difficulty of (a), we often float the sensitive mass in a fluid of specific gravity approximately the same as the mass element. How does such a system work?

(c) It is frequently desirable to improve the linearity and accuracy of the springs by using an electromagnet to provide the spring restoring force. Describe an appropriate feedback system.

(d) The crude pickoff of Fig. 4.9 is usually replaced by an electromagnetic pickoff, in which the mass element is magnetic and moves into a coil or a pair of coils to change the flux density for a given current. What would be the advantages of such a system?

The device described in this problem is essentially the accelerometer used in inertial navigation systems (the accelerometer is mounted on a gyro table, which is kept fixed in space orientation). In an inertial navigation system for a submarine, the ship must surface every few weeks to make stellar or sun fixes. Why?

12.12　If we wish to focus attention on a particular parameter K in a feedback system, we can redraw the block or signal flow diagram in the form shown in the figure: K is alone with feedback around it; there is a *leakage* transmission T_o describing the transfer function when $K = 0$. (This form is valid whenever the overall transfer function is a bilinear transformation in K—a condition whenever K is a normal circuit parameter.)

This canonic form of the system diagram permits us to prove the validity of the expression

$$S_K{}^T = \frac{1}{F}\left(1 - \frac{T_o}{T}\right)$$

where F is the return difference with respect to K—i.e., F is unity minus the loop gain through K. Prove this expression for the sensitivity and illustrate the use of the expression with the Problems 12.2 or 12.3.

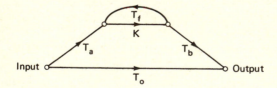

12.13　One of the most widely used control systems is that of a remotely controlled garage door opener. The basic system is shown in the diagram, with the various components described as follows:

Operational amplifier: a linear device described by

$$V_{\text{out}} = 4V_1 - 2V_2$$

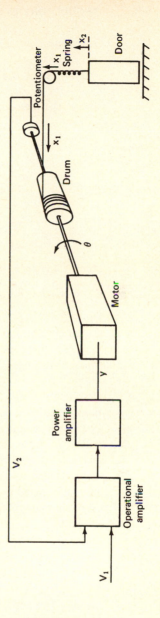

Power amplifier: a linear device with a unit-step response

$$y(t) = 0.4 - 0.4e^{-5t}$$

Motor: the speed-torque characteristic is shown in the diagram.

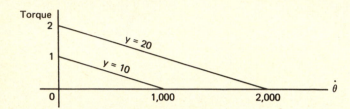

Potentiometer: a linear device with

$$V_2 = K_p \theta \quad \text{for all } \theta$$

Drum: with a radius of R. The moment of inertia of the combination of the drum and motor armature is J.

Spring: a linear spring with spring constant K.

Door: with a mass M and moving vertically (so gravity must be included).

Develop a signal flow diagram for the complete system. Show clearly all transfer functions and variables.

12.14 The block diagram shows a common antenna control system for a tracking radar. G_m represents the motor and the load, G_a the preamplifier, H_1 the feedback around the motor to linearize its performance, H_2 a compensation network, and everything to the left of the dashed line the radar which gives an output signal proportion to the error.

(a) Determine C/R

(b) If $H_1 H_2 = K$, show that the system is equivalent to two simpler systems in tandem. Show each of these systems in a block diagram.

(c) The $G_a - H_2$ loop can be wide-bandwidth and fast-response since it delivers essentially no power; the $G_m - H_1$ loop is inherently sluggish. Explain why z is the appropriate signal to use for computer prediction of the future target position, rather than c.

(d) What is the sensitivity of the C/R transfer function to changes in G_m (assume $H_1 H_2 = K$ still)?

(e) What is the sensitivity of Z/R to G_m, again with $H_1 H_2 = K$?

(f) As the target fades, the gain K tends to change and it is necessary to change H_2 automatically to ensure $H_1 H_2 = K$ at all times. If H_1 is a constant independent of frequency, and if K varies with e as shown in the figure, how might we realize the desired H_2?

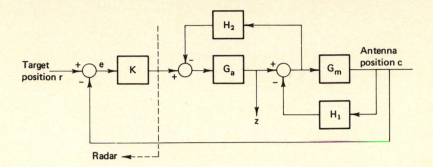

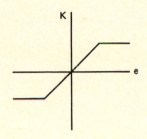

13 SAMPLING

Sampling is the process by which a signal varying continuously with time is measured or represented by its values at isolated, separated instants. The smooth curve of the signal $y(t)$ versus time is replaced by a sequence of values

$$\{y_n\} = y_0, y_1, y_2, \ldots$$

In this chapter, we will see that if we sample often enough, the *sampled signal* $\{y_n\}$ is entirely equivalent to the *analog signal* $y(t)$. If we want to determine the response of a system to the drive signal $y(t)$, we can use the sampled version instead as the input—often with the result that the analysis is much simpler. Furthermore, to communicate between two points, we can send either $y(t)$ or the sequence of numbers that are the sample values.

Thus, the possibility of sampling adds an entirely new dimension to system engineering.

13.1   WHAT IS SAMPLING?

A signal is ordinarily described by a graphical portrayal of its variation (Fig. 13.1, where the independent variable is time) or by a mathematical expression—for example,

$$y = 3 \sin(377t + 60°) \tag{13-1}$$

In either case, we say that the signal is a function of the continuous variable t; in other words, time t varies continuously.

In many engineering problems, a signal is actually known only at discrete instants of time. For example, a man driving an automobile tends to *sample* the various available signals. Occasionally he glances at the rearview mirror and observes the state of traffic behind him; he jiggles the steering wheel intermittently and observes the resultant change in the heading of the car in order to estimate the car-and-road dynamics; off and on he glances to the right or left to observe other traffic. In other words, he senses the signals intermittently—some of them perhaps once a second, others perhaps once every ten seconds, and so forth.

The original signals are functions of continuous time. For instance, the distance to the next car rearward exists at every value of t. There is no necessity, however, for the driver to observe this distance continuously. He can drive quite satisfactorily if he *samples* the signal, as long as he observes the signal often enough to detect any significant changes. Indeed, this possibility of sampling is essential if the driver is to be aware of the many different signals which are important to him.

This sampling phenomenon occurs in a wide variety of engineering systems; in the following paragraphs, we shall develop some of the basic characteristics of sampled signals.

Sequence of sample values

One way to describe a signal is to list a sequence of *sample values*—that is, a list of specific values of the signal at stated times. Table 13.1 for example, describes the temperature in New York City on November 4, 1969. From the standpoint of the average individual, the data of Table 13.1 (that is, the sample sequence) are clearly adequate to describe the temperature variation throughout that day. Even though we can only guess what the temperature was, for example, at 3:45 p.m., we are reasonably confident that it was about 63°; certainly it was not 97° or 18°. In this case, the sample sequence presents essentially all the information contained in the corresponding curve of temperature versus time.

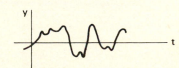

FIG. 13.1   A signal described graphically.

Table 13.1   New York City temperatures

Temp.		Temp.	
5 a.m.	– 58	1 p.m.	– 65
6 a.m.	– 58	2 p.m.	– 64
7 a.m.	– 58	3 p.m.	– 64
8 a.m.	– 59	4 p.m.	– 62
9 a.m.	– 60	5 p.m.	– 60
10 a.m.	– 61	6 p.m.	– 57
11 a.m.	– 62	7 p.m.	– 56
Noon	– 64	8 p.m.	– 54

Table 13.2 gives the heart rate of a man as measured every hour. In contrast to the temperature data, this set of sample values is of only limited usefulness. We certainly have no idea of the man's heart rate at 3:45 p.m.; it could well have been 150.

Why are hourly outside temperature data enough, while heart rate data taken every hour are of little value? A sequence of sample values is an adequate description of a signal only if the samples are taken frequently enough to show all significant changes of the signal. In the temperature example above, we know that the temperature does not vary wildly during a one-hour period; hence a sample an hour is adequate. The human heart rate, on the other hand, may go from 70 to 150 and back to 70 in just a few minutes; one sample an hour gives no picture of the signal variation.

Thus, a sequence of sample values is one possible way to describe a signal, provided that the samples are taken frequently enough. Subsequently, we shall consider the meaning of this phrase "frequently enough" in more quantitative terms. First, we consider the case of regular or periodic sampling.

Samples taken regularly

In many cases, we sample the signal at regularly spaced instants of time—for example, every millisecond or every hour. We call this *periodic sampling*; the *sampling period* is the time between successive samples, and the *sampling frequency* is the number of samples per second (or in any convenient unit of time). For example, the temperature readings of Table 13-1 constitute a sampled signal, with

Sampling period	T_s	= 1 hour	(13-2)
Sampling frequency	f_s	= 1 per hour	(13-3)

Table 13.2   Heart rate measured on the hour

7 a.m.	– 71	12 noon	– 78
8 a.m.	– 73	1 p.m.	– 88
9 a.m.	– 80	2 p.m.	– 70
10 a.m.	– 73	3 p.m.	– 75
11 a.m.	– 74	4 p.m.	– 83

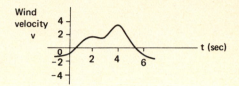

FIG. 13.2   Wind-velocity signal.

Figure 13.2 shows a different signal: the wind velocity as it varies with time. If we sample this signal every 0.5 seconds starting at $t = 0$, we obtain the sequence of values starting with

t	*Velocity v*
0	-1
0.5	-0.4
1	$+0.2$
1.5	$+1.4$
2	$+2$
2.5	$+1.9$

The signal is adequately represented by either the curve of Fig. 13.2 or this table of sample values. In this case,

$$f_s = 2 \text{ samples/sec} \qquad T_s = 0.5 \text{ sec} \tag{13-4}$$

Thus, sampling replaces a continuous or analog signal by a sequence of sample values—the amplitudes of the signal at the sampling instants. If we sample *regularly* (the most common situation in engineering), we can describe the rate of sampling by either the sampling period (T_s or also denoted T) or by the sampling frequency (f_s samples/sec, or we can use $\omega_s = 2\pi f_s$, the angular sampling frequency).

13.2 SAMPLING THEOREM

The sampling period T must be small enough to ensure that all significant variations of the signal are measured by the samples. For example, the T in Fig. 13.3 is clearly too large; the sharp dip in the signal is not even indicated by the sample sequence. While we can often estimate an appropriate T from the waveform of the signal, there is a famous theorem—the sampling theorem—which also helps in selecting T. The sampling theorem states:

> If the average sampling frequency is greater than twice the highest frequency component of the signal, the original signal can be recovered from the sample sequence.

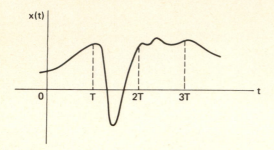

FIG. 13.3 Signal with sampling period *T* clearly too large.

As an example, a normal speech signal contains no significant components above 4000 Hz. Hence, samples taken at more than 8000 samples/second contain all the information of the original signal. We can recover the speech signal from this sample sequence.

Validity of the sampling theorem

In order to understand the logic underlying the sampling theorem, we consider a very special type of sampling, *impulse sampling*. In the last section, we viewed the sampled signal as simply a sequence of values: the signal at the sampling times. We might also define the sampling process as generation of a sequence of impulses at the sampling instants, with the area of each impulse equal to the original signal value at that time. In other words, we now define sampling as shown in Fig. 13.4. Sampling replaces the continuous signal by a train of impulses occurring at the sampling times.

Figure 13.4(c) shows that such a view of sampling can be described by an equation

$$x^*(t) = x(t) i(t) \tag{13-5}$$

The sampled signal is just the original signal multiplied by the train of unit impulses. From this definition, we can derive the sampling theorem.

The Laplace transform of $x^*(t)$ is

$$X^*(s) = \mathcal{L}[x(t) i(t)] \tag{13-6}$$

Now $i(t)$ is a periodic signal which can be expanded in a Fourier series

$$i(t) = \frac{1}{T} \sum_{n=-\infty}^{\infty} \exp(jn\omega_s t) \tag{13-7}$$

where ω_s is the angular sampling frequency. Substitution of Eq. (13-7) into (13-6) and interchange of the order of transformation and summation give

$$X^*(s) = \frac{1}{T} \sum_{n=-\infty}^{\infty} \mathcal{L}[x(t) \exp(jn\omega_s t)] \tag{13-8}$$

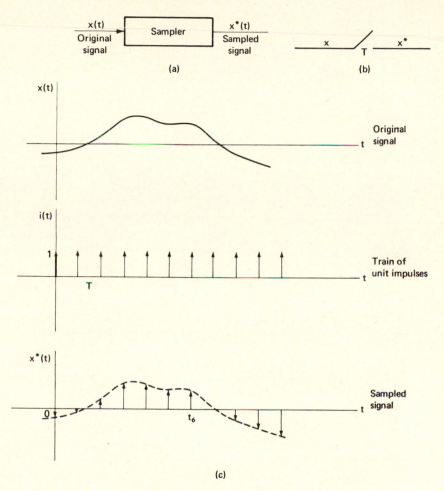

FIG. 13.4 Definition of impulse sampling. The area of the impulse of $x*(t)$ at $t_6 = 6T$ equals the amplitude of the original signal at that time—or $x(6T)$. (a) The sampling process; (b) the usual symbol for an impulse sampler in system diagram; (c) the sampling process $x*(t) = x(t)i(t)$.

The Laplace transform of a signal $x(t)$ multiplied by $\exp(jn\omega_s t)$ is just $X(s - jn\omega_s)$. Hence

$$X*(s) = \frac{1}{T} \sum_{n=-\infty}^{\infty} X(s - jn\omega_s) \qquad (13\text{-}9)$$

Equation (13-9) states that the Laplace transform of $x*(t)$ is just $1/T$ times a summation of terms: $X(s)$, $X(s)$ shifted by $j\omega_s$, $X(s)$ shifted by $-j\omega_s$, and so forth.

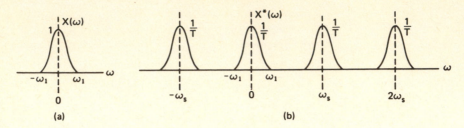

FIG. 13.5 Effect of the sampling process. (a) Spectrum of the original signal $x(t)$; (b) spectrum of the impulse-sampled signal $x*(t) = x(t) i(t)$.

Figure 13.5 shows the meaning of Eq. (13-9) in terms of the spectrum along the ω axis. The original $x(t)$ has a spectrum [part (a) of the figure] which spans from $-\omega_1$ to ω_1; that is, the frequency components lie within this range, and there is no energy outside this range. When we form $x*(t)$ by multiplying $x(t)$ by $i(t)$, we have a train of impulses of areas $x_0, x_1, x_2, x_3, \ldots$. The spectrum of this impulse train is the original spectrum (multiplied by $1/T$), plus this spectrum displaced by ω_s, plus the same spectrum displaced by $2\omega_s$, and so forth. The sampled signal has frequency components around $\omega = 0, \pm\omega_s, \pm2\omega_s, \ldots$*

Figure 13.5, which is a picture of Eq. 13-9, reveals two important facts:

(1) If $\omega_s > 2\omega_1$, there is no overlapping of the different parts of the spectrum in Fig. 13.5(b). Hence, the spectrum around $\omega = 0$ is a reproduction of the original $X(\omega)$. If we pass the $x*(t)$ signal through a filter with a gain of unity from $-\omega_s/2$ to $+\omega_s/2$ and a gain of zero elsewhere, the filter output is the original $x(t)$ signal (multiplied by the constant $1/T$). Since we can build a network which approximates such an ideal low-pass filter within any desired accuracy, we can recover the original signal from the impulse sampled form [$x(t)$ from $x*(t)$] if ω_s is greater than $2\omega_1$ (that is, if the sampling frequency is more than twice the highest frequency contained in the original signal.) (Fig. 13.6)

(2) If the sampling frequency is too low ($\omega_s < 2\omega_1$), we lose information in the sampling process. Figure 13.7 shows the situation. From 0 to ω_2, the spectrum of $X*(\omega)$ is the same as $X(\omega)$ except for the constant $1/T$ factor. From ω_2 out to ω_1, however, $X*(\omega)$ is the vector sum of two components, centered around $\omega = 0$ and $\omega = \omega_s$. The actual shape of this vector sum depends on the relative phases of the two parts, but it might have the form shown by the solid line in Fig. 13.7. In general, there is no filter or device now which permits recovery of the original signal from $x*(t)$. The sampling process irretrievably distorts the signal.

*In electrical engineering, we say that the impulse train $i(t)$ is amplitude-modulated by $x(t)$. The term *modulation* refers to the process by which the information in a signal is shifted to a different portion of the frequency spectrum—in this case, there are shifts to many different parts of the spectrum since $i(t)$ itself has components at $\omega = 0, \pm\omega_s, \pm2\omega_s, \ldots$

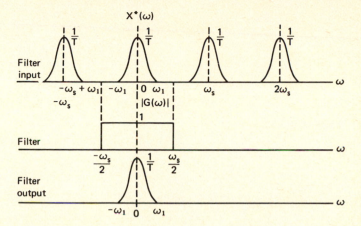

FIG. 13.6 Use of a low-pass filter to recover $x(t)$ from $x^*(t)$.

Thus, the sampling theorem states the minimum sampling frequency which can be used if the sampled signal is to contain all the information of the original signal. If we can anticipate the maximum frequency in the signal spectrum, we can select a sampling period T.*

13.3   IMPLICATIONS OF THE SAMPLING THEOREM

The sampling theorem is one of the important basic concepts of electrical engineering: it states essentially that we can work with the samples of a signal rather than the signal itself. We mention briefly some results of this possibility in this section.

Time multiplexing

Since there is no real difference between a signal $x(t)$ and the sequence of its sample values $(x_0, x_1, x_2, x_3, \ldots)$, we can *time share* a communication channel, computer, or control system among several different signals. In other words, the system might have three different inputs: $x(t)$, $y(t)$, and $w(t)$ in Fig. 13.8. Each is sampled every T seconds, but with the sampling instants staggered so that a different sample occurs every $T/3$ seconds and we move in rotation through $x(t)$, $y(t)$, and $w(t)$.

The information contained in each sample is carried by the amplitude of a short-duration pulse at the sampling time. Thus, the first pulse in the composite signal

---

*Actually, the sampling theorem specifies only the *average* sampling rate; there is no requirement that the sampling be at regular intervals. We can sample twice rapidly, then wait $2T$, for example. In most practical systems, we sample regularly; in a few cases, the system is designed to sample when the signal changes by a predetermined amount (then no samples are taken if the signal is just constant and nothing interesting is happening).

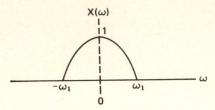

Original spectrum

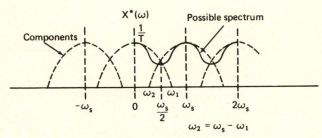

$$\omega_2 = \omega_s - \omega_1$$

Spectrum after sampling

FIG. 13.7 Sampling with ω_s too small.

has an amplitude equal to $x(0)$, the second pulse an amplitude of $y(T/3)$, and so forth. In this way, the composite signal has all the information contained in all three of the input signals.

The potential of this technique of time multiplexing or time sharing is illustrated by some simple calculations for telephone communication. Each speech signal contains frequency components to 4000 Hz; hence, we must sample at 8000 samples/sec or one sample every 125 μseconds. If the cable on which the signals are carried will transmit 1 μsec pulses with reasonable fidelity, we can send each message using a 1-μsec pulse every 125 μseconds. Hence, by proper sequencing, we can transmit 125 messages simultaneously on the one cable.* The economic implications are obvious: we need only one cable to carry 125 conversations from New York to Chicago, rather than 125.

The same principle is used in computation and control. In a process-control system, for example, a single computer is used to determine the optimum settings of many different controllers. The computer receives a variety of input signals; it performs one

---

*The terminal equipment is, of course, complex. At the transmitting end, we sample 125 different signals (conversations) in rotation and then add them to form the composite signal. At the receiving end, we separate all x pulses (one of every 125 incoming pulses) so $x(t)$ can be reconstructed and transmitted to the listener's telephone, then do the same with y and each of the 125 signals.

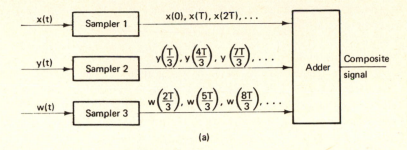

(a)

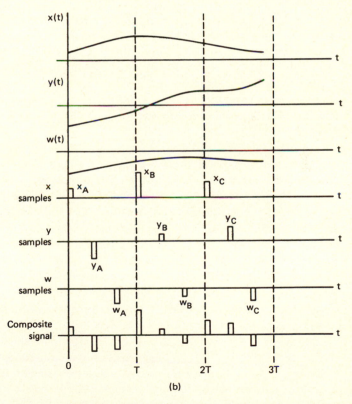

(b)

FIG. 13.8 Time multiplexing of three signals. (a) System for multiplexing; (b) signals.

calculation at a time in sequence, in each case using the appropriate part of the total incoming signal.

Human operations

At the beginning of this section it was indicated that the human being usually operates in a sampling mode, whether he is driving an automobile, reading, or performing other basic tasks. In reading, for example, we scan rapidly, observing groups of words or key words. When the information content is high (the frequency spectrum broad), we sample more frequently (more words) and the reading rate drops.

One of the basic problems in designing devices to permit a blind person to read a printed page involves a means of achieving this sampling characteristic. Reading machines for the blind, currently under development, usually scan each letter on the printed page—one letter at a time. There are two ways to inform the blind person what the letter is. In one scheme, the "reader's" finger is over an array of small reeds which can vibrate; if the letter h is being scanned, those reeds which form an h vibrate, and he "feels" the h. In the second scheme, a flying spot scanner electronically observes the h, information on the letter shape is sent to a computer which recognizes what the letter is by comparison with shapes stored in its memory, and a prerecorded voice then says h over a loudspeaker. The blind person hears the printed material being read letter by letter as he moves the scanner sensor down the printed line.

"Reading" rates as high as 100 words/minute have been reported with these systems. A significant disadvantage of this approach is that the blind "reader" must feel or listen to every letter; there is no way to skip letters or portions of the text when there is little content of interest.

Instrumentation

Measurement of physical variables is often accomplished with a sampling procedure, permitting the equipment to be simplified and, possibly, used for several different purposes. A strong motivation for sampling is the possibility of reducing the loading on the system being measured. For example, if we are measuring temperature very accurately within an enclosure, we must insert a thermocouple, thermistor, or some temperature-sensitive device. This device usually generates heat itself, hence affects the very temperature it is designed to measure.

If we use sampling and measure the temperature only during a short interval of time every T seconds, we can reduce significantly the *loading*—or the influence of the sensor on the temperature being measured. Consequently, the measurement can be far more accurate.

This loading phenomenon is similar to the effects of sampling public opinion on political issues. Indeed, the sampling theorem itself is analogous to the idea that, in public opinion polls, we need only query a relatively small fraction of the voters, the TV watchers, or the public. As long as the samples are taken frequently and regularly enough to avoid missing significant signal variations, the samples represent the total. Just as measurement of a physical variable tends to *load* the system, measurement of public opinion may influence that very opinion.

13.4 *z* TRANSFORM

We have mentioned three different types of periodic sampling in the preceding sections of this chapter:

(1) The signal $y(t)$ replaced by a sequence of sample values

$$y_0, y_1, y_2, y_3, \cdots$$

These y_j are just numbers, the values of $y(t)$ at $0, T, 2T, 3T$, and so on. The signal is represented by a sequence of numbers.

(2) Impulse sampling: $y(t)$ is represented by a train of impulses occurring at the sampling times and with the area of each impulse equal to the signal value at that time:

$$y^*(t) = y(0)\,\delta(t) + y(T)\,\delta(t - T) + y(2T)\,\delta(t - 2T) + \cdots \qquad (13\text{-}10)$$

where $\delta(t - jT)$ is the unit impulse occurring at jT.

(3) Pulse sampling: $y(t)$ is represented by a train of very short pulses occurring at the sampling times and with the amplitude of each pulse equal to the signal value at that sampling instant:

$$y_s(t) = y(0)\,p(t) + y(T)\,p(t - T) + y(2T)\,p(t - 2T) + \cdots \qquad (13\text{-}11)$$

where $p(t - jT)$ is a short pulse starting at $t = jT$.

These three, slightly different interpretations of the sampling operation can be represented as in Fig. 13.9.

Each of the three different views of sampling is useful, and indeed, we want to be able to go back and forth among them. The first interpretation (as a sequence of numbers) is the description used when we represent a signal in a digital computer or digital system. The second (impulse train) turns out to be convenient when we want to study sampling theoretically. The pulse-train interpretation describes the way we would actually sample with electronic equipment. The importance of these alternative views will become apparent in the remainder of this chapter.

Laplace transform of impulse sampling

For the moment, let us look at sampling as modulation of an impulse train—(2) above. Then the sampled signal $y^*(t)$ can be described mathematically as

$$y^*(t) = y(t)\,i(t) \qquad (13\text{-}12)$$

where $i(t)$ is a periodic train of unit impulses:

$$i(t) = \delta(t) + \delta(t - T) + \delta(t - 2T) + \cdots \qquad (13\text{-}13)$$

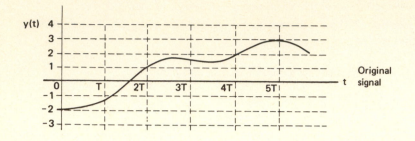

y(t) represented by y_n = -2, -1.2, 1.0, 1.6, 2.0, 3.0, . . . Sequence of values

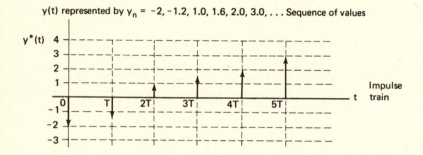

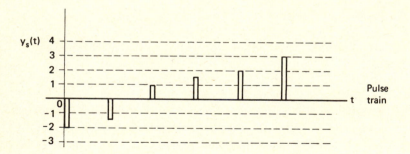

FIG. 13.9 Interpretations of sampling.

When we substitute this last equation into the preceding, we have

$$y^*(t) = y(t)\delta(t) + y(t)\delta(t - T) + y(t)\delta(t - 2T) + \cdots \qquad (13\text{-}14)$$

Each term is zero everywhere except at the instant when the impulse occurs. Hence $y(t)$ in each term can be replaced by its value at the moment of the impulse, or

$$y^*(t) = y(0)\delta(t) + y(T)\delta(t - T) + y(2T)\delta(t - 2T) + \cdots \qquad (13\text{-}15)$$

If we use our earlier notation that y_j is the value of $y(t)$ at jT, we have

$$y*(t) = y_0 \delta(t) + y_1 \delta(t - T) + y_2 \delta(t - 2T) + \cdots \tag{13-16}$$

This "derivation" has taken several steps, but all that Eq. (13-16) really says is that $y*(t)$ is a train of impulses at the sampling times, with the area of each impulse equal to the amplitude of $y(t)$ at that instant.

Since we often prefer to work in terms of Laplace transforms, we now transform Eq. (13-16) term by term:

$$Y*(s) = y_0 + y_1 e^{-Ts} + y_2 e^{-2Ts} + y_3 e^{-3Ts} + \cdots \tag{13-17}*$$

The complex frequency s appears in Eq. (13-17) only in the form e^{-Ts} or powers of that exponential. Hence, for simplicity in writing, we can define

$$\boxed{\frac{1}{z} = e^{-Ts}} \tag{13-18}$$

and Eq. (13-17) becomes

$$Y*(z) = y_0 + y_1 \frac{1}{z} + y_2 \frac{1}{z^2} + y_3 \frac{1}{z^3} + \cdots \tag{13-19}$$

The Laplace transform of $y*(t)$ is just a power series in $1/z$, with the coefficient of $1/z^n$ equal to the signal value at nT. This $Y*(z)$ is called the *z-transform* of the signal $y(t)$.†

---

*In the derivation of the sampling theorem, we also found $Y^*(s)$, but by first replacing $i(t)$ there by its Fourier series. The $Y^*(s)$ we find here certainly must equal our earlier form

$$Y^*(s) = \frac{1}{T} \sum_{n=-\infty}^{\infty} Y(s - jn\omega_s)$$

even though in appearance the two seem unrelated.

†At this point, one might logically ask: How ridiculous can engineers be? Why not define z as e^{-Ts}? Then $Y^*(z)$ would just be a power series in z, instead of $1/z$. Unfortunately, z transforms were of interest to mathematicians in the last century, long before engineers became interested in sampling in the late 1940s (after electronic circuits, to perform the sampling operation, and digital circuitry became available). Since extensive tables of z transforms were available, engineers followed the existing definition.

Meaning of z transform

Equation (13-18) states that $1/z$ is just a delay operator, since e^{-Ts} is the transform representation of a delay of T seconds. In other words, $1/z$ represents a delay of one sampling period, $1/z^2$ a delay of two periods, and $1/z^n$ a delay of n periods. In this sense, Eq. (13-19) is simply an alternate way of writing the table of sample values: the coefficient of $1/z^n$ is the sample value at a time of nT seconds.

In order to emphasize the inherent simplicity of the z transform, we can work a problem in the reverse direction. If we are given a z transform (for the sampling period specified as 2 msec)

$$Y(z) = 3 + \frac{4}{z} + \frac{1}{z^2} + \frac{0}{z^3} + \frac{-2}{z^4} + \frac{-2}{z^5} + \cdots \tag{13-20}$$

we know that the actual $y(t)$ passes through the points 3, 4, 1, 0, −2, −2, and so on. Furthermore, if the sample values were taken sufficiently frequently, the original $y(t)$ is a smooth curve through these points.

z transforms of simple functions

As defined above, the z transform is just an infinite power series in $1/z$, with the coefficient of $1/z^n$ being the signal value at nT. When the signal is a generalized exponential, the z transform can be written in closed form, which is demonstrated here by several examples (in each case, the signal is considered only for $t > 0$).

(a) Unit ramp function, $y(t) = t$. Then

$$Y(z) = \frac{T}{z} + \frac{2T}{z^2} + \frac{3T}{z^3} + \cdots + \frac{nT}{z^n} + \cdots \tag{13-21}$$

$$= T\left(\frac{1}{z} + \frac{2}{z^2} + \frac{3}{z^3} + \cdots\right) \tag{13-22}$$

Now, to write this series in closed form, we can consult an appropriate reference table in a math handbook. Alternatively, we can try to rewrite the series in a recognizable form. It turns out that we can manipulate as follows:

$$Y(z) = Tz\left(\frac{1}{z^2} + \frac{2}{z^3} + \frac{3}{z^4} + \cdots\right)$$

$$= Tz\left[-\frac{d}{dz}\left(\frac{1}{z} + \frac{1}{z^2} + \frac{1}{z^3} + \cdots\right)\right] \tag{13-23}$$

$$= Tz\left[-\frac{d}{dz}\left(\frac{1}{z-1}\right)\right]$$

$$Y(z) = \frac{Tz}{(z-1)^2}$$

(b) Unit step function, $y(t) = 1$. Here we must decide what the sample value is at $t = 0$. If we arbitrarily say that the step starts just before we sample,

$$Y(z) = 1 + \frac{1}{z} + \frac{1}{z^2} + \frac{1}{z^3} + \cdots \tag{13-24}$$

or

$$Y(z) = \frac{z}{z - 1} \tag{13-25}$$

(c) Simple exponential, $y = e^{-at}$. The direct substitution of $t = 0, T, 2T, \ldots$ gives the sample values from which we can write

$$Y(z) = 1 + \frac{e^{-aT}}{z} + \frac{e^{-2aT}}{z^2} + \cdots + \frac{e^{-naT}}{z^n} + \cdots \tag{13-26}$$

This is just a power series in e^{-aT}/z; hence

$$Y(z) = \frac{1}{1 - e^{-aT}/z}$$

or

$$Y(z) = \frac{z}{z - e^{-aT}} \tag{13-27}$$

Clearly, we could continue similar derivations for other common, generalized exponentials. Instead, Table 13.3 simply lists all the z transforms we shall have any occasion to use. The validity of any of these transforms can be checked by expanding $Y(z)$ in powers of $1/z$ (this is most easily accomplished by dividing the denominator polynomial into the numerator). The sample values given by $Y(z)$ can then be compared with the corresponding values of $y(t)$.

Summary comment

This section defines the z transform and develops z transform of generalized exponential signals. The z transform of a signal is really the Laplace transform of the impulse-sampled version of the signal. The z transform is often much easier to derive than the Laplace transform. If a signal $y(t)$ is given graphically or experimentally, the z transform is just

$$y_0 + \frac{y_1}{z} + \frac{y_2}{z^2} + \cdots$$

Table 13.3 Laplace and z transforms

Row	Column 1 Laplace transform	Column 2 Time function	Column 3 z transform	Column 4 Description of time function
a	1	$\delta(t)$	1	Impulse function at $t = 0$
b	e^{-nTs}	$\delta(t - nT)$	$\dfrac{1}{z^n}$	Impulse function at $t = nT$
c	$\dfrac{1}{1 - e^{-Ts}}$	$i(t)$	$\dfrac{z}{z - 1}$	Train of impulses at sampling instants
d	$\dfrac{1}{s}$	$u(t)$	$\dfrac{z}{z - 1}$	Step function
e	$\dfrac{1}{s^2}$	t	$\dfrac{Tz}{(z - 1)^2}$	Ramp function
f	$\dfrac{1}{s^3}$	$\dfrac{1}{2}t^2$	$\dfrac{1}{2}T^2\dfrac{z(z + 1)}{(z - 1)^3}$	Quadratic or acceleration function
g	$\dfrac{1}{s + a}$	e^{-at}	$\dfrac{z}{z - e^{-aT}}$	Exponential function
h	$\dfrac{a}{s^2 + a^2}$	$\sin at$	$\dfrac{z \sin aT}{z^2 - 2z \cos aT + 1}$	Sinusoidal function

(The sampling period T must be known if the z transform is to have meaning.) In the next sections, we consider how the z transform can be used in the analysis of sampled systems. In the next chapter, we show how z transforms are also useful in the study of ordinary systems which do not have sampling.

13.5 DISCRETE DATA PROCESSORS

The familiar RLC networks of electrical engineering "process" input signals. The input signal $x(t)$ in Fig. 13.10(a) drives the network; the result is the output $y(t)$. We often find it convenient to describe the relation between x and y in terms of the transforms

$$Y(s) = G(s)X(s) \tag{13-28}$$

FIG. 13.10 Continuous and discrete systems. (a) Conventional system, network described by $G(s)$ or $G(j\omega)$ or $g(t)$; (b) system processing discrete data, with system described by $G(z)$ or $g^*(t)$.

or the behavior for sinusoidal signals

$$Y(j\omega) = G(j\omega) X(j\omega) \tag{13-29}$$

In the time domain, the corresponding relation is the convolution integral.

$$y(t) = \int_0^t x(\tau - t) g(\tau) d\tau \tag{13-30}$$

If the input is a sequence of sample values or an impulse-sampled signal and the output has the same form, the system can be described in terms of z transforms instead of Laplace transforms. Indeed, the system of Fig. 13.10(b) turns out to be much simpler to analyze (and to design) than the continuous network.

Recursion equations

Just as the relation between $x(t)$ and $y(t)$ in Fig. 13.10(a) is described by a linear differential equation (when the system is linear), y_n is determined from x_n in the discrete-data system by a *linear recursion relation or equation*. The "present" value y_n is given as a linear combination of the earlier values of the response and the present and earlier values of the input. For example, we might have

$$y_n + 0.4y_{n-1} + 0.2y_{n-2} = 2x_n + 3x_{n-1} \tag{13-31}$$

Equation (13-31) is second order: three successive samples of y are involved. If the latest sample of y is y_n, and the equation also involves y_{n-j}, the order is j. Thus, the relation

$$y_n - 0.1y_{n-30} = x_{n-1} \tag{13-32}$$

is thirtieth order.

To solve a recursion equation of order j, we need to specify j starting values (possibly $y_0, y_1, \ldots, y_{j-1}$). Then the next value can be calculated algebraically from the equation. Each successive calculation gives one additional sample value of the response.

This brute-force solution procedure can be illustrated with Eq. (13-31) above, if we are given starting conditions* and the drive signal x:

$$y_0 = 0 \qquad y_1 = 1 \qquad x_n = \begin{cases} 0 & n < 0 \\ 1 & n \geq 0 \end{cases} \tag{13-33}$$

---

*We might be given y_{-1} and y_{-2} instead of y_0 and y_1. Then y_0 would first be calculated from the equation.

Then

$$y_2 = -0.4y_1 - 0.2y_0 + 2x_2 + 3x_1 = 4.6 \qquad (13\text{-}34)$$

$$y_3 = -0.4y_2 - 0.2y_1 + 2x_3 + 3x_2 = 2.96 \qquad (13\text{-}35)$$

and so on.

z transform of a recursion equation

Since Eq. (13-31) is a relation among sample values (or impulse samples), we might expect the x transform to be useful. We can interpret the recursion equation

$$y_n + 0.4y_{n-1} + 0.2y_{n-2} = 2x_n + 3x_{n-1} \qquad (13\text{-}36)$$

as a relation which states that at any particular sample time:

The y sample $+ 0.4\,(y$ sample one period ago)
$\qquad\qquad + 0.2\,(y$ sample two periods ago)
$\qquad = 2\,(x$ sample) $+ 3\,(x$ sample one period ago) $\qquad (13\text{-}37)$

Since a delay of one period corresponds to multiplying by $1/z$, this relation is equivalent to

$$Y(z) + \frac{0.4}{z}\,Y(z) + \frac{0.2}{z^2}\,Y(z) = 2X(z) + \frac{3}{z}\,X(z) \qquad (13\text{-}38)$$

Equation (13-38) is essentially the z transform of the recursion relation.

One modification must be made to take care of the specified starting values. If we recognize that $Y(z)$ is the power series in $1/z$, Eq. (13-38) is a shorthand notation for

$$\left(y_0 + \frac{y_1}{z} + \cdots + \frac{y_n}{z^n} + \cdots \right) + \frac{0.4}{z}\left(y_0 + \frac{y_1}{z} + \cdots + \frac{y_n}{z^n} + \cdots \right)$$

$$+ \frac{0.2}{z^2}\left(y_0 + \frac{y_1}{z} + \cdots + \frac{y_n}{z^n} + \cdots \right) \qquad (13\text{-}39)$$

$$= 2\left(x_0 + \frac{x_1}{z} + \cdots + \frac{x_n}{z^n} + \cdots \right) + \frac{3}{z}\left(x_0 + \frac{x_1}{z} + \cdots + \frac{x_n}{z^n} + \cdots \right)$$

If we equate the coefficients of $1/z^n$ on left and right sides, we find

$$y_n + 0.4y_{n-1} + 0.2y_{n-2} = 2x_n + 3x_{n-1} \qquad (13\text{-}40)$$

just the original recursion equation. For the $(1/z^0)$ and $1/z^1$ terms, we have

$$y_0 = 2x_0$$

$$y_1 + 0.4y_0 = 2x_1 + 3x_0$$

(13-41)

If we want to specify y_0 and y_1 arbitrarily (to represent initial energy storage in this second order system), we should remove the constant and $1/z$ terms in the transform equation (13-38), which then is properly written

$$Y(z) - y_0 - \frac{y_1}{z} + \frac{0.4}{z}[Y(z) - y_0] + \frac{0.2}{z^2}Y(z)$$

$$= 2\left[X(z) - x_0 - \frac{x_1}{z}\right] + \frac{3}{z}[X(z) - x_0] \quad (13\text{-}42)$$

(This subtraction is directly analogous to the Laplace transform of a second order differential equation when initial values of the response and its first derivative are given.)

Rules for finding z transform

Thus, given any linear recursion equation, we can find the z transform as follows. If the equation is in the form

$$y_n + \alpha y_{n-1} + \beta y_{n-2} + \cdots + \gamma y_{n-j} = Ax_n + Bx_{n-1} + \cdots + Cx_{n-k}$$

and if we arbitrarily specify the starting values $y_0, y_1, y_2, \ldots, y_{j-1}$, then the z transform is

$$\underbrace{\left[Y(z) - y_0 - \frac{y_1}{z} - \cdots - \frac{y_{j-1}}{z^{j-1}}\right]}_{\text{Transform of } y_n \text{ term}} + \underbrace{\frac{\alpha}{z}\left[Y(z) - y_0 - \cdots - \frac{y_{j-2}}{z^{j-2}}\right]}_{\text{Transform of } y_{n-1} \text{ term}} + \cdots$$

(13-43)

$$= \underbrace{A\left[X(z) - x_0 - \cdots - \frac{x_{j-1}}{z^{j-1}}\right]}_{\text{Transform of } x_n \text{ term}} + \cdots$$

This general formulation looks awesome mathematically, but it is actually easy in the simple cases we will consider, as illustrated by the two examples below.

(1) $\begin{cases} y_n + 0.8y_{n-1} = 0.2x_{n-1} \\ y_0 = 3 \end{cases}$ (13-44)

Then

$$\underbrace{Y(z) - 3}_{} + \frac{0.8}{z} Y(z) = \frac{0.2}{z} X(z)$$

The term over the brace represents the z transform of y_n when one starting condition is specified. Solving then gives

$$Y(z) = \frac{3 + (0.2/z) X(z)}{1 + 0.8/z} \tag{13-45}$$

(2) $\begin{cases} y_n + 0.3y_{n-1} + 0.3y_{n-2} = 0.2x_n + 0.2x_{n-1} \\ y_0 = 1 \qquad y_1 = 2 \end{cases}$ (13-46)

Now the system is second order, and we must worry about the $1/z^0$ and $1/z^1$ terms. Then

$$\left[Y(z) - 1 - \frac{2}{z} \right] + \frac{0.3}{z} \left[Y(z) - 1 \right] + \frac{0.3}{z^2} Y(z) = 0.2 \left[X(z) - x_0 - \frac{x_1}{z} \right]$$

$$+ \frac{0.2}{z} \left[X(z) - x_0 \right] \tag{13-47}$$

If we wish, this equation for $Y(z)$ can be solved by straightforward algebra.

Solution of a recursion equation by z transforms

The population of a closed island can be divided into two parts: children y and adults w. In any year n, the numbers of each are given by the relations

$$y_n = 0.95y_{n-1} + 0.1w_{n-1} \tag{13-48}$$

$$w_n = 0.95w_{n-1} + 0.05y_{n-1} \tag{13-49}$$

Equation (13-48) states that 95 percent of the children are alive and still children a year later; in addition, new children are equal to 10 percent of the adults. The second relation indicates 5 percent of the adults die and 5 percent of the children reach adulthood.

The system is second order (two coupled, first order systems); hence we need two starting values, which we take as

$$y_0 = 20$$

$$w_0 = 40$$
(13-50)

(measured, for example, in thousands). We now wish to solve for the adult population as a function of time.

Taking the z transform of our two equations gives

$$Y - 20 = \frac{0.95}{z} Y + \frac{0.1}{z} W$$

$$W - 40 = \frac{0.95}{z} W + \frac{0.05}{z} Y$$
(13-51)

Algebraic manipulation eliminates Y to give

$$W = \frac{40z^2 - 37z}{z^2 - 1.9z + 0.8975}$$
(13-52)

This W is the z transform of the adult population. It happens to be a ratio of polynomials rather than an infinite series in $1/z$. We can convert to the latter form simply by dividing the denominator into the numerator:

$$
\begin{array}{r}
40 \quad + \dfrac{39}{z} \quad + \dfrac{38.2}{z^2} \quad + \dfrac{37.678}{z^3} \quad + \cdots \\[2mm]
\hline
\end{array}
$$

$$z^2 - 1.9z + 0.8975 \, \overline{\smash{\big)}\, 40z^2 - 37z}$$

$$40z^2 - 76z + 35.9$$

$$39z - 35.9$$

$$39x - 74.1 + \frac{34.8025}{z}$$

$$38.2 - \frac{34.8025}{z}$$

$$38.2 - \frac{72.58}{z} \quad + \cdots$$

$$\frac{37.678}{z}$$

$$W(z) = 40 + \frac{39}{z} + \frac{38.2}{z^2} + \frac{37.678}{z^3} + \cdots$$
(13-53)

Thus, the adult population starts off at the value 40 (which we assumed). The next year it is 39, then 38.2, then 37.678, and so forth.

The solution requires comment. We started with the two coupled recursion equations, (13-48) and (13-49). We used the z transform to find by algebra the $W(z)$ in Eq. (13-52): the adult population summarized in a ratio of polynomials in z. Long division then gives as many values of the adult population as are required (or as we have the patience to find).

Just as the Laplace transform is useful because it reveals the global properties of the solution without tedious calculation, the z transform

$$W(z) = \frac{40z^2 - 37z}{z^2 - 1.9z + 0.8975} \tag{13-54}$$

can be interpreted without tedious long division and tabulation, or plotting many population values. In the next section, we want to consider such interpretations of $W(z)$.

13.6   WHAT CAN WE DEDUCE FROM $W(z)$?

We have found the z transform of the response of a system. We call this $W(z)$. It is known as a ratio of polynomials in z. What can we say about the sample values w_n?

Partial fraction expansion

First, we might make a partial fraction expansion of $W(z)$. It is slightly more convenient to expand W/z in partial fractions.* If all the poles of W are simple, this expansion contains terms of the form

$$\frac{W}{z} = \frac{k_0}{z} + \frac{k_1}{z - c} + \frac{k_2}{z - a - jb} + \frac{k_2^*}{z - a + jb} \tag{13-55}$$

There are three different, simple poles of interest that are possible: at the origin, at the real value c, and at the conjugate complex values $a \pm jb$.

Multiplication by z gives

$$W = k_0 + \frac{k_1 z}{z - c} + \frac{k_2 z}{z - a - jb} + \frac{k_2^* z}{z - a + jb} \tag{13-56}$$

The first term contributes only to w_0. The second term gives rise to a sequence of sample values determined by the series

---

*This apparently illogical approach results because we defined z as a time-advance factor, with $1/z$ the delay factor. We really want a partial fraction expansion in the variable $1/z$, with each term $k/(1/z + a)$. This corresponds to $kz/(1 + az)$. If we make the expansion of W/z, then multiply by z, the desired form is obtained.

$$\frac{k_1 z}{z - c} = \frac{k_1}{1 - c/z} = k_1 \left(1 + \frac{c}{z} + \frac{c^2}{z^2} + \frac{c^3}{z^3} + \cdots \right) \tag{13-57}$$

Thus, the $W(z)$ pole at $z = c$ represents an exponential term, where the nth sample is

$$k_1 c^n$$

This component of the response decreases exponentially if $|c| < 1$, grows exponentially if $|c| > 1$, and is constant if $|c| = 1$.

The conjugate complex poles similarly lead to exponentials. If the poles are written in polar form as $r \underline{/\pm\, \theta}$, the series expansion gives

$$k_2 \left(1 + \frac{r\underline{/\theta}}{z} + \frac{r^2\underline{/2\theta}}{z^2} + \cdots \right) + k_2^* \left(1 + \frac{r\underline{/-\theta}}{z} + \frac{r^2\underline{/-2\theta}}{z} + \cdots \right) \tag{13-58}$$

or

$$2 \operatorname{Re} \left[k_2 \left(1 + \frac{r\underline{/\theta}}{z} + \frac{r^2\underline{/2\theta}}{z^2} + \cdots \right) \right] \tag{13-59}$$

The nth sample is

$$2 \operatorname{Re}[k_2 r^n e^{jn\theta}] = 2 |k_2| r^n \cos(n\theta + \underline{/k_2}) \tag{13-60}$$

Again, if $r < 1$ the term decays with n; if $r > 1$, the magnitude increases with n; and if $r = 1$, we have a constant amplitude variation with n.

Stability

Thus, the poles of $W(z)$ show the stability situation at once. *The system is stable if all poles are inside the unit circle in the z plane.* Any pole outside the unit circle means instability. Poles on the unit circle represent the borderline between stability and instability and mean sustained sinusoidal oscillation (or a constant term).

These stability conditions could also have been derived from the relation between z and s:

$$z = e^{sT} \tag{13-61}$$

When the real part of s is positive, $|z| > 1$; the left half s plane corresponds to the interior of the unit circle in z.

Now let us return to our population example of the last section, where we found

$$W(z) = \frac{40z^2 - 37z}{z^2 - 1.9z + 0.8975} = \frac{40z^2 - 37z}{(z - 0.88)(z - 1.02)}$$

$$= 40 + \frac{39}{z} + \frac{38.2}{z^2} + \frac{37.678}{z^3} + \cdots$$

(13-62)

This is the expression for the annual size of the adult population. The first few terms of the series show that the population is decreasing, and we might predict an ultimate disappearance or stabilizing of adults. Actually, the pole at 1.02 indicates that the system is unstable. After the transient has died out, the population will grow indefinitely proportional to

$$(1.02)^n$$

The increase will be 2 percent per year (or a doubling about every 36 years).

If we return to the original model, Eqs. (13-48) and (13-49), we might ask: What birth rate b leads to zero population growth (ZPG)? The equations are

$$y_n = 0.95y_{n-1} + bw_{n-1}$$

$$w_n = 0.95w_{n-1} + 0.05y_{n-1}$$

(13-63)

The characterisitc polynomial in z is

$$(z - 0.95)(z - 0.95) - 0.05b$$

A constant, steady state population means a pole at $z = 1$:

$$(1 - 0.95)(1 - 0.95) - 0.05b = 0 \qquad \text{or} \qquad b = 0.05$$

(13-64)

The adult population holds steady (after the transient) if the number of new children each year equals 5 percent of the adult population: in other words, births equal deaths, as we knew intuitively. (The interesting feature is that doubling this birth rate only results in a 2 percent population growth per year.)

Growth and decay rates

The poles of $W(z)$ show the growth and decay rates. For instance, again in our population example, we have W poles at 0.88 and 1.02. The response has components

$$(0.88)^n \quad \text{and} \quad (1.02)^n$$

The transient dies at the rate of 12 percent per year; the growth term increases at 2 percent per year.

Response in closed form

From the partial fraction expansion of W/z, we can write the w_n in closed form. In our example,

$$\frac{W}{z} = \frac{40z - 37}{(z - 0.88)(z - 1.02)} = \frac{+12.9}{z - 0.88} + \frac{27.1}{z - 1.02} \tag{13-65}$$

$$W(z) = \frac{12.9z}{z - 0.88} + \frac{27.1z}{z - 1.02} \tag{13-66}$$

Since $z/(z - a)$ corresponds to $1 + a/z + a^2/z^2 + \cdots$, the coefficient of $1/z^n$ is

$$w_n = 12.9(0.88)^n + 27.1(1.02)^n \tag{13-67}$$

The population in any year (any desired value of n) can be calculated without the tedious process of long division until $1/z^n$ is reached.

In this sense, that Eq. (13-67) is a closed-form global solution, the z transform has exactly the same advantages as the Laplace transform.

Comment

In this section and those preceding, we discussed the solution of linear recursion equations: How to go from a given mathematical model (the equation plus required starting conditions) to a solution in the form of the sequence of output values or, through the z transform, to a closed-form expression for the output. Discussion in the two sections would appear to be primarily an algebraic exercise. In the next immediate section, we return to systems engineering with the question: How can these z transform techniques be used in system design? This emphasis on engineering applications is continued in the next chapter on numerical and digital simulation.

13.7 SYSTEM DESIGN EXAMPLE

Sampling is often used in instrumentation to decrease the loading effects of the measurement (suggested earlier in the chapter). In other cases, sampling is inherent in the system from the outset. For example, in an air traffic control system, a single radar set must be used to track many different airplanes. The radar may rotate continuously; then each target plane is observed for only a short interval during each antenna revolution. The signal representing target position is sampled.

The simplest form of a sampled data-control system is shown in Fig. 13.11. The error signal is sampled to give $e*(t)$. This sequence of sample values can be modified by a digital controller $D(z)$ to give a different sample sequence $m*(t)$. Our task is to design $D(z)$. The signal $m*$ drives a process G to generate the output c.

In approaching any feedback system design problem, we try first to find the overall transfer function $C(s)/R(s)$ as it depends on the separate components. In this case,

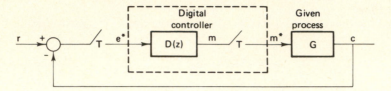

FIG. 13.11 A feedback system with sampling.

the signals e and m are already sampled; they are trains of impulses (or, equivalently, very short pulses). $c(t)$ is a continuous signal. We will find, however, that system analysis is much simpler if we look at $c(t)$ only at the sampling instants, every T seconds. In other words, we insert a hypothetical sampler at the output and consider $c*(t)$ as the output signal of interest (Fig. 13.12).*

Now we can write the relations

$$C(z) = G(z)M(z)$$
$$M(z) = D(z)E(z) \qquad\qquad (13\text{-}68)$$
$$E(z) = R(z) - C(z)$$

from which we obtain

$$C(z) = \frac{G(z)D(z)}{1 + G(z)D(z)} \qquad\qquad (13\text{-}69)$$

This equation is exactly parallel to that which we would have obtained for a system without sampling and involving the forward components $G(s)$ and $D(s)$.

In Eq. (13-69), what is the meaning of each term. $R(z)$ is the z transform of $r*(t)$:

$$R(z) = r_0 + \frac{r_1}{z} + \frac{r_2}{z^2} + \frac{r_3}{z^3} + \cdots \qquad\qquad (13\text{-}70)$$

If $r(t)$ is a sum of generalized exponentials, $R(z)$ can be written in closed form from the table of z transforms. For example, if $r(t)$ is a unit step function,

*As soon as this sampler is added, every signal is sampled. We can then describe the system in terms of z transforms, instead of a mixture of z and s relations.

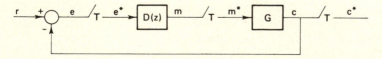

FIG. 13.12 Feedback system with sampler added at the output.

$$R(z) = \frac{z}{z-1} \tag{13-71}$$

Likewise, $C(z)$ is the z transform of the sampled output signal $c*(t)$.

$D(z)$ and $G(z)$ are z transfer functions

$$D(z) = \frac{M(z)}{E(z)} \qquad G(z) = \frac{C(z)}{M(z)}$$

In other words, $G(z)$ describes the process behavior when the input is a train of impulses and the output is sampled. Since the system is linear, superposition applies and we can use any convenient input $m*$ to measure $G(z)$. If $m*(t)$ is, for example, a unit impulse at $t = 0$ $[M(z) = 1]$, $c(t)$ is the unit impulse response of G, and $G(z)$ is just the z transform of $c*$.

Specific design example

We are to determine a network for $D(z)$ which meets the following specifications:

(1) Process is described by the transfer function $G(s) = 5/s^2$
(2) Input $r(t)$ is a unit step function.
(3) Output $c(t)$ ideally should follow or equal the input exactly. When the step is first applied, the output can not jump instantaneously, however, and we specify only that $c_n = 1$ for $n = 1, 2, 3, 4, \cdots$. That is, the output should equal exactly the input at every sample instant after the first.
(4) Sampling period is 2 seconds.
(5) Required $D(z)$ is to be realized by a network of delay elements and adders.

Solution

The solution is carried through in the following steps.

(1) First we find the required closed-loop z transfer function which meets the specifications on r and c. We are given

$$R(z) = \frac{z}{z-1} \quad \text{the } z \text{ transform of a unit step function} \tag{13-72}$$

$$C(z) = \frac{1}{z} + \frac{1}{z^2} + \frac{1}{z^3} + \cdots = \frac{1}{z-1} \tag{13-73}*$$

Hence

$$\frac{C(z)}{R(z)} = \frac{1}{z} \tag{13-74}$$

(the system represents a delay of T seconds).

---

*The c_0 value can be chosen arbitrarily. Since the process has inertia, the output cannot be moved instantaneously. Therefore, we recognize that c_0 will have to be zero if the signal applied to the process is finite.

(2) For the particular feedback configuration chosen, Eq. (13-69) relates the open-loop z transfer function $D(z)\,G(z)$ to the closed-loop $C(z)/R(z)$. If we solve this equation for $D(z)\,G(z)$, we obtain

$$D(z)\,G(z) \;=\; \frac{C(z)/R(z)}{1 - C(z)/R(z)} \tag{13-75}$$

Substitution of Eq. (13-74) yields

$$D(z)\,G(z) \;=\; \frac{1}{z - 1} \tag{13-76}$$

This is the open-loop system required to meet the performance specifications.

(3) $G(z)$ must be evaluated. The specifications state $G(s) = 5/s^2$. The table of z transforms indicates that, for $T = 2$,

$$G(z) \;=\; \frac{10z}{(z - 1)^2} \tag{13-77}$$

(4) The product $D(z)\,G(z)$ and $G(z)$ are both known; therefore

$$D(z) \;=\; \frac{z - 1}{10z} \tag{13-78}$$

This is the required controller transfer function.

(5) The network for this $D(z)$, shown in Fig. 13.13, uses a simple delay element; then the adder realizes

$$m_n \;=\; \frac{1}{10}(e_n - e_{n-1}) \tag{13-79}$$

Comment

The most striking feature of this design example is its triviality. The design of feedback systems with sampling is actually much simpler conceptually than without sampling. As we attempt to extend the above approach to more complex problems, a few difficulties arise.

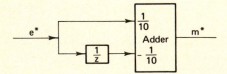

FIG. 13.13 Realization of $D(z)$.

First, it is often difficult to decide on simple specifications for $r(t)$ and $c(t)$. If $r(t)$ is a randomly varying signal, we might want $c(t)$ to be within 0.1 of r at as many sampling times as possible, or we might want to minimize the probability that the error exceeds a particular amount. In landing a plane on a runway (Fig. 13.14), we aim for a desired touchdown point. Errors of 50 feet either way are immaterial. There are significant disadvantages in an error which is more than 200 feet short; a major task in the control system should be to avoid such disaster. In all aspects of engineering design, we face the problem of how to interpret such specifications mathematically. Often we have to choose a $T(z)$ function rather arbitrarily, then test the resulting system in the laboratory or by simulation to be sure the real performance specifications are met.

Second, the calculation of $G(z)$ is often more tedious than implied by the above example. If $G(s)$ were a complex transfer function, for example, we would have to find the impulse response or the partial fraction expansion of $G(s)$ in order to use the table of z transforms.

Third, we must be sure that the $D(z)$ which we derive is causal. We cannot ask for a response m^* before the input e^* is applied. In other words, the numerator of $D(z)$ cannot be of higher degree than the denominator. In the above example, if we ask for the same output signal but with c_0 near unity instead of zero, we find the required

$$D(z) = \frac{(z+1)(z-1)^2}{-10z} \tag{13-80}$$

which is certainly noncausal and not physically realizable.

Finally, the $D(z)$ required in the example is

$$\frac{1}{10} - \frac{1}{10z}$$

This is so simple that the network can be found by inspection. In more interesting problems, the $D(z)$ comes out as a ratio of polynomials in z. We can still find a

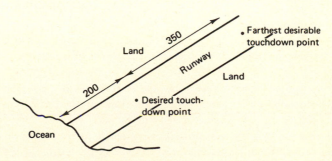

FIG. 13.14 Aircraft touchdown.

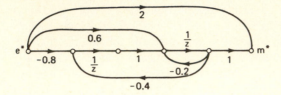

FIG. 13.15 Signal flow graph for

$$D(z) = 2 + \frac{0.6z - 0.8}{z^2 + 0.2z + 0.4}.$$

network of delay elements and ladders simply by using the same procedure employed for operational-amplifier realization of a transfer function in s. The integrator $1/s$ is replaced by the delay element $1/z$. For example, if

$$D(z) = \frac{2z^2 + z}{z^2 + 0.2z + 0.4} = 2 + \frac{0.6z - 0.8}{z^2 + 0.2z + 0.4} \tag{13-81}$$

one corresponding signal flow diagram is shown in Fig. 13.15, and the other canonic forms are also available.

13.8 CONCLUDING COMMENTS

This chapter focuses on the concept of sampling in systems engineering. The most important part is that devoted to the sampling theorem, where it can be seen that sampling need not mean the loss of information. If the rate of sampling is sufficiently high, the original continuous signal can be recovered from the sampled form within any desired accuracy.

As soon as the operational validity of sampling is established, we are interested in the analysis and design of sampled data systems—in this text, particularly the feedback configurations so common in instrumentation and control. The z transform is the most potent vehicle for such studies; the last section showed that the design of a sampled data-feedback system is often even simpler than an analogous problem with continuous data components.

Two features make sampled data systems particularly important. First, sampling permits time multiplexing—the sharing of equipment for several different tasks. The human being, as a control element in an overall system, uses sampling for just this purpose. While driving a car, the human operator controls speed and sidewise motion, observes traffic and pedestrian flow, navigates by interpreting road signs and directions, and monitors the environment for danger signals (sirens, potholes ahead, and so on). In addition, he may listen to the radio or cassette player, carry on a conversation, or (as one acquaintance of the author foolishly does) dictate letters into a portable dictating machine. The successful performance of these tasks in parallel depends on controlled sampling and time sharing.

Second, the sampled data-feedback system is a major step toward the use of a digital computer as a controller in a feedback system—called DDC (direct digital control). The $D(z)$ controller transfer function derived in the last section is essentially the program for a computer. The actual use of control computers involves not only sampling, but also quantizing the signal so the amplitudes of each sample are represented by a finite number of digits, rather than a pulse height or impulse area. Very often the error introduced by the quantization is negligible, however, and the DDC system operates exactly as the sampled data system.

When a digital computer is used to realize $D(z)$, the computer is typically occupied for only a small fraction of the time by the calculation of $m*(t)$ values (the process input signal). The same computer is also often used to control the startup and shutdown sequence of operations in the plant, to control other feedback loops, monitor particular warning signals, keep performance records, calculate economy of operation, and so on. In such a system, an important problem for the computer design and programmer is the establishment of priorities among these various functions.

APPENDIX—Buffalo Population Model

The z transform is useful in the analysis of linear difference or regression equations—for example, models in which a particular value of the signal is given in terms of earlier sample values. In this Appendix, we illustrate this use of the z transform by consideration of the buffalo population in the western United States during the nineteenth century: particularly, the size that population could have been if the government had adopted a rational policy for managing and preserving this great natural resource.

Actual history of the Buffalo

In 1830, the West was dominated by the buffalo (or American bison) to an extent probably unparalleled in the history of any one animal species. Estimates of the population range from 40 to over 100 million, but a reasonable compromise seems to be 60 million adults. Averaging 1000 pounds each, this represents 60 billion pounds of biomass, compared to today's human biomass in the total United States of less than 30 billion pounds.

The railroad appeared in 1830 and the rapid westward expansion in the United States began. By 1887, there were only 200 buffalo left. Animals were often killed in this slaughter for only the tongues or hides. An average of only 20 pounds of meat (of a possible 500) per buffalo was eaten. The peak was reached in 1872, when national heroes such as Buffalo Bill Cody led the killing of more than seven million, a total made possible by the lack of protective instincts and nonaggressiveness of the animals.

In less than 60 years, the absence of any sort of resource-management policy caused the destruction of a potential, major source of meat. A single buffalo could provide an entire meat supply for five people for a year. In the following paragraphs, we look at the basis for a resource-management policy which would permit maintaining the total population by controlling the number "harvested," or killed for food.

A population model for 1830

Before we can decide on a harvesting policy, we must know how the population rises or falls from natural causes. Unfortunately, most of the data on life expectancy and fertility of the buffalo come from recent studies of herds on well-protected land, where predation and disease are likely different from the open range. If we try to estimate parameters for an 1830 model, these assumptions seem reasonable:

(1) Buffalo reach maturity at age 2. (Actually, fertility increases at a slightly more advanced age, but we can call two-year-olds adults.)

(2) Five percent of the adults die each year. (Life expectancy seems to be about 30 years, but, of course, mortality is actually age-dependent.)

(3) Compared to the adult females alive at the beginning of year n, the number of *new* adult females in year $(n+2)$ is 12 percent. (This percentage depends on the female calves born and the infant mortality rate, which is always high for wild animals. The number of calves born actually depends on the ages of the adults, but for simplicity, we can assume that the 12 percent figure is constant. Calves are usually born in May, and we assume they become adults 19 months later.)

These data are obviously simplifications of the actual description of the population. Furthermore, we have focused on only the female population; the male population is described by a similar set of figures, except that the 12 percent figure of (3) becomes 14 percent of the adult *females* two years earlier (more male than female calves are born.)

When these data are accepted, we can construct a population model. If F_n is the adult female population at the start of year n and M_n the corresponding male population, we have

$$F_n = 0.95 F_{n-1} + 0.12 F_{n-2} \tag{A13-1}$$

$$M_n = 0.95 M_{n-1} + 0.14 F_{n-2} \tag{A13-2}$$

These two equations are merely succinct statements of the assumptions given above.

Analysis of the female population. The recursion equation (A13-1) requires two initial or starting conditions for solution. If 1830 is the start which we can call $n = 0$, we can use the adult female population in 1830 and 1831 — F_0 and F_1. F_2 is found from the equation; then substitution of F_2 and F_1 gives F_3; and so forth.

We can avoid the tedium of this year-by-year solution if we take the z transform of Eq. (A13-1). We then find

$$F(z) = \frac{F_0 z^2 + (F_1 - 0.95 F_0)z}{z^2 - 0.95z - 0.12} \tag{A13-3}$$

The poles of $F(z)$ occur at 1.063 and -0.113:

$$F(z) = \frac{F_0 z^2 + (F_1 - 0.95 F_0)z}{(z - 1.063)(z + 0.113)} \tag{A13-4}$$

The pole at 1.063 shows that the system is unstable (the pole is outside the unit circle). Indeed, once the transient disappears, the female population grows according to the expression

$$(1.063)^n$$

or at 6.3 percent per year. This would correspond to a doubling about every 12 years.*

Thus, Eq. (A13-4) reveals that the female population is given by an expression of the form

$$F_n = A(1.063)^n + B(-0.113)^n \tag{A13-5}$$

where A and B are determined by the initial conditions. The first term is the unstable exponential growth; the second term, the transient which dies out rapidly.

Male population.   The z transform of Eq. (A13-2) for the male population gives an $M(z)$ of the form

$$M(z) = \frac{\text{Numerator}}{(z - 1.063)(z + 0.113)(z - 0.95)} \tag{A13-6}$$

The response is again dominated by the unstable exponential growth $(1.063)^n$, and includes as well two transient terms: $(-0.113)^n$ and $(0.95)^n$. When the transient has died out, the male population increases percentagewise annually exactly as the female.

Harvesting policy

With no harvesting, the population increases 6.3 percent per year. Consequently, we can harvest an appropriate percentage each year and still keep the population constant. If the number of adult females harvested in year $n - 1$ is called H_{n-1}, Eq. (A13-1) for the female population becomes

$$F_n = 0.95 F_{n-1} + 0.12 F_{n-2} - H_{n-1} \tag{A13-7}$$

The simplest harvesting policy is one in which we kill, in any year, a certain percentage of the adult females alive at the beginning of that year, or

---

*In exponential growth or compound interest calculations, the time to double is approximately 72 divided by the annual percentage increase.

$$H_{n-1} = hF_{n-1} \tag{A13-8}$$

Then

$$F_n = (0.95 - h)F_{n-1} + 0.12F_{n-2} \tag{A13-9}$$

and the z transform equation (A13-3) becomes

$$F(z) = \frac{F_0 z^2 + [F_1 - (0.95 - h)F_0]z}{z^2 - (0.95 - h)z - 0.12} \tag{A13-10}$$

To obtain a constant, steady state value for the population, $F(z)$ should have a pole at $z = 1$. Substitution of $z = 1$ into the characteristic equation gives

$$1 - (0.95 - h) - 0.12 = 0$$

or

$$h = 0.07 \tag{A13-11}$$

We can kill 7 percent of the adult females each year and still maintain the population constant. With $h = 0.07$, $F(z)$ becomes

$$F(z) = \frac{F_0 z^2 + (F_1 - 0.88F_0)z}{(z + 0.12)(z - 1)} \tag{A13-12}$$

$$F(z) = \frac{(0.89F_1 + 0.11F_0)z}{z - 1} + \frac{0.89(F_0 - F_1)z}{z + 0.12} \tag{A13-13}$$

In other words,

$$F_n = (0.89F_1 + 0.11F_0) + 0.89(F_0 - F_1)(-0.12)^n \tag{A13-14}$$

The steady state population is the first term $(0.89F_1 + 0.11F_0)$; the transient is the second term, which dies out rapidly.

If there is a year's delay in completing the buffalo census, we should change our harvesting policy so that H_{n-1} depends on F_{n-2}—the number harvested in any year depends on the population at the beginning of the preceding year. Then the equations for the adult female model become

$$F_n = 0.95F_{n-1} + (0.12 - k)F_{n-2} \tag{A13-15}$$

$$F(z) = \frac{F_0 z^2 + (F_1 - 0.95F_0)z}{z^2 - 0.95z - (0.12 - k)} \tag{A13-16}$$

Forcing a pole of $F(z)$ at $z = 1$ specifies k:

$$1 - 0.95 - 0.12 + k = 0 \qquad k = 0.07 \tag{A13-17}$$

(The k is 7 percent, just as h was. When the population is constant, it makes no difference which year is used as a basis for determining the number to be harvested.) Figure A13.1 shows the simulation for both this case and the previous harvesting policy.

The resulting

$$F(z) = \frac{F_0 z^2 + (F_1 - 0.95F_0)z}{(z - 1)(z + 0.05)} \tag{A13-18}$$

$$F(z) = \frac{(0.95F_1 + 0.05F_0)z}{z - 1} + \frac{0.95(F_0 - F_1)z}{z + 0.05} \tag{A13-19}$$

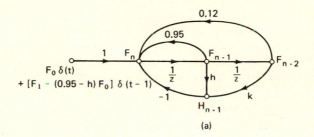

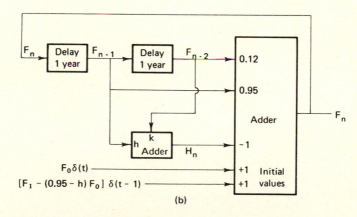

FIG. A13.1 Representation of the recursion equation for adult female population, F_n. The impulse inserted at $t = 1$ must be $[F_1 - (0.95 - h)F_0]$ in order to give a total F of F_1, since the $F_0 \delta(t)$ impulse has circulated through the delay network by $t = 1$. (a) Signal flow diagram; (b) block diagram.

and

$$F_n = (0.95F_1 + 0.05F_0) + 0.95(F_0 - F_1)(-0.05)^n \tag{A13-20}$$

Comparison of Eqs. (A13-14) and (A13-20) shows that the transient dies out much faster when the harvesting policy involves a year's delay. The steady state values differ slightly if F_1 and F_0 are not equal.

Other possible harvesting policies. We have considered only the two simplest harvesting policies, and indeed, only for the adult female population. If we were the Commissioner of Buffalo Resources, we would have to take into account many additional factors in determining a desirable policy. Some of these are:

(1) How do we regulate the number harvested? The simplest procedure (usually followed today in deer management) is to set the length of the hunting season. An alternative is to license each hunter to kill one female and anticipate the fraction who will be successful.

(2) In any case, we have to monitor the population annually to determine, as soon as possible, any deviation from the desired population level due to unexpected harvesting, drought, epidemic, and so forth.

(3) If the population is unexpectedly depleted for any reason, we have to select a policy which allows return to the desired level. The simplest policy is to stop all harvesting and allow the natural growth rate of 6.3 percent per year to restore the herd. Such a decision is likely to be totally impractical economically, however, in the interests of individuals who make their living from the buffalo industry. Consequently, we probably should adopt a reduced harvesting quota until the population has a chance to build up again. Analytically, we might represent this policy H_{n-1} as a constant plus a fraction of the female population over 20 million, where the constant is chosen to ensure stability of a minimum industry.

(4) There is no overwhelming reason for seeking a constant population. We might decide, for example, to try for an F_n which oscillates between 30 and 31 million (20 one year, 31 the next, 30 the next, and so on). Then the z transform $F(z)$ should have a denominator including the factor $(z^2 - 1)$. This can be realized in many ways; the simplest, analytically, is to select

$$H_{n-1} = 0.95F_{n-1} - 0.88F_{n-2} \tag{A13-21}$$

The corresponding $F(z)$ is

$$F(z) = \frac{F_0 z^2 + F_1 z}{z^2 - 1} \tag{A13-22}$$

and any oscillation established by F_0 and F_1 is continued indefintiely.

Comment

The buffalo example is important to illustrate the use of the z transform in obtaining greater understanding of the characteristics of a system under study. The recursion equations for F_n and M_n can be directly programmed on a computer. With an interactive terminal, the analyst can then try different harvesting policies and observe the results. Indeed, a harvesting policy can be adopted which is nonlinear and time variable, rather than the simple linear cases studied above.

The z-transform analysis has the advantage of giving us an understanding of the character of the solution—a global picture within the constraints imposed by the simplicity of the harvesting policies studied. Thus, the z-transform analysis complements the computer simulation.

Finally, the problem is even more important in a broader sense beyond electrical engineering. The intelligent management of our natural resources is clearly a major social challenge for the coming decades.

Questions:

(1) Equation (A13-3) states that the recursion equation

$$F_n = AF_{n-1} + BF_{n-2}$$

leads to the z transform

$$F_z = \frac{F_0 z^2 + (F_1 - AF_0)z}{z^2 - Az - B}$$

Show the validity of this transformation.

(2) Equation (A13-4) gives $F(z)$ for the case of no harvesting

$$F(z) = \frac{F_0 z^2 + (F_1 - 0.95F_0)z}{(z - 1.063)(z + 0.113)}$$

The interpretation of this $F(z)$ is that the adult female population grows exponentially as

$$(1.063)^n$$

after the transient dies out. This growth is represented by the denominator term $(z - 1.063)$. Since the numerator contains a linear factor, can we select F_0 and F_1 so that $(z - 1.063)$ is cancelled, with the result that the population has only a rapidly decaying transient?

(3) Our analysis winding up in Eq. (A13-11) shows that we can harvest 7 percent of the adult females alive at the beginning of that year. The growth rate without harvesting is 6.3 percent per year. Why is the allowable harvesting not then 6.3 percent?

(4) You are harvesting 7 percent of the adult females per year and the population is steady at 30 million. An epidemic in 1850 abruptly drops the population to 25 million adult females. You must allow harvesting of at least a million to keep the industry going. Determine a year-by-year policy which brings the population level back to its normal 30 million within a decade, then holds it constant thereafter.

PROBLEMS

13.1   A signal x with the frequency spectrum shown is sampled with the sampling frequency $\omega_s = 15$ rad/sec. Sketch the spectrum of the sampled signal. Can the original signal be perfectly recovered from the sampled version? Explain your answer. If not, can you suggest a method of approximate recovery and an estimate of the recovery error?

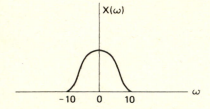

13.2   (a) A sound signal contains components at 120, 260, and 1100 Hz. What is the minimum sampling frequency?
(b) A signal $y(t)$ with the spectrum shown is sampled at 8 KHz. Sketch the spectrum of $y*(t)$.

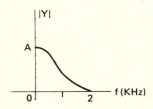

13.3   A moonquake vibration signal is measured by an instrument on the moon, and we wish to radio the data back to earth. The moonquake signal contains components at 0.1 Hz, 0.8 Hz, and 1.4 Hz.
(a) What condition should the sampling frequency satisfy, if we decide to transmit only very short, regular samples of the vibration signal?
(b) The original system designer had to guess at what the vibration signal-frequency components would be, and he chose a sampling frequency of 2 Hz. List the first 10 positive-frequency components of the sampled signal.
(c) Under the conditions of (b), which signal components can we expect to measure accurately at the ground station? What are the filter characteristics that should be used?

13.4   The problem concerns the transfer function

$$T(s) = \frac{200s^2(s + 10)}{(s + 1)^2(s^2 + 2s + 100)(s + 20)}$$

(a) Construct the asymptotic gain plot.

(b) Construct the smooth dB gain plot by adding appropriate corrections. For the quadratic factor, calculate only the correction at the resonant frequency and the bandwidth (for the half power points).

(c) Sketch the phase characteristic approximately.

(d) The input of this network is a signal with a flat spectrum (that is, a spectrum which is constant at all frequencies—we call such a spectrum *white*). We wish to sample the output. What is the minimum sampling frequency if we assume that, when the gain of T is below -40 dB, the signal energy is negligible?

13.5   A typical portion of a signal is shown in the diagram. Estimate the required sampling frequency. (Suggestion: In order to determine the maximum frequency contained in a given signal, we can find the most rapidly changing portion of the signal. If we can fit something like a half cycle of a sinusoid to that changing portion of the signal, the frequency of the sinusoid measures the maximum frequency present in the signal. This is admittedly an approximate approach, but it often gives a fairly valid answer.)

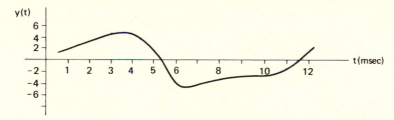

13.6   A sampled system is described by the recursion equation

$$y_n + 2y_{n-1} + 2y_{n-2} = x_n + x_{n-1}$$

where the sampling period is 0.1 sec.

(a) If the x_n are given for n of $0, 1, 2, \ldots$, what are the initial conditions required if we are to find the response?

(b) Is this a causal system? Why?

(c) If every x sample is unity (that is, x is a unit "step" function) and the system starts from rest, what are the first four samples of the response?

(d) What is $Y(z)/X(z)$?

(e) If this equation were an approximation of an analog continuous system, what constraint on the spectrum of x would follow from the sampling theorem?

13.7 (a) What is the definition of the z transform of a signal $y(t)$?

(b) What is the z tranform of the piecewise-linear periodic signal shown, if the sampling period is 0.2 sec?

(c) Can $Y(z)$ be found by first finding $Y(s)$, then consulting the table of z transforms? Show that the closed-form answer obtained in this way gives the same result as (b).

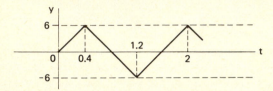

13.8 For the system shown, find $Y(z)$ in terms of $X(z)$, $D(z)$, and $G(z)$.

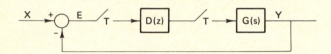

13.9 The z transform of a certain acceleration signal $a(t)$ is

$$A(z) = \frac{2.1}{z} + \frac{3.4}{z^2} + \frac{3.5}{z^3} + \frac{3}{z^4} + \frac{1}{z^5} + \frac{-1}{z^6} + \cdots$$

The sampling period is 2 msec, and we can assume the sampling frequency chosen was sufficiently large. Explain everything the above statements tell you about the signal $a(t)$.

13.10 There are two common ways to multiplex signals (send more than one message simultaneously through a communication system). We can sample each signal and time multiplex. Alternatively, we can frequency multiplex (shift the frequency of each message to a different portions of the frequency spectrum).

A communication system has a bandwidth of one MHz—in other words, we can transmit pulses of duration about one microsecond. Speech has a maximum frequency of 4 KHz. Which multiplexing scheme allows us to send more different speech signals simultaneously through this communication system?

13.11 The feedback control system shown has a

$$G(s) = \frac{K(1 - e^{-Ts})}{s^2(s + 10)}$$

T is chosen as 1/20 sec. Determine the z transfer function $L(z)$ of a computer controller which yields a $c(t)$ with the sample values

$$0, a, 1, 1, 1, 1, 1, \ldots$$

when $r(t)$ is a unit step function. Can a be chosen equal to unity? Could we choose the c sample at $t = 0$ equal to unity? Discuss this mathematically and explain in terms of physical principles.

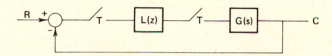

13.12   We are given data representing the value of a signal $y(t)$ every second in time (the sampling frequency is one sample per second). These values are called $y_1, y_2, y_3, \ldots, y_{4000}$ (we have 4000 values). The maximum frequency present in the signal is 1/5 Hz; hence the minimum sampling frequency is 2/5 sample/sec.

We wish to measure the spectrum of the signal below 1/100 Hz; hence, we would like to use only a sample every 50 seconds. If we sample at 1/50 Hz, however, we introduce aliasing errors, and the low-frequency components of y cannot be measured by using every 50th sample ($y_1, y_{51}, y_{101}, \ldots$).

Before we can use only every 50th sample, we must filter the data: that is, process $y_1, y_2, y_3, \ldots$ in such a way that the spectrum bandwidth is decreased without affecting the low-frequency components.

Our data processing consultant suggests that we can filter in a way illustrated by the following simplified example. We form a new sequence (of u's) according to the formulas:

$$u_3 = \frac{1}{3}(y_1 + y_2 + y_3)$$

$$u_4 = \frac{1}{3}(y_2 + y_3 + y_4)$$

$$u_5 = \frac{1}{3}(y_3 + y_4 + y_5)$$

In other words, each u is the average of the last three y's:

$$u_n = \frac{1}{3}(y_n + y_{n-1} + y_{n-2})$$

In terms of the z transforms

$$U(z) = \frac{1}{3} Y(z)\left(1 + \frac{1}{z} + \frac{1}{z^2}\right)$$

or the filter has the transfer function

$$\frac{U(z)}{Y(z)} = \frac{z^2 + z + 1}{3z^2}$$

Since $z = e^{j\omega T}$ (and here $T = 1$, the original sample period of the y data), sketch the magnitude of U/Y versus ω from $f = 0$ to $f = 1/2$ to illustrate that the system tends to cut out the high-frequency components as we go from y to u. (In order to determine the magnitude of the numerator of U/Y at a given ω, it is simple to represent each of the three terms by a vector and add these graphically. This rough calculation can be repeated for various ω over the band of interest.)

In an actual data-processing filter, we would utilize much heavier filtering (that is, averaging over a greater number of samples) in order to smooth the data before calculation of the low-frequency portions of the spectrum. This smoothing permits a reasonable calculation of the spectrum, without using vastly more data than we know are necessary for accuracy.

13.13 We wish to build a system realizing the z transfer function

$$\frac{Y(z)}{X(z)} = G(z) = \frac{2z(z + 1)}{(z - 1/2)(z - 3/4)}$$

The sampling period is 10 msec; the input is a train of sample pulses.

(a) Determine the recursion relation for y_n in terms of previous output samples and the input samples.

(b) Design a system using delay lines, adders, and coefficient multipliers.

(c) Make a partial fraction expansion of $G(z)/z$, multiply by z to obtain an expansion for $G(z)$. Find the corresponding $G(s)$ from a z-transform table and realize this $G(s)$ by an op-amp circuit.

13.14 A linear system described by $T(s)$ is stable if all poles lie within the left half of the s plane. Since

$$z = e^{sT}$$

stability of $T(z)$ requires that all poles lie within the unit circle of the z plane. Through specific examples, show the significance of $T(z)$ poles at

$$z = 1 \qquad z = -1 \qquad z = \pm j \qquad z = 1\underline{/\pm\theta} \qquad z = 2\underline{/\pm\theta}$$

14 SAMPLED AND DIGITAL SYSTEMS

Sampling is important for two reasons. In many systems, the signals are actually sampled—for example, by a human operator or by electronic equipment. In such cases, the determination of system performance requires that we use the z transform to represent the signals accurately.

Second, there are many continuous (nonsampled) systems for which analysis or design is enormously simplified if we replace the signals by their sampled form. In other words, we insert hypothetical samplers just to simplify the analysis.

In this chapter, we want to focus on the latter aspect of system engineering with two primary objectives:

(1) To show the possible simplification of system analysis through sampling.

(2) To develop the basic ideas of digital simulation—simulation (or system synthesis) using digital techniques rather than the op-amps of Chapters 6—10.

14.1 EQUIVALENCE OF CONTINUOUS AND SAMPLED SIGNALS

The sampling theorem of Chapter 13 shows that sampling need not discard any information in the signal. If the sampling frequency is sufficiently high, the sampled signal contains all the information of the original continuous signal.

The sampled signal is not, obviously, identical to the continuous signal, whether we consider the idealized impulse sampling or the practical sampling process (Fig. 14.1). If the δ in Fig. 14.1(c) is very short compared to the system time constants, the short pulses of $y_p(t)$ act as impulses of area $\delta y(t_n)$ for the sample at t_n. In other words, the two forms of sampling are equivalent except

$$y_p(t) = \delta y*(t)$$

For simplicity, then, we can restrict our consideration hereafter to the impulse sampling of Fig. 14.1 (b).

To what extent are $y*(t)$ and $y(t)$ equivalent? When can we insert a sampler in a system without changing significantly the response?

The gain of impulse sampling

In Chapter 13 we saw that the spectrum of the sampled signal $y*(t)$ is the original spectrum plus all the sidebands (the spectrum repeated every $\pm\omega_s$ rad/sec), with each component multiplied by $1/T$ as indicated in Fig. 14.2. Thus, the sampling process, even if we filter the sampled signal, gives a gain of $1/T$.

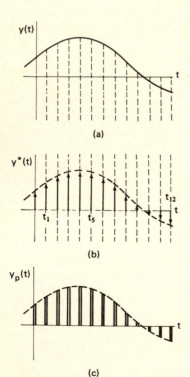

(a)

(b)

(c)

FIG. 14.1 The sampling operation. (a) Original continuous signal; (b) impulse-sampled form (each impulse of area equal to y at the sampling instant). $t_n = nT$ where T is the sampling period; (c) practical sampling (each pulse of width δ; during the pulse, y_p follows the original y).

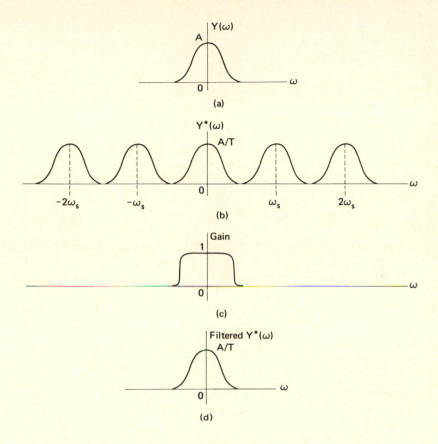

FIG. 14.2 Sampling inherently gives a gain of $1/T$. (a) Spectrum of original signal; (b) spectrum after impulse sampling; (c) ideal filter to convert sampled signal to original form (d) spectrum after filtering.

In other words, if the continuous and sampled signals are equivalent, we still have a gain of $1/T$ when we insert a sampler. If we sample often enough, we might then expect that the two systems of Fig. 14.3 would be equivalent.

The existence of this sampling gain $1/T$ can also be deduced from physical reasoning. If we sample so often that the signal does not change appreciably between samples (Fig. 14.4), the value of y is essentially y_1 over the interval T in Fig. 14.4. If the sampling period is much shorter than the system time constants, the effect of y on the output during the T interval is just the area under y during this time*—in other words,

---

*If a very short pulse is applied to a network or system, the shape of the pulse is immaterial. The contribution to the system output is determined only by the area under the pulse. This is the reason we can use the impulse in analysis to represent a very short pulse in real life.

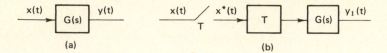

FIG. 14.3 Possible equivalent systems. (a) Original problem; (b) equivalent problem.

$$y_1 T$$

The impulse occurring at the middle of the interval has the area y_1; hence, the sampling process is equivalent to a gain $1/T$.

Equivalence of Fig. 14.3

With the gain T inserted as in Fig. 14.3(b), when are the two systems equivalent?

The answer is whenever the system involves low-pass filtering which essentially removes the sidebands in the manner of Fig. 14.2. In other words, if the $G(s)$ of Fig. 14.3 has a gain which falls with frequency so that the sidebands of $x*(t)$ produce negligible output, $y_1(t)$ is equivalent to $y(t)$.

If this low-pass filtering does not occur, $y_1(t)$ is the sum of low-frequency components (out to $\omega_s/2$), plus high-frequency components resulting from the sidebands of $x*(t)$.

In an actual case, we never have the ideal filtering shown in Fig. 14.2: a filter with zero gain beyond $\omega_s/2$. As long as $G(s)$ attenuates the high-frequency sideband components, however, $y_1(t)$ is the original $y(t)$ plus a small high-frequency signal superimposed. Merely by smoothing the graph of $y_1(t)$, Fig. 14.5, we often can obtain an accurate estimate of $y(t)$.

Summary

Thus, sampling can be artificially introduced to simplify a system analysis problem. The sampler is inserted with a tandem gain of T. The system response is unchanged if the sampler is followed by a subsystem which inserts reasonable low-pass filtering to attenuate the effects of the sidebands introduced in the sampling process.

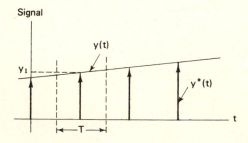

FIG. 14.4 Sampling at a high rate.

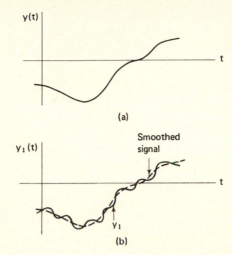

FIG. 14.5 Smoothing of $y_1(t)$ often gives an excellent approximation of $y(t)$. (a) Actual response y; (b) y_1 determined after sampling artificially introduced.

14.2 HOLD CIRCUITS

A common feedback system is shown in Fig. 14.6 (which was discussed in Chapter 13). The sampled error signal $e*$ is modified by the sampled data or digital controller $D(z)$ to generate the sample sequence $m*$. This signal in turn drives the process G to produce the output signal y.

Typically, the process might be a motor, gear train and load, or chemical or manufacturing equipment which is to be controlled. In most cases, it is not practical to drive this $G(s)$ with a train of very short pulses: if the pulses are large enough to deliver the required energy, they saturate the system (and the response is independent of pulse amplitude).

To avoid this difficulty, we commonly add a *simple* filter at the input of the process, a device to smooth out the signal $m*$. Such a smoothing device actually also removes much of the high-frequency content of $m*$, so that the system output y is a smooth, slowly varying signal.

Zero-order hold circuit

While any low-pass filter can be used, often the easiest characteristic to build is that of a hold circuit. The value of the sample is simply held until the next sampling instant (Fig. 14.7). If we recall that a practical sampler usually has a train of pulses of width δ and height equal to the sample values (rather than impulses), we recognize that the

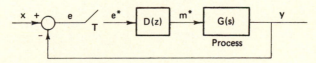

FIG. 14.6 Sampled data feedback system.

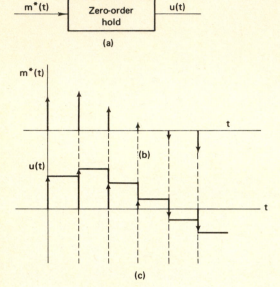

(a)

(b)

u(t)

(c)

FIG. 14.7 Operation of a zero-order hold circuit. (a) Hold circuit; (b) hold-circuit input; (c) hold-circuit output.

hold circuit merely extends the pulse of width δ to width T, so that each pulse lasts until the next sample arrives.

Since this hold circuit is linear and time-invariant, we can describe it by a transfer function. For the determination of that $H(s)$, we can apply any input signal and determine the resulting output. If the input is an impulse of area unity at $t = 0$, the output is a square pulse: the value unity from $t = 0$ to $t = T$. Hence

$$H(s) = \frac{\text{Output transform}}{\text{Input transform}} = \frac{(1/s)(1 - e^{-Ts})}{1} = \frac{1}{s}(1 - e^{-Ts}) \qquad (14\text{-}1)$$

With a hold circuit added, the feedback system of Fig. 14.6 takes the form shown in Fig. 14.8. If we use the analysis procedure developed in the last chapter,

$$\frac{Y(z)}{X(z)} = \frac{D(z)L(z)}{1 + D(z)L(z)} \qquad (14\text{-}2)$$

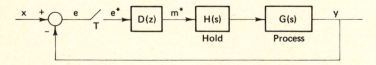

FIG. 14.8 Sampled data-feedback system with hold circuit.

where $L(z)$ is the z transform calculated from a partial-fraction expansion of $H(s)G(s)$. For example, if

$$G(s) = \frac{10}{s} \tag{14-3}$$

we have

$$H(s)G(s) = \frac{10}{s^2}(1 - e^{-Ts}) \tag{14-4}$$

The $(1 - e^{-Ts})$ term corresponds to $(1 - 1/z)$ since $1/z$ is a delay factor; the $10/s^2$ has a z transform given by

$$\frac{10\,Tz}{(z-1)^2}$$

so that

$$L(z) = \frac{10\,Tz}{(z-1)^2}\left(1 - \frac{1}{z}\right) = \frac{10\,T}{z-1} \tag{14-5}$$

[We have to form the $H(s)G(s)$ product before finding the z transfer function corresponding to the product.] Thus, the hold circuit effectively modifies the process to be controlled.

Other hold circuits

We have called H a "zero-order" hold circuit. The adjective in quotes refers to the constant value of the output between sample times; in this interval, the output is a polynomial of order zero. We might also let the output from t_n to t_{n+1} be a linear variation based on the change in the preceding sample interval, as illustrated in Fig. 14.9. Such a circuit is called a first-order hold; the output of this circuit tends to be smoother than the output of the zero-order hold. Actually, the latter (the zero-order hold) is so easy to build that it is used almost exclusively.

Hold circuit as filter

Equation (14-1) for the transfer function $H(s)$ of the zero-order hold circuit indicates that low-pass filtering is achieved. If we evaluate the gain $|H(j\omega)|$ versus frequency ω, we find

$$H(j\omega) = \frac{1 - e^{-j\omega T}}{j\omega} \tag{14-6}$$

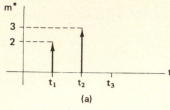

(a)

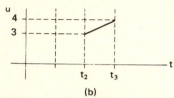

(b)

FIG. 14.9 Operation of first-order hold circuit. (a) Input at t_1 and t_2; (b) output from t_2 to next sample at t_3.

which can be written

$$
\begin{aligned}
H(j\omega) &= \exp\left[-j\omega\,\frac{T}{2}\right]\frac{\exp\left[j\omega\,(T/2)\right]\,-\,\exp\left[-j\omega\,(T/2\right]}{j\omega} \\
&= \exp\left[-j\omega\,\frac{T}{2}\right]T\,\frac{\sin\left(\omega T/2\right)}{\omega T/2}
\end{aligned}
\tag{14-7}
$$

Hence

$$
|H(j\omega)| = T\left[\frac{\sin\left(\omega T/2\right)}{\omega T/2}\right]
\tag{14-8}
$$

Figure 14.10, sketching the gain characteristic, shows the filtering or suppression of the high-frequency sidebands of the sampled input signal.

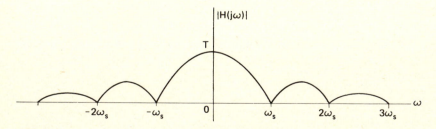

FIG. 14.10 Low-pass filtering achieved by the hold circuit.

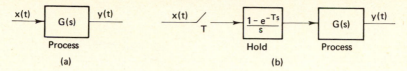

FIG. 14.11 Sampled data equivalent of a continuous system. (a) Continuous system; (b) sampled system.

Insertion of sampler hold-circuit combination

Let us return to the discussion of Sec. 14.1, where we questioned if a sampler could be inserted in a continuous system in order to simplify analysis. We found that the sampler insertion was permissible, provided a gain of T was added, and that the sampler was followed by a low-pass filter. Figure 14.10 indicates that a sampler hold-circuit combination is likely to satisfy both criteria (except in the rare case when the system following the hold circuit contains high gains at high frequencies).

Thus, the two systems of Fig. 14.11 are almost always equivalent, and we can analyze whichever is simpler.

14.3 ANALYSIS OF CONTINUOUS SYSTEMS

With the background of the preceding two sections, we are now ready to see how sampled techniques can be used to analyze continuous systems. We do so in terms of two examples, then continue the discussion somewhat farther in the problems at the end of the chapter.

First example

For the first example, we consider a problem which can also be solved by straightforward Laplace transform analysis. The system and input are shown in Fig. 14.12(a); we wish to find the response $y(t)$. The correct answer is

$$y(t) = \frac{1}{2} t^2 e^{-t} \qquad (14\text{-}9)$$

and is plotted in Fig. 14.13.

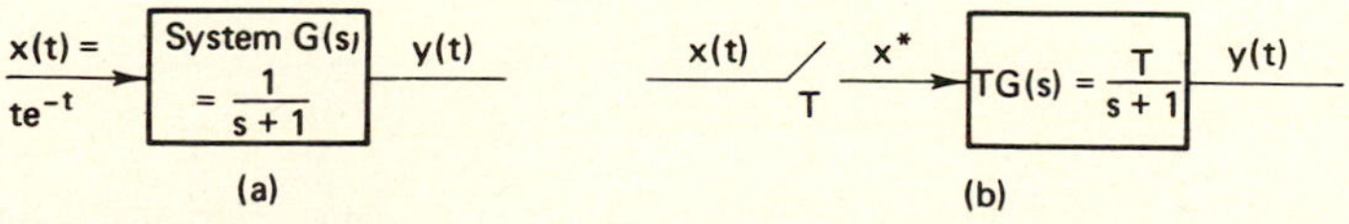

FIG. 14.12 System for first example. (a) Original system problem; (b) sampler inserted.

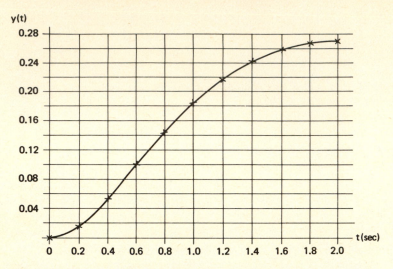

FIG. 14.13 Correct $y(t)$ and calculated sample values (shown by x). The accuracy of the z-transform calculations in this case is startlingly good; usually there is significant error, but the shape of the approximate output is generally correct.

In order to use z-transform or sampled techniques, we insert the sampler and gain T shown in Fig. 14.12(b). (A better approximation would include a hold circuit, but first we try the simpler case.) Now the sequences of sample values for $x(t)$ and $g(t)$ can, in this case, be found by substitution of the successive sampling times in the expressions for $x(t)$ and $g(t)$. If we choose a T of 0.2 sec (one-fifth of the system time constant), we find

$$X(z) = 0 + \frac{0.164}{z} + \frac{0.268}{z^2} + \frac{0.329}{z^3} + \frac{0.359}{z^4} + \frac{0.368}{z^5} + \frac{0.361}{z^6} + \frac{0.345}{z^7}$$
$$+ \frac{0.324}{z^8} + \frac{0.297}{z^9} + \frac{0.290}{z^{10}} + \cdots$$

$$(14\text{-}10)$$

$$G(z) = 0.5 + \frac{0.819}{z} + \frac{0.670}{z^2} + \frac{0.548}{z^3} + \frac{0.449}{z^4} + \frac{0.368}{z^5} + \frac{0.301}{z^6}$$
$$+ \frac{0.246}{z^7} + \frac{0.202}{z^8} + \frac{0.165}{z^9} + \frac{0.135}{z^{10}} + \cdots$$

$$(14\text{-}11)^*$$

*For the first sample g_0 we have used 0.5 rather than 1. The first sample is taken at $t = 0$—the time when $g(t)$ is not defined, since it is discontinuously changing from 0 to 1. Hence, we use the average of the two limits.

The sample values (every 0.2 sec) of $y(t)$ are then found by direct multiplication of the two series. This multiplication is a tedious process without a computer, but the result, after we add the gain $T = 0.2$, is

$$Y(z) = TG(z)X(z) = 0 + \frac{0.016}{z} + \frac{0.054}{z^2} + \frac{0.099}{z^3} + \frac{0.144}{z^4} + \frac{0.184}{z^5} + \cdots$$

$$(14\text{-}12)$$

The result is shown by the crosses in Fig. 14.13; clearly, the approximation in this case is essentially identical to the correct response.

Comment on first example

The above problem would never be worked with z transforms, when the correct Laplace transform solution is so simple. The example does illustrate, however, how systems can be analyzed when the input signal is given graphically (Fig. 14.14):

(1) A sampling period is chosen (much less than the important time constants of G and short enough so that x does not change appreciably during any T seconds).

(2) The sample values, $x^*(t)$ or $X(z)$, are determined form the given $x(t)$.

(3) The sample values, $g^*(t)$ or $G(z)$, are evaluated from a plot of $g(t)$, or from a z transform of each term in the partial fraction expansion of $G(s)$, or from measurement experimentally of the impulse response $g(t)$.

(4) $Y(z)$ is found as $TX(z)G(z)$.

(5) The smooth $y(t)$ is drawn through the sample values, $y^*(t)$ or $Y(z)$.

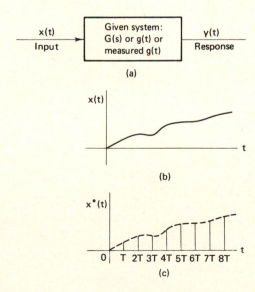

FIG. 14.14 System analysis when input is given graphically (commonly as a measured signal). (a) Problem is to determine y; (b) $x(t)$ given graphically; (c) once T is chosen, sample values $x^*(t)$ can be measured on graph of (b).

The exciting feature of this analysis is that we can find the system response to input signals which are given only graphically—signals which may not be sums of generalized exponentials (and hence may not have simple Laplace transforms).

Second example

For the second example, we consider again the problem shown in Fig. 14.14, but in this case the transfer function $G(s)$ is not given in factored form [so that a partial fraction expansion of $G(s)$ to evaluate $g(t)$ is not convenient]. For example, we might have

$$G(s) = \frac{2s + 4}{s^3 + 2s^2 + 3s + 4} \tag{14-13}$$

Can we find a suitable $G(z)$ without the partial fraction expansion and the term-by-term transformation from s to z?

If we divide through numerator and denominator by s^3 (the highest power of s), we obtain

$$G(s) = \frac{2/s^2 + 4/s^3}{1 + 2/s + 3/s^2 + 4/s^3} \tag{14-14}$$

Now each $1/s$, representing integration, is replaced by a z-operator which represents integration in the z or samples domain.

What should this z representation of $1/s$ be? If the input to the integrator is constant between samples, the integrator output changes each sample period by the preceding input multiplied by T. In other words (Fig. 14.15),

$$y_{n+1} - y_n = x_n T \tag{14-15}$$

In terms of z transforms

$$zY(z) - Y(z) = TX(z) \tag{14-16}$$

or the z transfer function for the integrator is

$$\frac{Y(z)}{X(z)} = \frac{T}{z - 1} \tag{14-17}*$$

*If we assume x varies linearly between sample values, we would obtain the trapezoidal rule for integration, with

$$\frac{Y(z)}{X(z)} = \frac{T}{2} \frac{z + 1}{z - 1}$$

replacing Eq. (14-17).

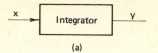

(a)

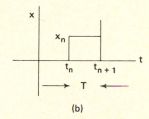

(b)

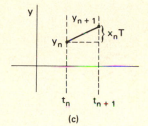

(c)

FIG. 14.15 Integration if x is essentially constant over each sampling interval. (a) y is the integral of x from 0 to t; (b) x is assumed constant from t_n to t_{n+1}; (c) y_{n+1} exceeds y_n by the value $x_n T$.

Thus, for each $1/s$ in $G(s)$, we substitute $T/(z - 1)$. Our example of Eq. (14-11) then becomes

$$G(z) = \frac{2[T/(z - 1)]^2 + 4[T/(z - 1)]^3}{1 + 2[T/(z - 1)] + 3[T/(z - 1)]^2 + 4[T/(z - 1)]^3} \qquad (14\text{-}18)$$

or, if we clear of fractions

$$G(z) = \frac{2T^2 z + (4T^3 - 2T^2)}{z^3 + (2T - 3)z^2 + (3T^2 - 4T + 3)z + (4T^3 - 3T^2 + 2T - 1)} \qquad (14\text{-}19)$$

This $G(z)$ can be written as an infinite series merely by dividing the denominator into the numerator by ordinary long division.

The $G(z)$ found by this procedure is an approximation. The $G(s)$ of Eq. (14-14) describes an integral equation relating x and y:

$$G(s) = \frac{Y(s)}{X(s)} = \frac{2/s^2 + 4/s^3}{1 + 2/s + 3/s^2 + 4/s^3} \qquad (14\text{-}20)$$

means

$$Y + \frac{2}{s} Y + \frac{3}{s^2} Y + \frac{4}{s^3} Y = \frac{2}{s^2} X + \frac{4}{s^3} X \tag{14-21}$$

or

$$y + 2 \int y + 3 \iint y + 4 \iiint y = 2 \iint x + 4 \iiint x \tag{14-22}$$

When we replace $1/s$ by $T/(z - 1)$, we assume that each function to be integrated is essentially constant over every sampling interval. [This is clearly an impossibility. If y is piecewise-constant, $\int_0^t y\,dt$ is piecewise-linear. In Eq. (14-22), $\iint y$ means the integral of $\int_0^t y\,dt$ or the integral of a piecewise-linear signal.] If T is chosen sufficiently small, however, the problem is minor, and this example illustrates one way to find $y^*(t)$ when $x(t)$ is given graphically and $G(s)$ is given as an unfactored transfer function.

Final comment

In this section, our purpose is merely to suggest a few of the ways that sampled data techniques can be used in the analysis of systems which are really continuous. We are really touching superficially on the subject of the numerical analysis of systems. Essentially, we try to find a recursion relation or difference equation relating the output sample values to those of the input (and also to the initial state, if the system is not inert).

The above discussion, perhaps, leaves a dominant impression that this sort of numerical analysis is incredibly tedious. Even after finding the $G(z)$ and $X(z)$ sample values, to determine $Y(z)$ we must multiply $X(z)$ by $G(z)$. Two features of such analysis are not apparent:

(1) The repetitive calculations involved are very simply programmed on a computer
(2) The analysis is often applicable directly to the study of nonlinear systems (which may not be open to any other type of evaluation).

These features alone establish the importance of the numerical analysis of systems.

14.4 DIGITAL SIMULATION

Throughout Chapter 13 and the preceding sections of this chapter, we discussed sampled data-system analysis and design. We noted that sampling may actually be present in a system, or that we may purposefully introduce a sampler to simplify the analysis. In either case, there is an analogy between the continuous and sampled data approaches:

	Continuous	*Sampled data*
Signals {	Continuous time signal $x(t)$	Sequence of samples $x*(t)$
	Laplace transform $X(s)$	z transform $X(z)$
	Differential equations	Recursion equations or difference equations*
Systems {	Impulse response $g(t)$	Weighting function or sampled impulse response $g*(t)$
	Transfer function $G(s)$	z transfer function $G(z)$
	Op-amp integrator simulation (Chapter 6)	Delay element simulation (Sec. 13.7)

Actually, we can carry the analogy in much more detail into the concepts of almost every idea of this book. For example, the state models of Chapter 7 have their analog in the set of first order difference or recursion relations describing the delay element simulation of Sec. 13.7 for $G(z)$. The multidimensional systems of Chapter 8 might be discussed equally well in terms of sampled data systems.

Throughout this discussion, we have considered sampled data systems—that is, systems where samplers convert the continuous signals to sequences of samples or very short pulses. The information is then carried in the amplitude of the various pulses. For many practical reasons, this way of conveying information is not advantageous.

Since the information about each sample (and hence about the original signal over an interval of T seconds) is concentrated in the amplitude of a very short pulse, any noise occurring during that pulse will cause error. The accuracy of the data transmission system depends on the accuracy with which the receiving equipment can read each pulse amplitude.

We can avoid these problems by taking two additional steps with the sampled signal:

(1) quantize each pulse amplitude, and
(2) represent the amplitude of each pulse in binary code.

By quantization, we mean rounding off the pulse amplitude to one of a finite number of levels. In voice communication, it is common to use 2^7 or 128 quantization levels; in other words, each pulse amplitude is assigned one of 128 allowable values. For example, if the pulses are always positive, the permissible values might be 0, 1, 2, 3, . . . , 127 (although there is no real reason for equal spacing; we can emphasize accuracy when signals are small by selecting 0, 1/4, 1/2, 3/4, 1, 2, 3, 4, . . . , 120, 121, 123, 125, and 127).

Once the quantization is completed, the signal can be coded. For instance, if 128 quantization levels are used, a seven-digit binary number suffices to describe the pulse amplitude. The measure of this amplitude can be transmitted by a sequence of seven

*A recursion equation describes y_n in terms of $y_{n-1}, y_{n-2}, \ldots$ and $x_n, x_{n-1}, \ldots$ (past values of the output and past and present values of the input). A difference equation involves a rearrangement of terms so that, instead of y_n, y_{n-1}, and so forth, we have y_n, $(y_n - y_{n-1})$—the first difference, $(y_n - y_{n-1}) - (y_{n-1} - y_{n-2})$—the second difference, and so forth.

Table 14.1 Summary of sampling

Concept	Meaning	Important results	Notation
(1) Sampling	The sampling theorem: a signal with maximum frequency f_m can be represented by a sample sequence taken at an average rate greater than $2f_m$	(a) From a given signal with a known or measurable spectrum, we can find the required sampling rate (b) If we must sample at f_s, we have to filter out all signal components beyond $f_s/2$ (important in data processing) (c) To recover $y(t)$ from $y*(t)$, we need an ideal low-pass filter with cutoff frequency $f_s/2$	f_s sampling frequency, $= 1/T_s$ (samples/sec) T_s period between samples (sec) ω_s angular sampling frequency, $= 2\pi f_s$ (rad/sec) f_m maximum frequency at which spectrum of $y(t)$ has significant components $y(t)$ a signal $y*(t)$ a sample version of the signal $y*(t) = y(t)\,i(t)$
(2) z transform	$Y(z)$ is the Laplace transform of $y(t)\,i(t)$ with e^{-sT} replaced by $1/z$	(a) $Y(z)$ can be found in the form of an infinite series merely by using the sample values: $$Y(z) = y_0 + \frac{y_1}{z} + \frac{y_2}{z^2} + \frac{y_3}{z^3} + \cdots$$ (b) Given $Y(z)$, sample values of $y(t)$ can be found from a power series expansion in $1/z$ (c) When $y(t)$ is the sum of generalized exponentials, $Y(z)$ can be found from a table of z transforms or the formula $$e^{-at} \longleftrightarrow \frac{z}{z - e^{-aT_s}}$$	$Y(z)$ z transform of $y(t)$ $y_0, y_1, y_2, \ldots$ successive sample values of $y(t)$ at $t = 0, T_s, \ldots$ $i(t)$ a periodic train of unit impulses, period of T_s, first at $t = 0$

Table 14.1 Summary of sampling (*Continued*)

Concept	Meaning	Important results	Notation
(3) z transfer function	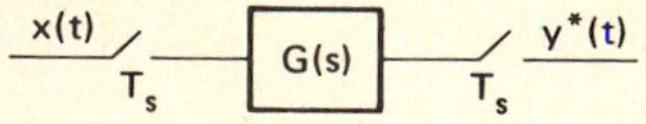 $\dfrac{x(t)}{\diagup T_s}$ — $\boxed{G(s)}$ — $\dfrac{y^*(t)}{\diagdown T_s}$ For the configuration shown above, $G(z) = \dfrac{Y(z)}{X(z)}$	(a) $G(z)$ can be found from a partial fraction expansion of $G(s)$ and use of a table of z transforms (b) For the system shown, we can find y_n (sampled values of y) from a known $x(t)$ and $G(s)$ (c) Even if the two samplers are not present, we can add them if we sample often enough—hence *any* system can be analyzed as a sampled system (d) If we know $x(t)$ samples and the desired samples of $y(t)$, we can find $G(z)$ as a numerical filter or transfer function	$G(s)$ The transfer function of a system y_n n th sample value of $y(t)$

507

on-off pulses (or ± pulses). The real advantage of this binary form is that extra pulses can then be added to correct automatically any one error in the sequence of seven signal pulses: the system automatically corrects itself for any error (cf. Problem 14.8 for an example).

When the sampling is supplemented by quantization and coding, the signal is digital in form. Processing of the signal is accomplished by a digital computer (possibly a special-purpose computer designed for the particular task). If we consider this three-part operation of sampling, quantization, and coding, we see that the sampling is the critical part from the standpoint of system analysis and design. If a reasonable number of allowable levels are used, the quantization merely introduces a small amount of "noise." In turn, the coding merely changes the form in which the signal appears.

Thus, when we discuss the analysis of sampled data systems, we are really considering digital systems. Or, from another viewpoint, digital simulation actually involves the realization of sampled data systems to represent the original components.

14.5 FINAL COMMENT

In these two chapters (13 and 14), we have taken an introductory look at sampling and sampled data systems. In recent years, with the availability of digital circuitry and instrumentation, a rapidly increasing percentage of feedback, instrumentation, and communication systems operate on sampled data. The problems at the end of the previous chapter (13) and this chapter (14) attempt to indicate some of this range of application, although details of analysis and design in many special cases have to be deferred to more specialized books.

Throughout these two chapters, the basic ideas are: the concept of sampling, the z transform of a signal, and the z transfer function of a system. There are summarized very briefly in Table 14.1.

PROBLEMS

14.1 The impulse response of a system is often called the *weighting function*. In this problem, we want to study the role of the impulse response in determining the system output; one goal is to find out where the term weighting function comes from. In other words, what is "weighted?"

We consider the system shown in the figure. We are given the sampling period T and the sample values of $x(t)$ and $g(t)$:

$x(t)$ sample values: $x_0, x_1, x_2, x_3, \ldots$
$g(t)$ sample values: $g_0, g_1, g_2, g_3, \ldots$

(a) Determine the sample values of the response—that is, y_0, y_1, y_2, ...—by using the z transforms. In other words, write $X(z)$, $G(z)$, and then find $Y(z)$.

(b) Now consider one particular sample value of y, for example, y_4. Write the expression for y_4. Notice that y_4 depends on the input at the same time (x_4), the input one sample period ago (x_3), the input two samples ago (x_2), and so on. The input j samples ago is weighted or multiplied by what? In other words, g_7, for example, measures the importance of the input $7T$ ago in determining the output. In this sense, $g(t)$ is a *weighting function*.

(c) Suppose that the $g(t)$ sample values are

$$0, 0, 1, 2, 0, -1, -1, 0, 0, \ldots \quad \text{(all 0 thereafter)}$$

and T is 2 msec.

(1) A step function is applied as the input. How long does the transient last?

(2) What is the memory of the system—that is, how long does the system remember the input?

(3) What is the transportation lag in the system? In other words, after an input is applied, what length of time passes before the response starts?

14.2 A zero-order hold circuit, as described in this chapter, simply maintains the sample value until the next sampling instant. A first-order hold circuit gives a linearly varying prediction: the output at the nth sampling instant is the $(n-1)$st sample value plus the difference from the $(n-2)$nd to the $(n-1)$st samples.

Sketch the gain versus frequency for each of these two hold circuits and compare the two filtering characteristics achieved.

14.3 A network with the transfer function

$$G(s) = \frac{3}{s+1}$$

is driven by a continuous $x(t)$ with sample values every 0.2 sec of

$$0, 1, 3, 1, 0, 0, \ldots$$

(All later values are zero.)

(a) Determine the sample values of the response y. (Should g_0 be called 3 or 1.5?)

(b) If $x(t)$ is piecewise-linear (linear between sample values), compare the answer to (a) with the correct y sample values.

(c) How is the error observed in (b) modified if the sampling period is halved?

14.4 We desire to investigate various z operators corresponding to integration $(1/s)$.

(a) If the integrator input is a train of impulses, the z operator appropriate for integration can be determined. Since the operation is linear, we can use any allowable input. If we choose a simple unit impulse at $t = 0$ as the input, the output of the integrator is a unit step. The sample values (and hence z transforms) of input and output can now be written. The z transfer function is the ratio of these transforms. Complete this determination.

(b) Repeat (a) when the input is a piecewise-constant signal between the sampling instants.

(c) Repeat (a) when the input is a piecewise-linear signal between the sampling instants.

(d) We now have three different integrating operators in z; how would you decide which to use in an actual system study?

14.5 (a) For the transfer function

$$T(s) = \frac{2s + 4}{s^2 + 2s + 3}$$

determine an op-amp realization and the corresponding signal flow diagram.

(b) Now replace each integrator $1/s$ by the z operator found in Problem 14.4(c). Find the corresponding $T(z)$.

(c) Determine a "digital simultation" by realizing $T(z)$ using delay lines instead of integrators as the memory elements.

(d) Describe briefly alternative procedures for realizing a z-domain simulation of the system.

14.6 A second order system is described by the state model

$$\dot{x}_1 = -2x_1 + x_2 + 2u$$
$$\dot{x}_2 = -3x_1 + 4u$$
$$y = x_1$$

The state model is a particularly convenient form to convert to a sampling realization, since we can use the approximation for each of the state variables that

$$\dot{x} = \frac{x_{n+1} - x_n}{T}$$

(an approximation which comes from the definition of the derivative when T is small).

(a) Determine the recursion relations for x_{1n} and x_{2n}.

(b) Take the z transform of the results of (a).

(c) Simulate the system in terms of delay elements, adders, and coefficient multipliers.

(d) Find $Y(z)/U(z)$.

(e) Comment on a comparison of this problem and the results with Problem 14.5.

(f) Outline a general procedure for determining a z-domain simulation from a state model.

14.7 It is often convenient to think of epidemic models in terms of a sampling operation. Each sample instant corresponds to a contact between two people in the total population. If the contact is between an infected person and a susceptible person, the susceptible man becomes infected.

In this problem, we wish to derive a model for the spread of an epidemic. At any "sampling" time (that is, contact time), there are three signals or three types of people:

i_n the number of infected at nT
s_n the number of susceptible at nT
r_n the number of recovered at nT

Here

$$i_n + s_n + r_n = N$$

where N is the total population.

We must decide how people recover (move from i to r). We might assume each man in i remains for a fixed period of time, then moves to r. Alternatively, we can say that there is one transfer from i to r each time the two people in contact are both i or are i and r.*

(a) Determine recursion-equation models for the system under each of the two assumptions in the last paragraph. (Assume that each two-person contact every T units of time involves two people selected *at random* from the total population.)

(b) Comment on the method of analysis, and carry the second model through a few analysis steps to show the detailed procedure when we start from given initial conditions.

14.8 As mentioned in the chapter, a major advantage of digital systems for information processing stems from the possibility of using automatic correction of any single error in a packet of binary digits. This problem demonstrates such capability by a simple example.

We wish to transmit a four-digit binary number—for example,

1 0 1 1

We construct three overlapping circles ($A, B,$ and C) and mark the seven regions defined as $a, b, c, \ldots, g$, as indicated in the first figure. In other words, a is the region common to all three circles, b to circles A and B, and so on.

Now we insert the four digits of our message in regions $a, b, c,$ and d. Then in e we place a 0 or a 1, whichever is required to make the number of 1s inside

*This latter hypothesis is particularly useful in rumor models. Then r is the group no longer spreading the rumor (presumably, people who hear the rumor often enough eventually tire of repeating it).

circle *A even*—in this case, there are two 1s already, so we place a zero in *e*. Analogously, we place a 0 in *f*, a 1 in *g*.

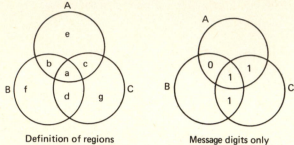

Definition of regions   Message digits only

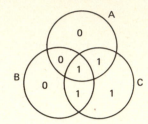

Message plus error-correcting digits

The signal to be transmitted is now

a b c d e f g

or, in our case,

1 0 1 1 0 0 1

The first four digits constitute the message, the next three the basis for error correction.

When the signal is "received," any single error in the packet of seven digits can be corrected. The receiver is, of course, aware of the original coding basis. At the receiver, we know that each of the subpackets

a, b, c, e   *a, b, d, f*   *a, c, d, g*

must have an even number of 1s.

(a) Prove that, if any one digit is received incorrectly, that digit can be uniquely determined.

(b) For an example, suppose we receive 1001001 instead of the transmitted message above. How does the receiver detect the error in the third digit?

(c) If we are willing to use four (rather than three) error-correcting digits, how many message digits can be sent?

(d) Repeat (c) for five error-correcting digits.

(e) As we increase the packet size, we increase the probability of two errors occurring during transmission. Show, by an example, that the system cannot correct two errors.

The procedure described in this problem is known as the Hamming code.

15 STABILITY

The homeward-bound commuter leaves his city office at 5 p.m. to drive to his suburban home 30 miles away. For the first 10 miles, the bumper-to-bumper traffic crawls along at an average speed of 5 miles per hour, even though a few cars leave the stream at each exit from the throughway. Just as his frustration and exasperation are approaching their limits, he passes another exit where, again, a very small percentage of the cars leave. Now traffic flow suddenly speeds up to 35 mph.

What has happened? Why the great difference in average speed with a very small change in the density of cars on the throughway? Is this phenomenon sufficiently predictable to be used for traffic control?

In engineering terms, the departure of a few cars has converted the system from *instability* to *stability*. When the density of cars is too high, the system is unstable and average speed is very low. Reduction of the density results in a stable system.

And, indeed, the phenomenon is used to guide traffic control. In many modern systems, sensors measure the number of cars on a throughway. On the basis of these measurements, entrance of additional cars is controlled. For example, at an entrance ramp, a red light holds cars trying to enter for varying periods of time in order to control the rate at which cars are added to the stream. Thus, when we attempt to enter

513

the throughway, a red light may hold us for perhaps 14 seconds, even though the density of cars visible just ahead is small. Sensors farther downstream have determined that density there is approaching the instability level.

Throughout many aspects of engineering, a primary goal of system design is stability. For example, in the design of a racing yacht, we have to compromise between speed and stability; the boat should be as fast as possible within the constraint that it must be stable in the seas. In the design of a servomechanism for remote control of the position of an object, we want the fastest and most accurate possible response, subject to the condition that the system must be stable (it must not oscillate after a sudden input).

What does the term *stability* mean? In the following sections, we shall try to define stability in mathematical terms, so that we can decide definitely whether a given system is stable or unstable. In more general terms, a system is stable if it is under control—that is, if the responses are following (at least approximately) the input signals. An unstable system is one where we have lost control: the output varies out of control of the input. In an unstable system, the output may grow without bound, it may oscillate indefinitely at a constant amplitude.

System stability in the above sense is a fundamental engineering concept. In this chapter and the next, we want to consider different manifestations of this concept, and ways we can test a system to determine whether it is stable or unstable.

15.1 STABILITY OF LINEAR SYSTEMS

If we first consider linear time-invariant systems, we find that the definition of stability is straightforward. These systems can be described either by transfer functions or by sets of linear differential equations. In the simple case with one input and one output, the notation we use is described in Fig. 15.1.

Definition

Such a linear system is stable if every element of the state transition matrix approaches zero as $t \rightarrow \infty$. In mathematical terms

$$\lim_{t \to \infty} \phi_{ij}(t) = 0 \tag{15-1}$$

for all i and j.*

*Actually, we say that a system satisfying this condition is *asymptotically stable,* the adverb "asymptotically" referring to the behavior as $t \rightarrow \infty$.

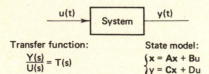

Transfer function:
$$\frac{Y(s)}{U(s)} = T(s)$$

State model:
$$\begin{cases} \dot{x} = Ax + Bu \\ y = Cx + Du \end{cases}$$

FIG. 15.1 System with one input and one output.

There is one obvious implication of this definition. If the input is zero (or remains at zero after a certain time t_1), the output eventually decays to zero. No input results ultimately in no output.

Stability from transfer function

In many cases, evaluation of the elements ϕ_{ij} of the state transition matrix is unnecessarily tedious, and we would like a stability criterion in terms of the transfer function $T(s)$. In *most* cases, the system is stable if

All poles of $T(s)$ lie within the left half plane.

In other words, we can determine stability by finding the poles of the system function $T(s)$. If these poles are all inside the left half plane, the corresponding terms in the natural response are all exponentially decaying, or have envelopes which decay exponentially.

Why do we say "in *most* cases" above? What are the exceptions? The problem is complicated by the fact that, in the calculation of $T(s)$ for a particular output y, there may be zeros of the system characteristic polynomial (poles of T) which are cancelled by numerator factors in T. When looking at the $T(s)$ function, we may not see certain modes of the system.

Figure 15.2 shows two examples of the special cases which may occur. In (a), even if $T = G_1 G_2$ is stable, the state variable x_2 grows without bound (with a component e^{at}). In (b), the pole at a is cancelled by a zero in the transfer function Y/U, but if there is any non-zero initial condition x_3, the output y is unbounded (because Y/X_3 has a pole at $s = a$).

Such "special" cases occur in practice, and the engineer must always be aware of the possibility. In the vast majority of situations, however, the transfer function exhibits all the natural frequencies of the system. Under these circumstances, we can determine system stability by examination of the poles of $T(s)$. In our concern for complete generality, we should not be misled by the amazing complexity of occasional special cases.

Borderline stability

If the transfer function has all poles within the left half plane, the system is stable. If at least one pole lies within the right half plane, the system is unstable. The borderline between stability and instability occurs with poles exactly on the $j\omega$ axis.

If at least one pole on the $j\omega$ axis is multiple (of order two or greater), the system is clearly unstable. The transfer function then has a form such as:

$$T_1(s) = \frac{p(s)}{s^2 q(s)} \quad \text{with a response term } At$$

$$T_2(s) = \frac{p(s)}{(s^2 + \beta^2)^2 q(s)} \quad \text{with a response term } Bt \sin(\beta t + \psi) \tag{15-2}$$

In both cases, the output grows without bound.

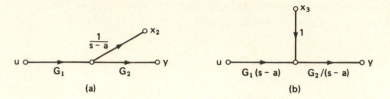

FIG. 15.2 Two cases in which $T = Y/U$ does not reveal system instability.

If all poles on the $j\omega$ axis are simple, the corresponding terms in the response are

$$C \qquad D \sin(\beta t + \theta)$$

for poles at the origin and $\pm j\beta$, respectively. Then the system output approaches a constant or a sinusoidal oscillation of constant amplitude. Such a system is clearly on the borderline between stability and instability; by our earlier definition, the system is unstable (the response does not approach zero as $t \to \infty$). In engineering practice, such systems are often called stable, although the practice varies according to the desired use of the system.

One final comment: the above discussion assumes that the degree of the denominator of $T(s)$ is at least as great as the degree of the numerator. If the numerator degree exceeds the denominator by q, there is a pole of order q at infinity. A partial fraction expansion then yields the form

$$T(s) = \underbrace{As^q + Bs^{q-1} + \cdots + D} + T_1(s)$$

where $T_1(s)$ is regular (degree of denominator greater than of numerator). The part over the brace can be treated separately; it corresponds to an output component which depends on the first q derivatives of the input. Whether we call such a system stable or unstable is, again, a matter of personal preference.

Summary

In the vast majority of systems, we can say the system is stable if the transfer function $T(s)$ possesses poles only within the left half of the s plane.

15.2 ROUTH TEST

The Routh test is a mechanistic procedure to determine, from a given polynomial in s, the number of zeros in the right half of the s plane. Thus, if we use the characteristic polynomial of a system, the Routh test can be used to determine stability: the system is stable if the characteristic polynomial has no zeros in the right half plane.

The impressive feature of the Routh test is that the information about the zeros is obtained directly from the coefficients of the polynomial. No factoring is required.

Procedure

The polynomial is written in descending powers of s, with no term omitted (one or more of the coefficients may be zero):

$$p(s) = a_0 s^n + a_1 s^{n-1} + a_2 s^{n-2} + a_3 s^{n-3} + a_4 s^{n-4}$$

$$+ \cdots + a_{n-1} s + a_n \tag{15-3}$$

Now we form the first two rows (the s^n and the s^{n-1} rows) of an array:

$$
\begin{array}{c|ccccc}
s^n & a_0 & a_2 & a_4 & a_6 & \cdots \\
s^{n-1} & a_1 & a_3 & a_5 & a_7 & \cdots
\end{array}
$$

The first row uses a_0 and every other coefficient thereafter; the second row, the other coefficients.

We now form the next (or s^{n-2}) row according to the following rules:

$$
\begin{array}{c|ccccc}
s^n & a_0 & a_2 & a_4 & a_6 & \cdots \\
s^{n-1} & a_1 & a_3 & a_5 & a_7 & \cdots \\
s^{n-2} & b_1 & b_2 & b_3 & b_4 & \cdots
\end{array}
$$

$$b_1 = \frac{a_1 a_2 - a_3 a_0}{a_1} \tag{15-4}$$

$$b_2 = \frac{a_1 a_4 - a_5 a_0}{a_1} \tag{15-5}$$

$$b_3 = \frac{a_1 a_6 - a_7 a_0}{a_1} \tag{15-6}$$

In other words, b_j is formed form the two rows above: we take the two leftmost entries and the two in the $(j + 1)$ st column—for b_2, we take

We multiply in direction I, subtract the product in direction II, and divide by a_1.

As soon as the third row is completed, we use rows 2 and 3 in the same way to form row 4.

This process is continued until $n + 1$ rows have been formed (the s^0 row is found).

The theorem now states: the number of right half plane zeros of the polynomial equals the number of sign changes as we go down the first column of the array.

Numerical example

To illustrate the procedure, we investigate the polynomial

$$p(s) = s^4 + 2s^3 + 6s^2 + 10s + 7 \tag{15-7}$$

We proceed in the following steps.

(1) The first two rows are formed directly from the given polynomial.

$$
\begin{array}{c|ccc}
s^4 & 1 & 6 & 7 \\
s^3 & 2 & 10 &
\end{array}
$$

(2) The third row has entries

$$\frac{2 \times 6 - 10 \times 1}{2} = 1 \qquad \frac{2 \times 7 - 0 \times 1}{2} = 7$$

(There is no third term in the second row, so we take this term as zero, which we do not bother to write in—zeros are omitted at the end of a row.) The array is now

$$
\begin{array}{c|ccc}
s^4 & 1 & 6 & 7 \\
s^3 & 2 & 10 & \\
s^2 & 1 & 7 &
\end{array}
$$

(3) From the last two rows, we form the next row

$$
\begin{array}{c|ccc}
s^4 & 1 & 6 & 7 \\
s^3 & 2 & 10 & \\
s^2 & 1 & 7 & \\
s^1 & -4 & &
\end{array}
$$

(4) Finally, we form the s^0 row: a single entry $(-4 \times 7 - 0 \times 1)/-4 = 7$, and the array is complete:

s^4	1	6	7
s^3	2	10	
s^2	1	7	
s^1	−4		
s^0	7		

(5) There are two changes of sign in the first column (from 1 to −4 and from −4 to +7); hence

$$p(s) = s^4 + 2s^3 + 6s^2 + 10s + 7$$

has *two* of its four zeros in the right half of the s plane.

A second example
The polynomial

$$q(s) = s^5 + s^4 + 6s^3 + 4s^2 + 5s + 4 \tag{15-8}$$

provides a second example in which we show only the work actually necessary:

s^5	1	6	5
s^4	1	4	4
s^3	2	1	
s^2	$\dfrac{7}{2}$	4	
s^1	$-\dfrac{9}{7}$		
s^0	4		

Therefore, $q(s)$ has two right half plane zeros.

Form of the array
In order to avoid silly errors (for example, omission of a term), we should note that the array has a generally triangular shape. The bottom two rows (s^0 and s^1) have one

entry each, the next two rows up have two entries each, the next two have three entries each, and so forth. Thus, the array has the form

$$
\begin{array}{|cccccc}
\times & \times & \times & \times & \times & \times \\
\times & \times & \times & \times & \times \\
\times & \times & \times & \times & \times \\
\times & \times & \times & \times \\
\times & \times & \times & \times \\
\times & \times & \times \\
\times & \times & \times \\
\times & \times \\
\times & \times \\
\times \\
\times
\end{array}
$$

Multiplication of a row by a positive constant. Any row can be multiplied throughout by any desired positive constant just before a new row is calculated. This possibility often allows us to avoid annoyingly large or small numbers.

The polynomial

$$ s^5 + 4s^4 + 11s^3 + 16s^2 + 5s + 8 $$

provides an illustration. The first two rows are

$$
\begin{array}{c|ccc}
s^5 & 1 & 11 & 5 \\
s^4 & 4 & 16 & 8
\end{array}
$$

Before finding the s^3 row, we can multiply the s^4 row by 1/4; we then proceed to complete the array.

s^5	1	11	5
s^4	1	4	2
s^3	7	3	
s^2	$\dfrac{25}{7}$	2	
s^1	$-\dfrac{23}{25}$		
s^0	2		

The constant by which we multiply (or divide) must be positive.

First special case: two successive rows proportional

The above procedure for formulating the Routh array breaks down when two successive rows are proportional, as illustrated by the following example:

$$p(s) = s^7 + 4s^6 + 5s^5 + 5s^4 + 6s^3 + 9s^2 + 8s + 2 \qquad (15\text{-}9)$$

If we proceed to form the array in the usual fashion, we obtain

s^7	1	5	6	8
s^6	4	5	9	2
s^5	$\dfrac{15}{4}$	$\dfrac{15}{4}$	$\dfrac{30}{4}$	
s^4	1	1	2	
s^3	0	0		

The entire s^3 row consists of zeros; hence, we cannot form the s^2 row since each term is 0/0. The problem arises because the two rows above the all-zero row are proportional—if we divide the s^5 row by 15/4, both the s^5 and s^4 rows are

1   1   2

In other words, when two consecutive rows are proportional, the array cannot be continued.

When we have two proportional rows, the second of these is a factor of the original polynomial. In the above example,

$$s^4 + s^2 + 2$$

is a factor of the original polynomial, $p(s)$ of Eq. (15-9).

We can continue the array by differentiating this factor. In our particular example, the derivative is

$$4s^3 + 2s$$

Hence, the s^3 row of the array is

4   2

and the complete array is

s^7	1	5	6	8
s^6	4	5	9	2
s^5	$\frac{15}{4}$	$\frac{15}{4}$	$\frac{30}{4}$	
s^4	1	1	2	
s^3	4	2		
s^2	$\frac{1}{2}$	2		
s^1	-14			
s^0	2			

The $p(s)$ has two zeros in the right-half plane; furthermore, these are zeros of

$$s^4 + s^2 + 2$$

since the changes of sign occur after the two proportional rows.

There is the phenomenon of two proportional rows whenever the original polynomial has a factor which is purely even. Figure 15.3 shows a few of the zero patterns which lead to such an even factor; in general, the zeros are not only conjugate complex, but also symmetrically located around the $j\omega$ axis.

If differentiation of the second proportional row results in a subsequent portion of the table with no sign changes in the first column, all zeros of the even factor must be on the $j\omega$ axis (since there are none in the right half plane).

Second special case: zero in first entry of a row

The only other special case arises when the first entry in a row is zero. An example is

$$s^6 + 3s^5 + 2s^4 + 6s^3 + 3s^2 + 6s + 3$$

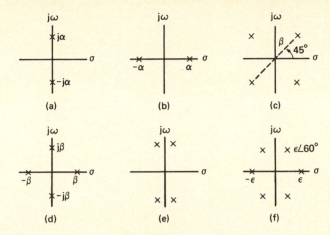

FIG. 15.3 Some zero configurations resulting in even factors in the polynomials. (a) $s^2 + \alpha^2$; (b) $s^2 - \alpha^2$; (c) $s^4 + \beta^4$; (d) $s^4 - \beta^4$; (e) $s^4 + \gamma s^2 + \delta$; (f) $s^6 - \epsilon^6$.

The beginning of the array is

$$
\begin{array}{c|cccc}
s^6 & 1 & 2 & 3 & 3 \\
s^5 & 3 & 6 & 6 \\
s^4 & 0 & 1 & 3
\end{array}
$$

If we now attempt to form the s^3 row, the entries are

$$
-\frac{3}{0} \qquad -\frac{9}{0}
$$

and we are in trouble.

The difficulty can be circumvented if we realize that the polynomial coefficients can be changed slightly without changing the number of zeros in the right half plane (we assume there are no zeros on the $j\omega$ axis). In our example, we might try working with a new polynomial

$$
s^6 + 3s^5 + 2.1s^4 + 6s^3 + 3s^2 + 6s + 3
$$

The array is then

s^6	1	2.1	3	3	
s^5	1	2	2		◄——— After dividing by $+3$
s^4	1	10	30		◄——— After multiplying by $+10$
s^3	-2	-7			◄——— After dividing by $+4$
s^2	13	60			◄——— After multiplying by $+2$
s^1	$+\dfrac{22}{13}$				
s^0	60				

Hence, the altered polynomial has two zeros in the right half plane, and it is reasonable to assume that the original polynomial has the same.

Actually, we cannot be sure of what constitutes a "slight change" in one coefficient. To avoid this question, we return to the original partial table

s^6	1	2	3	3
s^5	3	6	6	
s^4	0	1	3	

Instead of changing the 2 to 2.1 in the first row, we change the zero in the s^4 row to ϵ, a very small positive number. Then we can continue the array:

s^6	1	2	3	3
s^5	3	6	6	
s^4	ϵ	1	3	
s^3	$-\dfrac{3}{\epsilon}$	$-\dfrac{9}{\epsilon}$		
s^2	$1 - 3\epsilon$	3		
s^1	$\dfrac{27}{1 - 3\epsilon}$			
s^0	3			

Thus, there are two changes of sign in the first column; the original polynomial has two zeros in the right half plane.

When a zero appears in the first entry of a row, a second approach is useful in almost all cases.* In the original polynomial $p(s)$, we replace s by $1/s$; then $p(s)$ becomes

$$\frac{1}{s^n} q(s)$$

where $q(s)$ is simply $p(s)$ with the coefficients reversed in order. The right half plane zeros of p also lead to right half plane zeros of q, so we can test q rather than p.

The procedure is illustrated by

$$p(s) = s^5 + 2s^4 + 3s^3 + 6s^2 + 8s + 5 \qquad (15\text{-}10)$$

The start of the array is

$$
\begin{array}{c|ccc}
s^5 & 1 & 3 & 8 \\[4pt]
s^4 & 2 & 6 & 5 \\[4pt]
s^3 & 0 & \dfrac{11}{2} &
\end{array}
$$

Since there is a zero starting the s^3 row, we start the problem over again with consideration of

$$q(s) = 5s^5 + 8s^4 + 6s^3 + 3s^2 + 2s + 1 \qquad (15\text{-}11)$$

(q is written by inspection of p since we only reverse coefficients). The Routh array for q is

$$
\begin{array}{c|ccc}
s^5 & 5 & 6 & 2 \\[6pt]
s^4 & 8 & 3 & 1 \\[6pt]
s^3 & 33 & 11 & \quad\longleftarrow\text{After multiplication by 8} \\[6pt]
s^2 & 1 & 3 & \quad\longleftarrow\text{After multiplication by 3} \\[6pt]
s^1 & -88 & & \\[6pt]
s^0 & 3 & &
\end{array}
$$

*It happens not to work with the preceding example.

Hence, $q(s)$ has two zeros in the right half plane. The same statement can be made for the original $p(s)$.

Stability test

Often we use the Routh test on the denominator of a transfer function (the characteristic polynomial of a system). If we are interested only in whether the system is stable or not, we often do not want to know the number of zeros in the right half plane. Rather, we only seek a yes-or-no answer to the question: Are there any zeros in the right half plane?

In such circumstances, the work associated with the Routh test can be simplified occasionally because of this fact: a change of sign *anywhere* in the array ultimately results in a change of sign in the first column. Thus, as soon as we encounter a minus sign, we can terminate the test if our only interest is in the stability-instability question.

The simplification is illustrated by two examples:

$$s^7 + 2s^6 + 5s^5 + 4s^4 + 5s^3 + 12s^2 + 8s + 4 \qquad s^7 + 2s^6 + 5s^5 + 4s^4 + 6s^3 + 12s^2 + 8s + 4$$

s^7	1	5	5	8
s^6	2	4	12	4
s^5	3	-1	6	

s^7	1	5	6	8
s^6	2	4	12	4
s^5	3	0	6	

The polynomial has one or more zeros in the right half plane since we have a minus sign in the array.

A zero anywhere in the area where the complete array would normally have an entry also means system instability, unless we have an entire row of zeros (the first special case above).*

Thus, the Routh test is a mechanistic procedure for determination of system stability, directly from the coefficients of the characteristic polynomial. It is not necessary to factor the polynomial.

15.3 MORE ON THE ROUTH TEST

A story about the origin of the Routh test is interesting, whether true or not. Routh and Maxwell originally met as classmates at Cambridge University, where they competed vigorously (Routh graduated first in his group, just ahead of Maxwell in 1854). Even though they went their separate ways, the bitter professional competition continued.

In the late 1870s, Maxwell remarked (in a published paper) that determining stability from the coefficients of the characterisitc polynomial (without factoring) was, unfortunately, apparently an insoluble problem. After reading this, Routh

*This is the reason that we can say: any polynomial with missing terms (excluding the first and last) has RHP zeros unless the polynomial is even or odd (every other term missing).

worked diligently for three years to develop the Routh test, essentially as described above. He presented it with the opening remarks, "It has recently come to my attention that my good friend James Clerk Maxwell has had difficulty with a rather trivial problem. . . ."

Regardless of the motivation, Routh's contribution stands as a remarkable achievement. We can examine the stability of a system without factoring the characteristic polynomial (which may be of a very high degree). The simplicity of the Routh array also allows use of the test in several other forms.

Stability as a function of a parameter

The Routh test is frequently useful to determine rapidly the value of a parameter at which the system changes from stability to instability. The array is formed with the parameter left unspecified (for example, indicated by K). Inspection of the first column then allows us to find the critical value of K.

The complexity of the analysis depends on which of the polynomial coefficients depend on K. For example, if K appears only in the constant term, the calculation is relatively simple:

$$p(s) = s^4 + 2s^3 + 4s^2 + 6s + K \qquad (15\text{-}12)$$

The array is formed:

$$
\begin{array}{c|ccc}
s^4 & 1 & 4 & K \\
s^3 & 2 & 6 & \\
s^2 & 1 & K & \\
s^1 & 6-2K & & \\
s^0 & K & &
\end{array}
$$

For the system to be stable (no polynomial zeros in the right half plane), every entry in the first column must be positive. Hence

$$K > 0 \qquad 6 - 2K > 0$$

or

$$0 < K < 3 \qquad (15\text{-}13)$$

When $K = 0$ or $K = 3$, the system is on the borderline of instability. In the former case ($K = 0$), $p(s)$ has a zero at $s = 0$. In the latter case ($K = 3$), the s^3 and s^2

rows are proportional; hence

$$s^2 + 3$$

is a factor of $p(s)$, or there are zeros at

$$s = \pm j\sqrt{3} \tag{15-14}$$

The system is oscillatory at $\sqrt{3}$ rad/sec.

When the parameter appears in one of the other coefficients of the polynomial or in more than one coefficient, the analysis is likely to be more complex. The following example illustrates the difficulty:

$$p(s) = s^4 + 2s^3 + Ks^2 + 6s + 3 \tag{15-15}$$

The array is:

s^4	1	K	3
s^3	1	3	← After dividing by $+2$
s^2	$K - 3$	3	
s^1	$\dfrac{3K - 12}{K - 3}$		
s^0	3		

For stability, we require

$$K - 3 > 0 \qquad\qquad 3\,\frac{K - 4}{K - 3} > 0$$

or

$$K > 3 \qquad\qquad K > 4 \tag{15-16}$$

Hence, $K > 4$ is necessary; the borderline case is $K = 4$, with the frequency of oscillation equal to $\sqrt{3}$ rad/sec.

We can complicate the same polynomial by inserting K in two terms:

$$p(s) = s^4 + 2s^3 + Ks^2 + 6s + (3 + K) \tag{15-17}$$

The array is

$$
\begin{array}{c|ccc}
s^4 & 1 & K & 3+K \\
s^3 & 1 & 3 & \longleftarrow \text{After dividing by 2} \\
s^2 & K-3 & K+3 & \\
s^1 & \dfrac{2K-12}{K-3} & & \\
s^0 & K+3 & &
\end{array}
$$

For stability,

$$
K > 3 \qquad \frac{2K-12}{K-3} > 0 \qquad K+3 > 0
$$

or

$$
K > 6 \tag{15-18}
$$

If we go to higher degree polynomials, the inequalities to make all first column entries positive are of higher degree.

A final example illustrates a more interesting application. Here the analysis is so messy, we shall not bother to complete the array:

$$
p(s) = s^5 + 40s^4 + 500s^3 + (1200 + K)s^2
$$
$$
+ (1010 + 20K)s + (400 + 100K) \tag{15-19}
$$

If we start the array, we obtain

$$
\begin{array}{c|ccc}
s^5 & 1 & 500 & 1010 + 20K \\[2ex]
s^4 & 1 & 30 + \dfrac{K}{40} & 10 + \dfrac{5K}{2} \quad \longleftarrow \text{After division by 40} \\[3ex]
s^3 & 470 - \dfrac{K}{40} & 1000 + \dfrac{35K}{2} & \\[3ex]
s^2 & \dfrac{13100 - \dfrac{13K}{2} - \dfrac{K^2}{1600}}{470 - \dfrac{K}{40}} & 10 + \dfrac{5K}{2} &
\end{array}
$$

The next entry is a cubic polynomial divided by a quadratic.

This example illustrates the possible complexity. Actually, if we carried out the calculations, we would find that stability varies with K according to the diagram

Stable	Unstable	Stable	Unstable	
0	K_1	K_2	K_3	K

The system is stable for $0 < K < K_1$ and again $K_2 < K < K_3$. As K increases (Fig. 15.4), a pair of conjugate complex zeros of $p(s)$ moves into the right half plane, then back into the left half plane, and then finally into the right half plane. There are three values of K at which the system is on the borderline between stability and instability.*

Routh test in root-locus plotting

In Chapter 5, we discussed the root locus method for describing the dependence of system dynamics on a particular parameter. We cited a series of rules which allow construction of the root loci from a given open-loop transfer function. To this list, we can now add an additional technique: the Routh test for determining the value of the parameter when the loci cross the $j\omega$ axis (and the ω values at which this crossing occurs).†

The system of Fig. 15.5 illustrates the combination of the root locus techniques with the Routh test. From the basic rules of Chapter 5, part (b) of the figure is constructed. Before inserting the loci moving into the right half plane, we can use the Routh test to find the crossing points.

Another application of the Routh test

The real part of the poles of a transfer function determine the time constants associated with the free response. In order to be sure that the free response dies out in

---

*In order to realize small sensitivity, we often wish to make K as large as possible. If we operate with K between K_2 and K_3, the system is stable, but we have the unusual situation that a reduction in K causes instability. Such a system is termed *conditionally stable.*

†The root loci can, themselves, be used as a form of stability analysis to determine the values of a parameter which lead to instability.

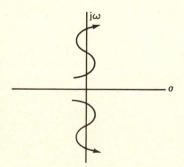

FIG. 15.4 Migration of zeros of $p(s)$, Eq. (15-19), as K increases.

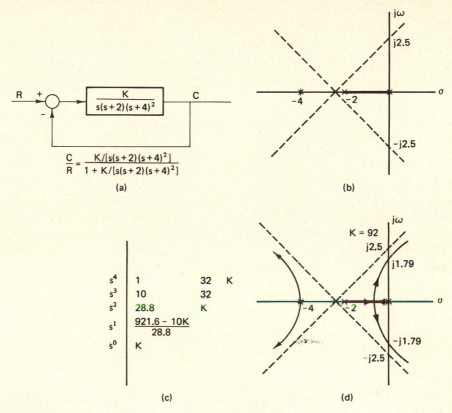

FIG. 15.5 Routh test used in constructing root loci. (a) System with the parameter K; (b) asymptotes and real axis portion of loci shown; (c) Routh array; for zeros on $j\omega$ axis, $K = 92$, $s = \pm j\sqrt{3.2} = \pm j1.79$; (d) Root loci sketched after Routh test.

A seconds, we must have all time constants less than $A/4$.* Hence, all poles must be to the left of the line

$$\sigma = -\frac{4}{A} \tag{15-20}$$

The Routh test can be used to check for this situation. We simply move all polynomial zeros to the right by $4/A$; this yields a new polynomial to which we apply the regular test.

As an example, we have the characteristic polynomial

---

*There may be poles of $T(s)$ which do not show up in the system response because they are very close to zeros. (The residues in these poles are then so small that the corresponding response terms are insignificant.)

$$p(s) = s^3 + 18s^2 + 108s + 216 + K \tag{15-21}$$

We wish to select K as large as possible, subject to the constraint that no time constant exceed 0.5 sec. Hence, all zeros of $p(s)$ should be to the left of $\sigma = -2$.

First we move all zeros rightward by 2 units. This change can be effected by replacing s in $p(s)$ by $z - 2$:*

$$p(z) = (z - 2)^3 + 18(z - 2)^2 + 108(z - 2) + 216 + K$$

$$p(z) = z^3 - 6z^2 + 12z - 8 + 18z^2 - 72z + 72 + 108z \tag{15-22}$$
$$- 216 + 216 + K$$

$$p(z) = z^3 + 12z^2 + 48z + 64 + K \tag{15-23}$$

Now the array for $p(z)$ is:

z^3	1	48
z^2	12	$64 + K$
z^1	$512 - K$	⟵ After multiplication by 12
z^0	$64 + K$	

Therefore, $K = 512$ results in a $p(z)$ with zeros on the imaginary axis at $\pm j\sqrt{48}$. Hence, $K = 512$ gives a $p(s)$ with zeros at

$$-2 \pm j\sqrt{48}$$

and this K of 512 is the maximum allowable value, subject to the constraint that all time constants be at least 0.5 second.

15.4 STABILITY WITH TWO PARAMETERS

When there are two parameters, each of which may vary over a specified range, the investigation of stability becomes far more tedious. We might hope that, in the two-dimensional parameter space, we could investigate only the boundaries of the region. For example, if the two parameters are α and β and each varies from 0 to 1 during operation, we could investigate stability for

---

*The zeros of a polynomial can be shifted horizontally by Horner's method, involving repeated divisions. In our simple example, it is equally easy to substitute $z - 2$ for s.

$$\alpha = 1 \qquad\qquad 0 < \beta < 1$$
$$\alpha = 0 \qquad\qquad 0 < \beta < 1$$

$$\beta = 1 \qquad\qquad 0 < \alpha < 1$$
$$\beta = 0 \qquad\qquad 0 < \alpha < 1$$

Unfortunately, this "boundary" investigation is not enough, as the following example shows.

In the configuration of Fig. 15.6, which involves only three simple negative feedback paths and two frequency-sensitive transfer functions, the parameters α and β represent the gains of active devices (transistors, transducers, motors, and so forth). The normal operating value of each parameter is unity.

According to the usual feedback system theory, analysis of the characteristics of such a system normally involves:

1. Determination of the overall transfer function $C(s)/R(s)$, either in factored form, in order to place in evidence the natural modes of the system, or in terms of gain and phase characteristics

2. Investigation of the effects of decreases in α and β from the normal design values of unity (in order to ascertain the effects of transistor aging, saturation, variation in power-supply voltages, and so forth). It is instructive to make such tests for this particular system.

Direct application of Mason's reduction theorem yields

$$\frac{C}{R}(s) = \frac{\alpha\beta G_1 G_2}{1 + \alpha G_2 + \beta G_2 + \alpha\beta G_1 G_2} \tag{15-24}$$

Substitution of specific functions for G_1 and G_2 and clearing of fractions give

$$\frac{C}{R}(s) = \frac{2\alpha\beta}{s^2 + (1 + \alpha + \beta)s^2 + (1 + \alpha + \beta)s + (0.5625 + 3\alpha + 3\beta + 2\alpha\beta)} \tag{15-25}$$

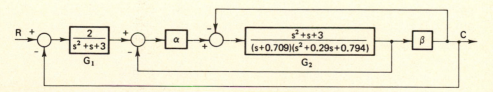

FIG. 15.6 Feedback configuration.

When α and β both assume the normal values of unity

$$\frac{C}{R}(s) = \frac{2}{s^3 + 3s^2 + 3s + 8.5625} \tag{15-26}$$

which can be written in factored form as

$$\frac{C}{R}(s) = \frac{2}{(s + 2.964)(s^2 + 0.036s + 2.9)} \tag{15-27}$$

The system is stable, even though the relative stability would not ordinarily be satisfactory for control applications, with the ζ of the complex pole pair equal to approximately 0.01 (the Q of this pole pair is equal to approximately 50). The light damping is not, however, an essential feature of the system, and we might have readily determined another system with better damping and with the characteristics noted below.

In order to study the effects of variations in α and β, we should make stability tests for the following conditions:

$$\left.\begin{array}{l} \beta = 0, \alpha \text{ varying from 0 to 1} \\ \alpha = 1, \beta \text{ varying from 0 to 1} \end{array}\right\} \quad \text{or} \quad \left\{\begin{array}{l} \alpha = 0, \beta \text{ varying from 0 to 1} \\ \beta = 1, \alpha \text{ varying from 0 to 1} \end{array}\right.$$

The particular system here is unchanged if β and α are interchanged, with the result that the two procedures yield identical results at each step. Hence, we first check stability with $\beta = 0$, α varying from 0 to 1. The characteristic polynomial becomes [from Eq. (15-25)]

$$s^3 + (1 + \alpha)s^2 + (1 + \alpha)s + (0.5625 + 3\alpha)$$

Because the polynomial is of low degree, the Routh test is simple:

$$
\begin{array}{c|cc}
s^3 & 1 & (1 + \alpha) \\[2mm]
s^2 & (1 + \alpha) & (0.5625 + 3\alpha) \\[2mm]
s^1 & \dfrac{0.4375 - \alpha + \alpha^2}{1 + \alpha} & \\[4mm]
s^0 & (0.5625 + 3\alpha) &
\end{array}
$$

The numerator of the s^1 row is positive for all α in the interval $0 \le \alpha \le 1$ (also, for all real α), and the system *is never unstable*. Indeed, for α and β each near zero, the relative stability is appreciably better than for $\alpha = \beta = 1$:

$$\left.\frac{C}{R}\right|_{\substack{\alpha \to 0 \\ \beta \to 0}} = \frac{2\alpha\beta}{(s + 0.709)(s^2 + 0.29s + 0.794)} \quad \left\{\begin{array}{l} \zeta \simeq 0.16 \\ \omega_n \simeq 0.89 \end{array}\right\} \tag{15-28}$$

The second part of the test requires that α be set equal to unity and the stability determined as a function of β; the characteristic polynomial is

$$s^3 + (2 + \beta)s^2 + (2 + \beta)s + (3.5625 + 5\beta)$$

and the Routh array is

s^3	1	$(2 + \beta)$
s^2	$(2 + \beta)$	$(3.5625 + 5\beta)$
s^1	$\dfrac{0.4375 - \beta + \beta^2}{3.5625 + 5\beta}$	
s^0	$(3.5625 + 5\beta)$	

Again, the system is stable for all real β.

If absolute stability only is of interest, we might hope to terminate the stability investigation at this point. The system is stable when

$$
\begin{array}{ll}
\alpha = 0 & 0 \le \beta \le 1 \\
\alpha = 1 & 0 \le \beta \le 1 \\
\beta = 0 & 0 \le \alpha \le 1 \\
\beta = 1 & 0 \le \alpha \le 1
\end{array}
\tag{15-29}
$$

It is revealing, however, to carry the analysis one step further and investigate the stability in the $\alpha - \beta$ plane.

We have tested the stability along the boundary of the region of interest in the $\alpha - \beta$ plane—along the four straight-line segments shown in Fig. 15.7 and described by Eq. (15-29). In the present case, involving only the third order characteristic polynomial, it is not difficult to extend the tests to study the stability boundary in the entire $\alpha - \beta$ plane (although in more practical cases it would be a more tedious task, particularly with more than the two parameters used here). The Routh array in general becomes [from Eq. (15-25)]

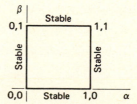

FIG. 15.7 Interesting region of $\alpha - \beta$ plane.

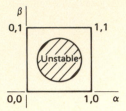

FIG. 15.8 Region of instability.

$$
\begin{array}{c|cc}
s^3 & 1 & (1 + \alpha + \beta) \\[4pt]
s^2 & (1 + \alpha + \beta) & (0.5625 + 3\alpha + 3\beta + 2\alpha\beta) \\[4pt]
s^1 & \dfrac{0.4375 - \alpha - \beta + \alpha^2 + \beta^2}{1 + \alpha + \beta} & \\[8pt]
s^0 & (0.5625 + 3\alpha + 3\beta + 2\alpha\beta) &
\end{array}
$$

When the numerator of the s^1 term is zero, the system is on the boundary between instability and stability. In this particular example, setting the numerator equal to zero yields, after slight rearrangement of terms,

$$
(\alpha - 1/2)^2 + (\beta - 1/2)^2 = 1/16 \tag{15-30}
$$

the equation of a circle of radius $1/4$ and centered at $\alpha = \beta = 1/2$. Hence, the system is *unstable* whenever α and β assume values within the shaded region of Fig. 15.8.

The remarkable aspect of this example is not the existence of the interior region of instability in Fig. 15.8, but rather the fact that the region exists for such a simple configuration and for transfer functions which seem to possess no unusual features. The situation leads to the discouraging thought that, in more realistic and complex systems, stability must be checked for each parameter assuming every possible, distinctly different value (that is, over the entire region of interest in the multidimensional parameter space).

This example is not, in itself, important. It does demonstrate, however, that stability analysis even for relatively simple linear systems can be a difficult task. Yet the study of stability is central in the analysis and design of feedback systems. The engineer must investigate the possibility of system instability over the entire range of possible operating conditions.

15.5 NONLINEAR SYSTEMS

The subject of stability is not as straightforward as, perhaps, is implied in the preceding sections. When a system is unstable, we know from experience that the output never increases exponentially or with an exponential envelope. Instead, the output usually grows into a constant amplitude oscillation, often with a waveform which bears no resemblance to a sinusoid. The discrepancy from our neat theory of the preceding sections results because the system is nonlinear. The nonlinearity is particularly apparent when the signal amplitudes within the system become large. Thus, a system which is linear and unstable leads to signals which grow until the nonlinearities become important.

What dynamic behavior can be observed in nonlinear systems that would not be found in linear systems? This is a very expansive subject, since there are so many different types of nonlinearities. We can list a few, however, which are particularly important in stability studies.

Equilibrium points

The nonlinear system may possess many different equilibrium points—states in which the system is in equilibrium with no tendency to move unless driven. For the linear system made up of passive elements (no batteries or sources, electrically), the only equilibrium point is with all signals zero.

In the network of Fig. 15.9, for example, the two state equations, in terms of the state variables e_C and i_L, are

$$\left. \begin{aligned} C\dot{e}_C &= -\frac{1}{R_1} e_C - i_L + \frac{1}{R_1} e_1 \\ L\dot{i}_L &= e_C - R_2 i_L \end{aligned} \right\} \qquad (15\text{-}31)$$

When the drive (e_1) is zero, the equilibrium point (defined by $\dot{e}_C$ and $\dot{i}_L$ both zero) requires

$$e_C = 0 \quad \text{and} \quad i_L = 0$$

In other words, in this stable linear system, removal of the drive results in both state variables decaying toward zero. Indeed, all voltages and currents within the system decay toward zero.

The system might be unstable, although still linear. For example, if $L = C = 1$, $R_1 = 1/2$, and $R_2 = -1$ in Fig. 15.9, the characteristic polynomial is

$$s^2 + s - 1$$

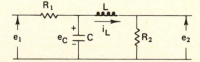

FIG. 15.9 Simple electrical network.

The state equations are

$$\left.\begin{array}{l} \dot{e}_C = -2e_C - i_L + 2e_1 \\ \dot{i}_L = e_C + i_L \end{array}\right\}$$ (15-32)

The equilibrium point is still $e_C = i_L = 0$. As long as e_C and i_L are both zero, the system stays at this point (this is the meaning of "equilibrium"). If there is a slight disturbance (such as noise) which moves the state slightly, however, the system then moves farther and farther away from the equilibrium point. The system is unstable.

Thus, in a linear system, one difference between stability and instability is observed in the response to a disturbance away from the equilibrium point.

The *nonlinear* system shown in Fig. 15.10 possesses *three* equilibrium points (at $x = -1, -2$, and -3). The blocks labelled "Mult." yield an output which is the product of the two input signals. When the circuit is connected by closing the switch, the resulting x may be any of the three values. Which value of x results depends on the initial conditions. In addition, the system may be totally unstable, with the signal x growing without bound (until the multipliers or adder saturate).

In a nonlinear system, then, we may have a number of equilibrium points. Furthermore, each equilibrium point may be stable or unstable. If it is stable, a disturbance slightly away from the equilibrium point results in further motion back toward equilibrium. If it is unstable, a slight disturbance is followed by motion away from that equilibrium. In this sense, the system usually behaves like a linear system in the immediate vicinity of the equilibrium point.

Limit cycles

When a linear system is unstable, the output grows without bound, once started. In the system with two state variables (for example, e_C and i_L for Fig. 15.9), we can

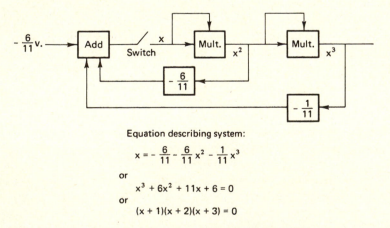

Equation describing system:

$$x = -\frac{6}{11} - \frac{6}{11}x^2 - \frac{1}{11}x^3$$

or

$$x^3 + 6x^2 + 11x + 6 = 0$$

or

$$(x + 1)(x + 2)(x + 3) = 0$$

FIG. 15.10 A system for finding the roots of a cubic equation.

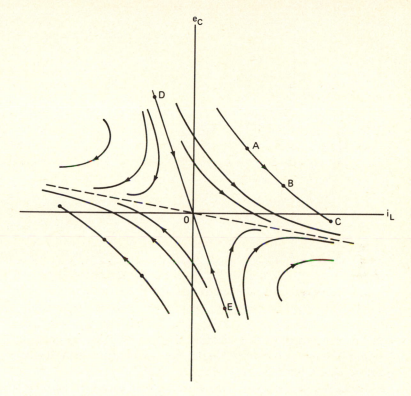

FIG. 15.11 Phase portrait for Fig. 15.9 with $L = C = 1$, $R_1 = 1/2$, $R_2 = -1$.

portray this situation by a *phase portrait*: a plot of the way the two state variables change from any starting conditions. Figure 5.11 shows the phase portrait for the electric network of Fig. 15.9 (with the drive $e_1 = 0$). When the initial conditions start the system at point A, the values of e_C and i_L change with time along the path passing through A. That is, the system moves toward B, then C, and so on, as time progresses. Time itself is not shown on the phase portrait.

In the phase portrait of Fig. 15.11, all paths eventually lead to unbounded values of e_C and i_L [except for initial values along the line DE, which correspond to the special values resulting in a numerator for $E_C(s)$ and $I_L(s)$ which cancels the unstable denominator factor]. Except for these particular values of $e_C(0)$ and $i_L(0)$, any initial conditions result in unbounded system response.

In the phase portrait of a nonlinear system, a very different phenomenon may occur. Here we may find that the paths approach neither an equilibrium point nor infinity, but rather a *limit cycle*: a closed path in the phase portrait. Figure 15.12 displays a *stable* limit cycle $ABCD$—stable because it is approached by paths just inside and just

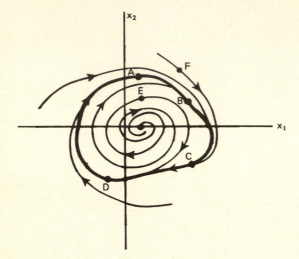

FIG. 15.12   A stable limit cycle.

outside. With initial conditions at either E or F, for example, the system performance approaches a repetitive, constant amplitude oscillation around the $ABCD$ closed contour. The behavior shown in Fig. 15.12 corresponds to a nonlinear system which can exhibit a constant finite oscillation.

We can also have an unstable limit cycle: a closed path, with the paths on either side diverging away from the limit cycle. As long as the system operates on the limit cycle, the output is periodic. Any small disturbance results eventually in operation at an equilibrium point, another limit cycle, or toward infinity.

The existence of a limit cycle is clearly a most important physical property of the system. Unfortunately, the construction of the phase portrait (in order to detect limit cycles) is generally a tedious procedure even for second order systems (with only two state variables, so the portrait can be shown in two dimensions). When the system is higher order, the construction of the phase portrait and its portrayal are hopelessly complex tasks. Thus, in stability analysis of nonlinear systems, we would like a simple test for the presence of a limit cycle. The most useful engineering test is based on the describing-function concept, discussed at the end of the next chapter.

Phenomena in driven nonlinear systems

Equilibrium points and limit cycles discussed above are in terms of undriven systems—systems responding to initial conditions, but with no continually changing drive signal. The phenomena observed when a nonlinear system is driven are exceedingly complex and beyond the scope of this book. We only mention some of the possibilities:

(1) Multivalued responses, depending on the way the drive signal is applied. For example, a sinusoidal input of a particular amplitude and frequency may give very

different outputs when the frequency is increased or decreased toward its desired value.

(2) Outputs which are high harmonics of the input (the phenomenon which is used to multiply the frequency of a sinusoidal drive signal).

(3) Outputs which are subharmonics of the input, with the output frequency 1/2, 1/3, . . . of the input.

15.6 LYAPUNOV STABILITY

There really are two stability questions in the study of nonlinear systems. In most cases, we want to be sure that the system is stable, or we want to determine the value of a particular parameter to ensure stability. In some cases, we want to investigate an instability (particularly a limit cycle) to evaluate the frequency and amplitude of the oscillation.

When we discuss nonlinear systems, care must be taken to define stability mathematically, simply because of the wide variety of dynamic phenomena which may occur. In this section, we introduce such a definition, then give a useful stability criterion which was first presented by A. M. Lyapunov, the Russian mathematician, at the beginning of the twentieth century. In the past 15 years, the Lyapunov criterion has been extensively applied, first in the Soviet Union and later in the United States, in the design of feedback systems.

Definition of stability

In the definition of stability, it is convenient to use the state model. For a nonlinear time-invariant system which is not driven, the state model has the general form

$$\dot{x} = f(x) \tag{15-33}$$

Here $f(x)$ represents a column matrix of nonlinear functions of the state variables. For example, in a second order system we might have

$$\left.\begin{array}{l} \dot{x}_1 = -2x_1 + 3x_2 + x_1 x_2 \\ \dot{x}_2 = +x_1 - 2x_2 - x_2^2 \end{array}\right\} \tag{15-34}$$

or the f functions might be given graphically rather than algebraically.

Furthermore, we are interested in the stability of an equilibrium point x_e—that is, a point where all derivatives of state variables are simultaneously zero, or where

$$f(x_e) = 0 \tag{15-35}$$

The stability question is: If the system sits at x_e, there is no motion. Now the system is disturbed slightly; we move slightly away from the equilibrium point. What happens? Do we remain *near* the equilibrium point? Do we return to the equilibrium point? Clearly, the latter response seems a stronger type of stability than the former.

Stability in the sense of Lyapunov

An equilibrium point x_e is stable in the sense of Lyapunov if, given an $\epsilon > 0$, there exists a $\delta > 0$ so that when the initial condition x_0 satisfy

$$||x_0 - x_e|| < \delta \tag{15-36}*$$

then

$$||x(t) - x_e|| < \epsilon \quad \text{for} \quad t \geq 0 \tag{15-37}$$

In other words, we are given a distance ϵ from the equilibrium point; the response must stay within this distance for all positive time. We can then find a second distance δ so that when the initial conditions are anywhere within this distance of x_e, the response stays within ϵ of x_e. The meaning is shown in Fig. 15.13 for the second order system. Given the larger circle, we can find the smaller circle so that any initial conditions within the shaded circle result in response always staying within the outer circle.

---

*Here $||x_0 - x_e||$ is the norm of $x_0 - x_e$—that is, the distance in the n-dimensional space, where n is the system order. If the system is second order, $||x_0 - x_e||$ is just

$$\sqrt{(x_{10} - x_{1e})^2 + (x_{20} - x_{2e})^2}$$

the distance in the plane from the equilibrium point (x_{1e}, x_{2e}) to the initial condition (x_{10}, x_{20}).

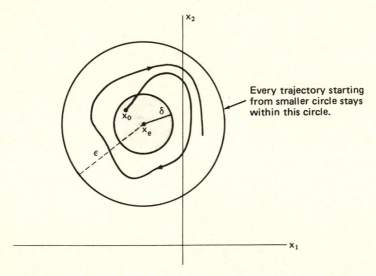

FIG. 15.13 Stability in the sense of Lyapunov. Given ϵ, we can find δ greater than zero.

In a sense, this stability definition does not seem particularly strong; we do not even require that the system return to its equilibrium state after a small disturbance. Actually, the criterion (or definition) really prohibits the response after a disturbance from moving toward infinity, another isolated limit cycle, or another equilibrium point. Thus, if we were considering a linear system, stability in the sense of Lyapunov would guarantee all system poles inside the left half plane (or at best, simple poles only on the imaginary axis).

Asymptotic stability

Usually we are more interested in the tendency of the system to return to the equilibrium point after a disturbance: an idea embraced in the term asymptotic stability. The equilibrium point x_e is asymptotically stable if it is stable *and* if

$$\lim_{t \to \infty} x(t) = x_e \tag{15-38}$$

for $||x_0 - x_e|| < \delta$.

In other words, asymptotic stability adds to the idea of stability the further requirement that every path starting within the smaller circle of Fig. 15.13 must ultimately approach the equilibrium point. Any disturbance within δ from the equilibrium point results in a system response which not only stays within the ϵ circle (since the system is stable), but also finally settles back to the original equilibrium.

Clearly, the linear system with a transfer function having poles within the left-half plane is asymptotically stable. Indeed, such a system is asymptotically stable *in the large* (meaning the return to equilibrium occurs for *any* initial conditions).

Lyapunov function

In order to establish asymptotic stability, Lyapunov presented the following theorem. We consider a nonlinear (or linear) system with an equilibrium point at the origin—in other words, a system described by the equations

$$\dot{x} = f(x) \tag{15-39}$$

$$f(0) = 0 \tag{15-40}$$

The system is asymptotically stable in the large if we can find a function $V(x)$, called the Lyapunov function, which has the following properties

$$
\left.
\begin{array}{lll}
(1) & V(x) > 0 & \text{for} \quad x \neq 0 \\
& V(x) = 0 & \text{for} \quad x = 0 \\
(2) & \dot{V}(x) < 0 & \text{for} \quad x \neq 0 \\
& \dot{V}(x) = 0 & \text{for} \quad x = 0 \\
(3) & V(x) \to \infty & \text{as} \quad ||x|| \to \infty
\end{array}
\right\}
\tag{15-41}
$$

In other words, the Lyapunov function is positive everywhere (except zero at x_e), its derivative is negative everywhere except at the equilibrium point. The Lyapunov function is a generalized concept of the energy stored in the system. If $V(x)$ is positive everywhere, it is an energy type of function. If this stored energy is decreasing everywhere, the system is continually settling back toward equilibrium, and the system is asymptotically stable.

The theorem states that we can prove asymptotic stability by finding *any* function $V(x)$ satisfying these conditions; the $V(x)$ need not be the actual energy stored in the system (it may be impossible to prove that that energy is always decreasing).

Example of a linear system

The linear system of Fig. 15.14 provides an example of the Lyapunov stability criterion. (We know this system is stable since the characteristic polynomial is $s^2 + 3s + 5$.) To demonstrate stability, we proceed as follows:

(1) We are interested in only the undriven system, so $u = 0$ or we neglect the two branches from u.

(2) The state model is (with $u = 0$):

$$\left.\begin{aligned} \dot{x}_1 &= -3x_1 + x_2 \\ \dot{x}_2 &= -5x_1 \end{aligned}\right\} \tag{15-42}$$

The equilibrium point is the origin $x_1 = x_2 = 0$

(3) We now try to find an appropriate Lyapunov function satisfying Eqs. (15-41). As the first trial, we might select

$$V(x) = x_1{}^2 + x_2{}^2 \tag{15-43}$$

This V certainly is positive definite (positive for any x_1 and x_2 except zero at the origin).

(4) We next evaluate the time derivative $\dot{V}(x)$:

$$\dot{V} = 2x_1\dot{x}_1 + 2x_2\dot{x}_2$$

Substitution of the values of $\dot{x}_1$ and $\dot{x}_2$ from the state model gives

$$\begin{aligned} \dot{V} &= 2x_1(-3x_1 + x_2) + 2x_2(-5x_1) \\ \dot{V} &= -6x_1{}^2 - 8x_1x_2 \end{aligned} \tag{15-44}$$

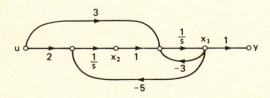

FIG. 15.14 Second order linear system.

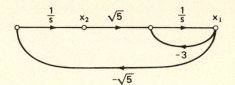

FIG. 15.15 State model giving same characteristic polynomial as Fig. 15.14.

Unfortunately, this V is not negative everywhere away from equilibrium. For example, if $x_1 = 1$ and $x_2 = -1$, $\dot{V}$ is $+2$. Hence our $V(x)$ is not a suitable Lyapunov function. This does not mean the system is unstable; it only means we have not proved stability.

(5) In order to prove the stability of the system, we must try a different $V(x)$. Alternatively, we can transform to a new set of state variables—and it turns out this is a somewhat simpler approach. We simply realize the same characteristic polynomial in a different signal flow diagram (that shown in Fig. 15.15). Then

$$\left.\begin{array}{l} \dot{x}_1 = -3x_1 + \sqrt{5}x_2 \\ \dot{x}_2 = -\sqrt{5}x_1 \end{array}\right\} \tag{15-45}$$

(6) Now $V(x)$ is chosen as

$$V(x) = x_1^2 + x_2^2 \tag{15-46}$$

Then

$$\dot{V} = 2x_1(-3x_1 + \sqrt{5}x_2) + 2x_2(-\sqrt{5}x_1) \tag{15-47}$$

or

$$\dot{V} = -6x_1^2 \tag{15-48}$$

This $V(x)$ satisfies Eqs. (15-41); the system is asymptotically stable in the large.

There are several important comments on the above example. First, the existence of a Lyapunov function satisfying Eqs. (15-41) proves asymptotic stability. Our inability to find such a function may result from system instability, but it may also be the consequence of our failure to guess an appropriate form. In the study of nonlinear systems, there is no known basic procedure for finding a Lyapunov function when it exists (although the literature on nonlinear control includes many procedures which are often successful).

In the study of linear systems, when a Lyapunov function exists it can always be found by the following procedure:

(a) We realize the desired characteristic polynomial by a state model in the general form shown in Fig. 15.16.*

(b) The Lyapunov function is then

$$V(\mathbf{x}) = x_1^2 + x_2^2 + \cdots + x_n^2 \tag{15-49}$$

which always leads to

$$\dot{V}(\mathbf{x}) = -a x_1^2 \tag{15-50}$$

Hence, the linear system is asymptotically stable in the large if (and only if) the signal flow diagram of Fig. 15.16 can be determined with the parameters a, b, c, d, and e all positive.

Relation to Routh test

It is interesting that the coefficients (a, b, c, d, e) in Fig. 15.16 are equal to the ratios of coefficients in the first column of the Routh array—a fact which is apparent if we take a specific example, such as the characteristic polynomial

$$s^4 + 4s^3 + 8s^2 + 4s + 2$$

The corresponding Routh array is

s^4	1	8	2
s^3	4	4	
s^2	7	2	
s^1	$\dfrac{20}{7}$		
s^0	2		

The corresponding signal flow diagram (Fig. 15.17) is developed to realize the characteristic polynomial

$$1 + \frac{4}{s} + \frac{8}{s^2} + \frac{4}{s^3} + \frac{2}{s^4}$$

by identification with the determinant of the diagram

---

*There are non-touching loops in Fig. 15.16. Hence, the evaluation of a, b, c, d, and e requires solving simultaneous algebraic equations, as indicated in Prob. 15.15 at the end of the chapter.

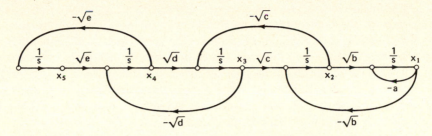

FIG. 15.16 Form of the Lyapunov state model for a fifth order system. In general, there is feedback from x_1 to $\dot{x}_1$, and then from x_j to $\dot{x}_{j+1}$ for all j.

$$\Delta = 1 + \frac{a}{s} + \frac{b + c + d}{s^2} + \frac{a}{s}\left(\frac{c + d}{s^2}\right) + \frac{bd}{s^4} \qquad (15\text{-}51)$$

Equating of like coefficients in the two preceding expressions gives:

$$\left.\begin{array}{llll} a = 4 & & & a = 4 \\ a(c + d) = 4 & (c + d) = 1 & & b = 7 \\ b + c + d = 8 & b = 7 & & c = 5/7 \\ bd = 2 & d = 2/7 & c = 5/7 & d = 2/7 \end{array}\right\} \qquad (15\text{-}52)$$

Thus, the parameters are

$$\left.\begin{aligned} a &= \frac{\text{First entry in } s^3 \text{ row}}{\text{First entry in } s^4 \text{ row}} \\[2mm] b &= \frac{\text{First entry in } s^2 \text{ row}}{\text{First entry in } s^4 \text{ row}} \\[2mm] c &= \frac{\text{First entry in } s^1 \text{ row}}{\text{First entry in } s^3 \text{ row}} \\[2mm] d &= \frac{\text{First entry in } s^0 \text{ row}}{\text{First entry in } s^2 \text{ row}} \end{aligned}\right\} \qquad (15\text{-}53)$$

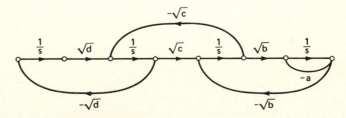

FIG. 15.17 Stability flow diagram.

For linear systems, the Lyapunov and Routh tests are essentially equivalent. No sign change in the first column of the Routh array guarantees that the parameters in the Lyapunov signal flow diagram are positive (the coefficients in the state model are real). Indeed, we can use the Lyapunov theorem to prove the validity of the Routh test.*

The power of the Lyapunov stability criterion is its applicability to a broad range of nonlinear systems. A very few of these applications are illustrated in the problems at the end of the chapter; other illustrations are left to subsequent courses in feedback system design and nonlinear control systems.

15.7 FINAL REMARKS

In this chapter, we have introduced the engineering concept of stability, considered in detail the mechanics of the Routh test for linear systems, and finally indicated some of the complexities encountered in stability studies of nonlinear systems. Stability analysis is continued in the next chapter, devoted to the other important technique for testing: the Nyquist criterion.

In the analysis and design of man-made systems, the easiest definition of instability is that the system runs away, out of control. The unstable linear system exhibits a response which grows without bound. When we try to apply these stability-analysis ideas to social systems or systems involving man, we find that instability (or absence of control) is often much more difficult to describe mathematically. In the commuter-traffic flow example mentioned at the beginning of the chapter, for example, no signal is unbounded (not even the time required to reach home; the commuter always gets there eventually). As another example, Problem 15.14 discusses epidemic models. Here instability means that a significant fraction of the population ultimately has the disease, but eventually the epidemic dies out.

Thus, the instability we have discussed in this chapter is mathematically a relatively simple manifestation of the loss of control, the very special case when the system response grows without bound.

PROBLEMS

15.1　We wish to study the relationship between a manufacturer of tape recorders and a retail outlet, with particular attention on the effects of different pricing policies by the manufacturer and different ordering policies of the retailer. The retailer places his order every Tuesday for a quantity that can be sold in a week. Every Thursday the manufacturer announces the new price for the following week.

---

*This may seem ridiculous, since we have not proved the Lyapunov criterion. However, for linear systems when we can choose the Lyapunov function as $x_1{}^2 + x_2{}^2 + \cdots + x_n{}^2$, this V is just an energy function and stability is apparent. If the system is unstable, we have not proved that the flow diagram with positive real coefficients does not exist.

There are two strategies: the manufacturer's price versus last order strategy, and the retailer's order versus last price strategy. Investigate graphically the stability of such a system when

(a) Retailer's strategy is linear with increasing order at lower price, manufacturer's strategy is linear with lower price as quantity sold increases.
(b) Same linearity but with each slope separately positive, then both slopes positive.
(c) Nonlinear relationships

By a study of the various possible strategies, can you deduce a stability criterion? Under what conditions does a limit-cycle oscillation occur?

15.2 Which of the following polynomials are Hurwitz (that is, all zeros within the left half plane)? If a polynomial is not, how many of its zeros lie in the right half of the s plane?

(a) $s^4 + 4s^3 + 6s^2 + 7s + 2$

(b) $s^5 + 2s^4 + 3s^3 + 8s^2 + 4s + 6$

(c) $s^{12} + 4s^{11} + 6s^{10} + 4s^9 + 3s^8 + 8s^7 + 12s^6$

$$+ 8s^5 + 3s^4 + 4s^3 + 6s^2 + 4s + 1$$

15.3 A linear system has the characteristic polynomial

$$s^5 + s^4 + 4s^3 + 2s^2 + Ks + 1$$

We want to adjust K so the system is stable. Is this possible? If so, for what values of K?

15.4 A system is described by the transfer function

$$T(s) = \frac{s^2 + s + 8}{s^7 + 2s^6 + 8s^5 + 14s^4 + 20s^3 + 16s^2 + 5s + 4}$$

Is the system stable? If not, how many poles are there in the right half plane?

15.5 In the system transfer function

$$G(s) = \frac{K(s + 1)}{s^5 + 4s^4 + 12s^3 + 36s^2 + K(s + 1)}$$

we wish to select K as large as possible, but with the system still stable. Determine the marginal K. What is the oscillation frequency if K assumes this value?

15.6   We are concerned with

$$G(s) = \frac{1}{s(s^5 + s^4 + 6s^3 + 4s^2 + 12s + 4)}$$

in the system shown. As K is increased from zero toward very large values, describe what happens when $K = 8$. Is the system stable or unstable? Discuss. What happens with larger values of K? What happens with a smaller K? How would you describe what is happening in terms of motion of the poles of Y/X?

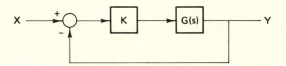

15.7   A system characteristic polynomial is

$$s^8 + 8s^7 + 28s^6 + 56s^5 + 70s^4 + 56s^3 + 28s^2 + 8s + 21 - 12K$$

(a) Determine the values of K (positive and negative) which cause zeros directly on the imaginary axis.

(b) Since there are two values of K, there are two different frequencies at which oscillation may occur. Indicate clearly how you would determine these frequencies (that is, what equation would be solved).

(c) Notice that the characteristic polynomial is

$$(s + 1)^8 + (20 - 12K)$$

We can visualize that this model comes from a feedback system. The Bode plot of the open-loop gain indicates clearly the conditions under which the system is on the borderline of instability. How does this approach compare to the Routh test in simplicity, in susceptibility to arithmetic error?

15.8   (This problem is to be worked only on December 31, 1972). A feedback system depends on two parameters (call them Y and P), one transducer sensitivity (which we shall call H), and an amplifier of gain A. The characteristic polynomial for the multiloop structure is

$$q(s) + Hs^8 + \left[\frac{s(s + 4)}{H} + 1\right] [As^6 + Y + Ps^2(s^2 + 1)]$$

where

$$q(s) = s^{10} + 4s^9 + \frac{1}{4}s^8 + s^7 + \frac{9}{4}s^6 + 9s^5$$

$$+ \frac{7}{4}s^4 + 7s^3 + \frac{3}{4}s^2 + 3s$$

Determine the s^8 and s^7 rows of the Routh array. The result demonstrates conclusively that there is at least one practical application of the Routh test.

15.9   A dynamic system is described by the state equations

$$\dot{x}_1 = x_1 - 2x_2$$
$$\dot{x}_2 = x_1 + 4x_2$$

Using the energy function $V = x_1^2 + x_2^2$, investigate the stability by the Lyapunov method.

15.10   A very famous class of electronic oscillators is described by a nonlinear differential equation of the form

$$\frac{d^2y}{dt^2} - 2(1 - y^2)\frac{dy}{dt} + y = x$$

where y is the response and x is the input. Explain why this system is unstable for small values of y, and stable for large values of y. How can we, therefore, deduce that the response will show a limit cycle or oscillatory value, even if x is non-zero only momentarily?

15.11   A linear dynamical system is described by the differential equation

$$\frac{d^3y}{dt^3} + (a + 1)\frac{d^2y}{dt^2} + (a + b - 1)\frac{dy}{dt} + (b - 1)y = f(t)$$

where a and b are adjustable gains. Determine the range of a and b for which the system is stable. Sketch the region of admissible values on the b-a plane.

15.12   A relay is shown in the sketch. When there is no current through the coil of the electromagnet, the flapper rests against the right stop. When the relay is energized by a current of more than 2, the flapper rests against the left stop and the contacts (not shown) are closed. The curves of force versus displacement of the flapper are also shown.

(a) With no current through the coil, show that the system has a stable equilibrium point for $x = 0$.

(b) Indicate in detail, how to calculate the time required for the relay to close after a current of 2 is suddenly applied through the coil. For this calculation, we can assume that the flapper arm has a mass M, but that the friction (for example, due to air resistance) is negligible.

(c) With $i = 2$, where is the equilibrium point and is it stable?

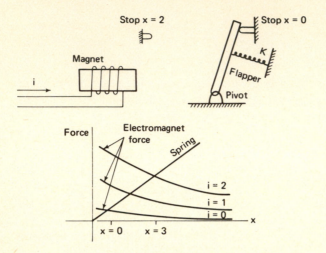

15.13   Consider the number of rabbits and foxes in a controlled environment. The number of rabbits is x_1 and, if left alone, would grow indefinitely (until the food supply were exhausted), so that

$$\dot{x}_1 = kx_1$$

However, with rabbit-eating foxes present

$$\dot{x}_1 = kx_1 - ax_2$$

where x_2 is the fox population. If the foxes must have rabbits to exist,

$$\dot{x}_2 = -hx_2 + bx_1$$

The experiment starts with a population of $x_1(0)$ rabbits and $x_2(0)$ foxes. Assume a, b, h, and k are positive numbers.

(a) Draw a signal flow graph for this system and indicate clearly the initial conditions. Terms involving s must be in the form $1/s$.

(b) Let the output y be the number of rabbits x_1. Determine the functions $G(s)$ and $H(s)$ so that

$$Y = G(s)\,x_1(0) + H(s)\,x_2(0)$$

(c) Let $h = 3$, $k = 1$, $a = b = 2$. Is the system stable? What will be the number of rabbits as t tends to infinity?

(d) Generalize the result of (c). Use Routh's test to determine conditions on the parameters a, b, h, and k so that the system is stable.

(e) If $b/h = k/a$, determine an expression for the number of rabbits as t tends to infinity. Explain your result (for example, in terms of the poles).

15.14 An epidemic model is described in terms of the following signals: N, the total population; s, the number of people susceptible to the disease; i, the number infected at any time; and r, the number who have recovered or are immune. A common epidemic model assumes that the slope of r is B times i, and that the rate of newly infected is A times the product si.

(a) Explain the logic behind this model.

(b) Indicate an appropriate analog computer program, including initial conditions.

(c) What devices might be used to realize the signal multiplier?

(d) How would you define a stability criterion in this problem? Explain in terms of examples.

(e) How might the system of equations be linearized about a particular operating point? Carry through this linearization for the system with $N = 100, A = 0.01$, and $B = 0.75$, about the operating point

$$(i, s, r) = (10, 80, 10)$$

Is the linearized system stable? Is the actual system stable according to your criterion of (d)? Is the linearized system useful?

Notice that this problem illustrates the problem when we attempt to extend the definition and concept of stability to nonelectrical systems. In an epidemic analysis, the number of infected clearly cannot approach infinity (it can never exceed the total population).

15.15 An important model in biology describes the dynamics of the populations of two species, one of which is a predator on the other, the prey. The classical model for this situation is the Lotka-Volterra equations, with B, K, and D positive:

$$x' = Bx - Kxy$$
$$y' = Kxy - Dy$$

(a) Describe the reason for each term in the model. Which is the predator, which is the prey? What assumptions are made which are counter to common sense? If these assumptions are to be valid, what can we say about the region within which the model is useful?

(b) Where are the equilibrium points?

(c) From a linearized model in the vicinity of each of these equilibrium points, what can we say about stability? Describe the time behavior of x and y for various initial states near these equilibrium points.

(d) Discuss the model: its value, its limitations, and ways it might be improved.

16 NYQUIST
STABILITY CRITERION

Stability is usually a system requirement. As we develop the detailed design of a complex feedback system, we frequently pause to make sure that the individual subsystem (or the final system) is stable. Often we arrive at the point depicted in Fig. 16.1.

We have designed and built the system $G(j\omega)$. We are ready to connect the feedback around G—the path through the sensor H. Before the system is actually built and connected, we want to determine if it will be stable. In other words, from the known characteristics of G and H, can we determine the stability of the closed-loop system?

If $G(s)$ is known as a transfer function and H is a constant, we can use the Routh test on the characteristic polynomial. The overall system function is

$$\frac{C}{R}(s) = \frac{G}{1 + GH} \tag{16-1}$$

and the characteristic polynomial is just the numerator of $(1 + GH)$.

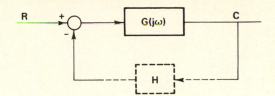

FIG. 16.1 $G(j\omega)$ is determined and the loop is to be closed through the feedback sensor H.

In many cases, however, $G(s)$ is not known as a ratio of polynomials in s. Perhaps the physical laws underlying the operation of G are not known.* Perhaps we cannot measure all the parameters determining the transfer function coefficients.

In many such cases, we can measure the frequency response $HG(j\omega)$—the gain and phase variations with ω. Can we determine the stability of the closed-loop system from $HG(j\omega)$, the frequency characteristics of the open-loop system? The answer is yes; the procedure is the Nyquist stability test, which is the central topic of this chapter.

16.1 HISTORICAL BACKGROUND

Until the late 1920s, the answer to our question seemed obvious. We visualize the loop broken at AA' in Fig. 16.2. At point A, a sinusoidal signal is inserted at a frequency ω_1. This signal passes through G and H, then is reversed in sign (shifted $180°$ in phase) from B to A'.

If the phase shift from A all the way around the loop to A' is $360°$, the signal at A' is in phase with that at A. If the A' signal is *larger* than that at A, we should be able to close the loop and remove the drive signal. The oscillation would grow as the signal traveled around and around the loop. In other words, the stability criterion could be stated:

> The system is unstable if, at any frequency where the total loop phase shift is $360°$ (or $0°$ or $720°$, and so forth), the gain is greater than unity.

This was called the Barkhausen stability criterion.

In the late 1920s, Black, of Bell Telephone Laboratories, found experimentally that he could build feedback amplifiers that violated the above stability criterion. The amplifiers had a gain greater than unity at a frequency where the loop phase shift was zero, and still were stable. This laboratory discovery led directly to the work by Nyquist which established a theoretical stability criterion.

We should note that the incorrect Barkhausen criterion does make sense, if a feedback system is viewed as a signal traveling around and around the loop. If we have a signal of amplitude 2 at A in Fig. 16.2 and this causes an in-phase signal of 6 at A', we would think that we could apply the A' signal to A to generate first 18 at A', then

---

*An example is the complex system which determines missile stability during the thrust phase, when fuel may be sloshing around the partially empty tanks.

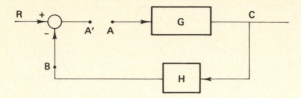

FIG. 16.2 Closed-loop system with loop broken at AA'.

54, and so on, until the response blows up completely. The fallacy in such an argument is not at all obvious. Apparently we just must not think of signals as "traveling around a loop." If the system is stable and a sinusoid is applied as R at a frequency where

$$GH = -3 \qquad (H = 1, \ G = -3) \tag{16-2}$$

the resulting overall transfer function is

$$T = \frac{G}{1 + GH} = \frac{-3}{1 - 3} = \frac{3}{2} \tag{16-3}$$

An R signal of amplitude 4 causes an output C of 6 and a signal at A of C/G or -2. In the steady state, equilibrium is achieved; no signal circulates indefinitely around the loop.

16.2 NYQUIST CRITERION

The Nyquist stability criterion determines the stability of the closed-loop system in terms of the frequency characteristics of the open-loop system. In Fig. 16.3, if we close the loop by connecting A to A',

$$T(s) = \frac{G(s)}{1 + G(s)H(s)} \tag{16-4}$$

Stability depends on the zeros of $1 + G(s)H(s)$ or the points at which

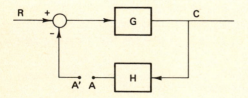

FIG. 16.3 Definition of terms in statement of Nyquist criterion.

$$G(s)H(s) = -1 \tag{16-5}$$

The system is stable only if all roots of Eq. (16-5) lie within the left half of the s plane. The Nyquist criterion states:

$[G(s)H(s)]$ is plotted as the contour Γ_{GH} in the GH plane as s varies along the contour γ_s shown in Fig. 16.4. Then

$$N = P_T - P_{GH} \tag{16-6}$$

where $\begin{cases} N \text{ is the number of clockwise encirclements of the point } (-1,0) \text{ in the} \\ \quad GH \text{ plane by } \Gamma_{GH}, \\ P_T \text{ is the number of poles of } T(s) \text{ inside the right half } s \text{ plane, and} \\ P_{GH} \text{ is the number of poles of } GH \text{ in the right half of the } s \text{ plane.} \end{cases}$

It so often happens that, in an attempt to make a perfectly general statement, we end up with a theorem which sounds discouragingly complex. Actually, in almost all practical situations, the theorem is quite simple and straightforward, as we shall now attempt to show.

The contour γ_s

To make the test, we allow s to vary along a large semicircle enclosing the right half of the s plane. As a specific example, we might know that the motor and amplifier represented by the G block give essentially no gain at frequencies higher than 10 rad/sec. Then, certainly, all interesting behavior of GH is below 30 rad/sec, or

$$|s| < 30 \tag{16-7}$$

For larger $|s|$, $|GH|$ is effectively zero. Accordingly, the right half of the s plane can be enclosed by a semicircle of radius 30. We have defined γ_s.

Now, to let s vary along this contour, we simply pick a sequence of values, some of which might be

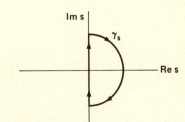

FIG. 16.4 γ_s contour includes the $j\omega$ axis and a semi-circle large enough to enclose any interesting variation of GH inside the right-half plane.

$$s = j0 \qquad\qquad s = 30 \angle{-80}°$$

$$s = j2 \qquad\qquad s = -j30$$

$$s = j4 \qquad\qquad s = -j6$$

$$s = j30 \qquad\qquad s = -j2$$

$$s = 30 \angle{80}° \qquad\qquad s = j0$$

$$s = 30 \angle{0}°$$

The contour Γ_{GH}

As s takes on this succession of values around γ_s (Fig. 16.5), we determine for each s the corresponding value of GH. These values of GH are plotted in the GH plane to generate the contour Γ_{GH}. In the mathematical sense, Γ_{GH} is the mapping of γ_s.

Figure 16.6 shows a possible pair of contours, in the s and GH planes. When $s = 0$, GH has the value $+2$. As s moves up the $j\omega$ axis, GH changes in values, spiraling clockwise through $-180°$. By the time we reach point A on γ_s, GH is essentially zero (at the origin of the GH plane). All the way around the semicircle in γ_s, $|s|$ is large, and GH stays essentially at the origin. In other words, the semicircle in γ_s corresponds to the vicinity of the origin for Γ_{GH}. As we leave B up toward the origin, Γ_{GH} now starts out from the origin and spirals toward $+2$.

Along the Γ_{GH} contour, we can mark a few values of s (Fig. 16.6) in order to show which part of the Γ_{GH} contour corresponds to which part of γ_s. Each value of s corresponds to a single point on the Γ_{GH} contour.

Criterion for Fig. 16.6

For the particular example shown in Fig. 16.6, there are no encirclements of the -1 point (the point on the plane where $GH = -1$). As we travel around Γ_{GH} in the direction of the arrows (which correspond to the arrows on γ_s), we are always to the right of the -1 point. We can see by inspection that there is no encirclement. Consequently, the Nyquist criterion

$$N = P_T - P_{GH} \tag{16-8}$$

gives, in this case,

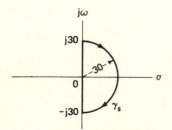

FIG. 16.5 Contour in s plane.

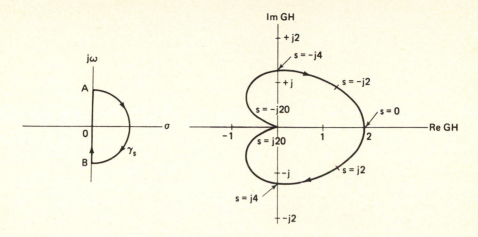

FIG. 16.6 Typical γ_s and Γ_{GH} contours.

$$0 = P_T - P_{GH} \tag{16-9}$$

P_T (the number of RHP poles of T) equals P_{GH} (the number of RHP poles of GH).*

In other words, the criterion only tells us how the stability of T compares with that of GH. In this example, if the open-loop GH is stable ($P_{GH} = 0$), the closed-loop system is also stable. If GH is unstable, T is also.

16.3 SIMPLIFICATIONS USUALLY VALID

The Nyquist test, as described in the preceding section, seems discouragingly complicated. We have to construct a polar plot of GH as s varies around the entire γ_s contour. Fortunately, there are usually simplifications.

Semicircle of γ_s

Our one example showed that the entire semicircle of the γ_s contour maps into the region very close to the origin of the GH plane. Usually, if GH describes a practical set of components, the gain falls off to zero at very high frequencies. Then the semicircle in γ_s can be neglected since the corresponding part of Γ_{GH} is near $GH = 0$, and we are only interested in the encirclements of $GH = -1$.

Negative imaginary axis of γ_s

If the semicircle of γ_s is of no concern, we need only let s vary along the entire $j\omega$ axis. Here, again, we can simplify. If GH describes a real system, its coefficients as a transfer function in s are real. Then $GH(-j\beta)$ is just the complex conjugate of $GH(j\beta)$ for any value of β. In other words, if we plot Γ_{GH} for s along the *positive* imaginary

*RHP is the abbreviation for right-half plane (where the reference is to the s plane).

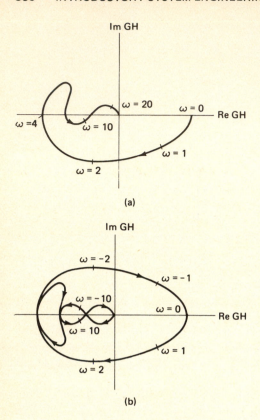

(a)

(b)

FIG. 16.7 We need plot only Γ_{GH} for s positive imaginary. (a) Γ_{GH} when $s = +j\omega$, ω varying from 0 to $+\infty$; (b) complete Γ_{GH} contour—curve of (a) plus its conjugate.

axis, we can complete the plot by adding the conjugate curve. Figure 16.7 shows one example.

This simplification is particularly important (although obvious) when we measure experimentally the gain and phase characteristics of GH to collect the data for the Γ_{GH} plot (the Nyquist plot). In the system of Fig. 16.8, we can measure the Nyquist plot by using a sinusoidal signal of variable frequency for either R or E as the input, with the signal at A as the response. Such an experiment might yield the curve of Fig. 16.7(a). We start with ω at or near zero and then, step-by-step, increase the frequency until the gain has dropped essentially to zero—in this example, $\omega = 20$. The complete Nyquist plot can then be drawn by adding the conjugate curve.

Indeed, usually we do not bother to draw the conjugate curve (for ω negative). The

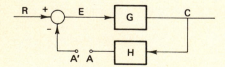

FIG. 16.8 Feedback system with loop opened.

mapping of the positive $j\omega$ axis is loosely called the Nyquist plot of the open-loop transfer function GH; to apply the theorem, we visualize the complete plot.

Nyquist equation

The criterion is centered on the equation

$$N = P_T - P_{GH} \tag{16-10}$$

The number of clockwise encirclements of the -1 point equals the number of RHP poles of T minus the number of RHP poles of GH.

In many cases, GH is known to be stable ($P_{GH} = 0$). If we can measure the Nyquist plot experimentally, GH *must* be stable (otherwise the output at A in Fig. 16.8 would grow without bound). If we draw the Nyquist plot theoretically by substituting values of $j\omega$ into $GH(s)$ or by constructing the Bode plots and then converting to polar form, we know very often that the open-loop system is stable. [We may have $GH(s)$ in factored form, or we can use the Routh test on the denominator.]

If GH is indeed stable, then the Nyquist-criterion equation reduces to

$$N = P_T \tag{16-11}$$

The number of clockwise encirclements of the -1 point by the Nyquist diagram is the number of RHP poles of T.*

16.4 EXAMPLES OF NYQUIST PLOTS

Each of the following examples refers to the single-loop configuration of Fig. 16.9. Here, instead of the H feedback block we considered earlier, we use unity feedback, and insert a variable gain K in the forward path. Then the overall system function is

$$T(s) = \frac{C(s)}{R(s)} = \frac{KG(s)}{1 + KG(s)} \tag{16-12}$$

and stability depends on the encirclements of the point

$$KG = -1 \tag{16-13}$$

(the point where the denominator of T is zero).

---

*In this case, of GH stable, there cannot be a net number of counterclockwise encirclements (N negative) of the -1 point by the Nyquist plot. If GH is unstable, there may be such a negative value of N—for example, if T is stable and GH has 2 poles in the RHP, $N = -2$.

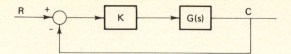

FIG. 16.9   Single-loop system for first examples.

A stable system

Figure 16.10 shows the first example: the Nyquist plot for $KG_1(s)$, drawn with the value of K set at unity. In the preceding section it was suggested that the plot be shown only for positive ω. The arrow indicates the direction of increasing ω.

What can we tell about the system from this plot?

(1) If $G_1(s)$ itself is stable, the closed-loop system is stable for any positive K (no matter how large we make K—how much the plot is magnified—, the -1 point is never encircled).

(2) The plot incidentally suggests that $G_1(s)$ is a constant (2) at $s = 0$, and that the degree of the denominator is 2 greater than that of the numerator (since the phase shift at high frequencies is $-180°$).

A system that can be made unstable

Figure 16.11 is a more interesting Nyquist diagram, again referring to the single-loop feedback system of Fig. 16.9. This plot tells us that, if G_2 itself is stable,*

(1) With $K = 1$, the system is stable.

---

*If G_2 were unstable, the closed-loop system would be unstable for any positive K. As K increases through 3, two *more* poles of T move from the left to the right half s plane.

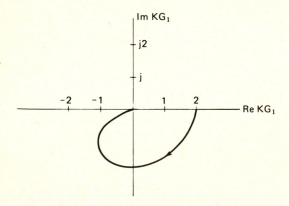

FIG. 16.10   Nyquist plot for $KG_1(s)$, drawn for $K = 1$.

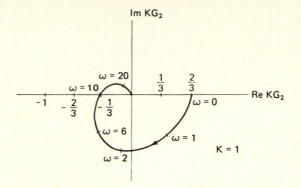

FIG. 16.11 Nyquist plot for second example, $KG_2(s)$.

(2) As K is increased, the curve is simply magnified—for example, at any ω, the phase is unchanged but the magnitude is multiplied by K. Therefore, the system is unstable for $K > 3$. Indeed, when $K > 3$, $T(s)$ has two poles in the right half of the s plane.

(3) When $K = 3$, the Nyquist plot passes through the -1 point at $\omega = 10$ and $\omega = -10$. Hence, for $K = 3$, $T(s)$ has poles on the $j\omega$ axis at $s = \pm j\,10$. The system is on the borderline between stability and instability.

A more complex system

An even more interesting Nyquist plot is shown in Fig. 16.12. Here the phase angle varies from $0°$ to $-290°$, then comes back to $-170°$, then goes more negative toward $-360°$ at high frequencies.

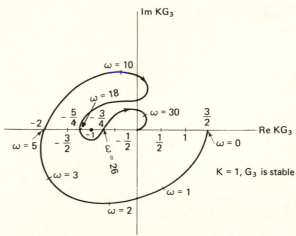

FIG. 16.12 A Nyquist plot which shows a system stable for a range of values of K.

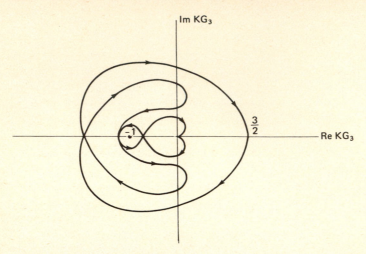

FIG. 16.13   Complete contour for KG_3.

The interesting feature of Fig. 16.12 is that the system *is stable*; the -1 point is *not* encircled by the contour shown in the figure.

In order to see that there is no encirclement, we can work with either Fig. 16.12 or the complete contour of Fig. 16.13. In either case, place your left forefinger on the -1 point and hereafter do not move it. Now place the right forefinger on the $+3/2$ point for $\omega = 0$. Move the right forefinger along the contour in the direction of the arrow. When you have traced the complete contour, your two forefingers are back exactly in their starting positions. Your right arm is *not* wound around your left. If there were

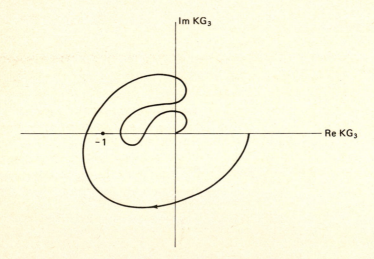

FIG. 16.14   Plot of Fig. 16.12 when K slightly less than $4/5$.

two clockwise encirclements, your right arm would be wound twice clockwise around your left by the time you had traced the total path of Fig. 16.13. (If such arm winding is not physiologically possible, we can use two pencils tied together by a string. One pencil is held at −1, the other traces the contour, and we count at the end the number of times the string is wound around the stationary pencil.)

To return to Fig. 16.12, we see that the system is stable for $K = 1$. As K is increased or decreased, the plot expands or collapses. Hence, when $K = 4/5$, the $\omega = 18$ point passes through −1. For slightly smaller K, the plot has the appearance of Fig. 16.14. This system is unstable. As K decreases further, it reaches the value 1/2 at which the locus again passes through the −1 point.

If we analyze the plot of Fig. 16.12 in detail, we then find:

$$
\begin{cases}
0 < K < 1/2 & \text{System stable} \\
K = 1/2 & \text{Borderline, } T \text{ poles } s = \pm j5 \\
1/2 < K < 4/5 & \text{Unstable} \\
K = 4/5 & \text{Borderline, } T \text{ poles } s = \pm j18 \\
4/5 < K < 4/3 & \text{Stable} \\
K = 4/3 & \text{Borderline, } T \text{ poles } s = \pm j26 \\
K > 4/3 & \text{Unstable}
\end{cases}
$$

In a root locus plot, as K is increased, two poles move into the right-half plane at $K = 1/2$, these same poles move back into the left-half plane at $K = 4/5$, and two poles move into the right-half plane finally at $K = 4/3$. The crossings of the $j\omega$ axis are, respectively, at $\pm j5$, $\pm j18$, and $\pm j26$.

Conditionally stable system

When the above example of Fig. 16.12 operates at $K = 1$, the system is stable and behaves like any normal stable system. In most stable feedback systems, a continual increase in gain eventually causes instability; this is also true here: when the gain is increased beyond 4/3, the system becomes unstable.

This system, however, has another important feature: when the gain is *reduced* (to below 4/5), instability appears. We call such a system *conditionally stable*—meaning it is stable only as long as the gain is kept up.*

Inspection of Fig. 16.12 reveals further that this is exactly the type of system which puzzled Black, and motivated the development of the Nyquist criterion in the 1920s.

---

*An interesting problem arose in the pretransistor days, when vacuum tubes had to warm up when the system was first turned on. During this period, the gain gradually increased from zero to its normal value (in our example, unity). If the system were unstable for low gains, components could be damaged by the oscillations during this warmup time. If we are to use a conditionally stable system, we cannot tolerate too great a drop in gain from any cause.

The gain all the way around the loop is $-KG_3$. At the frequency of 5 rad/sec, this gains is +2: the gain around the loop is greater than unity, yet the system is stable.

Thus, conditionally stable systems are those for which the Nyquist criterion was developed. A surprisingly high percentage of real feedback control systems, which appear in both industrial automation and other applications, are conditionally stable. Often the system starts out with a Nyquist plot such as that depicted by the solid line in Fig. 16.15. In the process of system design, we add a network to $G_3(s)$ which changes the shape of the Nyquist plot to that shown by the dotted lines. In other words, we try to find an additional network to insert which will steer the curve around the -1 point in order to stabilize the system. The design of such networks is a detailed aspect of feedback system technology, covered in texts on control system and feedback amplifier design.

16.5 PROOF OF THE NYQUIST CRITERION

The rigorous mathematical proof of the Nyquist stability criterion is not of direct interest, if we are primarily concerned with the engineering problems of feedback system design. Appreciation of the applicability of the criterion can be enhanced, however, if we consider briefly a heuristic explanation of why the criterion is valid—the purpose of this section.

For stability analysis, we want to investigate the number of right-half plane (*RHP*) zeros of

$$1 + G(s)$$

We first select a contour, γ_s in Fig. 16.16, which encloses the "entire" right-half plane. As s varies around this γ_s, what happens to the behavior of $[1 + G(s)]$? How does this behavior depend on the number of zeros of $[1 + G(s)]$ inside the contour?

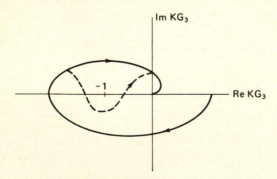

FIG. 16.15 Effect of design of a feedback system.

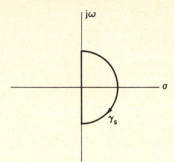

FIG. 16.16   The contour γ_s.

$(1 + G)$ is a ratio of polynomials in s.* Hence $(1 + G)$ can be written in terms of its zeros and poles, for example, as

$$(1 + G) = K \frac{(s - z_1)(s - z_2)}{(s - p_1)(s - p_2)(s - p_3)} \tag{16-14}$$

where the zeros are z_1 and z_2, and the poles are p_1, p_2, and p_3. These zeros and poles are either inside γ_s (in the *RHP*) or outside γ_s (in the left-half plane). What happens to $(1 + G)$ in each case as s traverses the complete γ_s?

Figure 16.17 shows a zero in the left-half plane at z_1. When s is at point A, the term $(s - z_1)$ is just the vector α. As s moves to B, $(s - z_1)$ becomes the vector β. In other words, this factor $(s - z_1)$ varies continuously through the values α, β, γ, and δ. The magnitude grows, then wanes; the phase grows, wanes, and grows again. By the time we have gone completely around γ_s and are back again at A, the term $(s - z_1)$ has returned to its *original* value.

The situation is different when the zero is inside the contour γ_s, or in the right-half plane (Fig. 16.18). Now the factor $(s - z_1)$ takes on the succession of vector values

---

*We continue to consider here only transfer functions which are ratios of polynomials in s. If the system has a pure time delay (a component of e^{-Ts}), we have to modify this discussion.

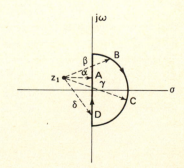

FIG. 16.17   *LHP* zero.

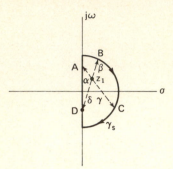

FIG. 16.18 *RHP* zero.

shown as α, β, γ, δ, and α again. The magnitude of this term returns to its original value as we complete traversal of γ_s. The phase, however, changes by $-360°$ in a complete traversal of γ_s. In Fig. 16.18, the phase of α is $150°$; then

$$
\left.\begin{array}{l}
\angle \beta = 75° \\
\angle \gamma = -60° \\
\angle \delta = -120° \\
\angle \alpha = -210°
\end{array}\right\} \tag{16-15}
$$

For *each* zero of $(1 + G)$ inside the contour, the phase of $(1 + G)$ changes by $-360°$ as we traverse the contour once. If it were a pole of $(1 + G)$ rather than a zero inside γ_s, the phase change would be $+360°$. In summary,

As we traverse γ_s once, we give the total change in the angle of $(1 + G)$ the symbol θ in degrees. Then

$$
-\frac{\theta}{360°} = Z - P \tag{16-16}
$$

where Z is the number of *RHP* zeros of $(1 + G)$, P is the number of *RHP* poles of $(1 + G)$.

If $Z = 2$ and $P = 0$ ($1 + G$ has two zeros in the *RHP* and no pole), θ is $-720°$. As we go all the way around γ_s, the angle of $(1 + G)$ changes by $-720°$.

Now what does such an angle change mean about the polar plot of $(1 + G)$—the contour Γ_{1+G}? If the angle of $(1 + G)$ changes by $-720°$, the polar plot of $(1 + G)$ encircles the origin of the $(1 + G)$ plane twice in a clockwise direction. If we generalize to any number of *RHP* zeros or poles of $(1 + G)$, we find

$$
N = Z - P \tag{16-17}
$$

The number N of clockwise encirclements of the origin by the $(1 + G)$ polar plot equals the number Z of RHP zeros of $(1 + G)$ minus the number P of RHP poles of $(1 + G)$.

Essentially, we have now reached the Nyquist criterion. If, instead of plotting $(1 + G)$, we choose to plot G, we are interested in the encirclements of the -1 point in the G plane, since

$$(1 + G) = 0 \qquad \text{means} \qquad G = -1 \tag{16-18}$$

Hence, as s traverses γ_s, the number of clockwise encirclements of $G = -1$ by Γ_G equals the number of RHP zeros of $(1 + G)$ minus the number of RHP poles of $(1 + G)$. In the earlier notation of Sec. 16.2,

$$N = P_T - P_G \tag{16.19}$$

where P_T is the number of RHP poles of $T(s)$ [or zeros of $(1 + G)$ or $G = -1$ points], and P_G is the number of RHP poles of G [or of $(1 + G)$].

Thus, the Nyquist stability criterion follows directly from the behavior of the phase of a transfer function as we let s vary on a closed path surrounding zeros and poles of that function.

16.6 EXAMPLE OF THE NYQUIST TEST

The Nyquist test uses the open-loop frequency characteristics to predict system stability before the loop is closed. The system of Fig. 16.19 provides a simple example. Here we know $G(s)$ to be

$$G(s) = \frac{K}{(s + 10)^3} \tag{16-20}$$

We want to determine the effects of K on the system stability.

(1) We first plot the gain and phase characteristics, shown in Fig. 16.20. We arbitrarily select $K = 1000$ for this plot, with the result that $G(0) = 1$ or 0 dB. Later, we will change K by scaling the gain up or down.

In this example, we started from the transfer function $G(s)$ in factored form. In many cases, $G(s)$ is given in unfactored form, for example, as

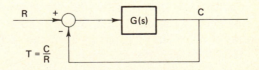

FIG. 16.19 Single-loop feedback system.

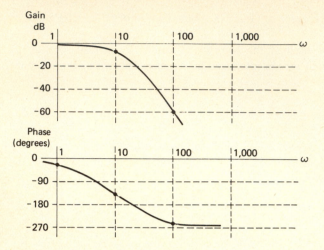

FIG. 16.20 Gain and phase characteristics for

$$G(s) = \frac{1000}{(s + 10)^3} \;.$$

$$G(s) = \frac{1000}{s^3 + 30s^2 + 300s + 1000} \tag{16-21}$$

Then it may be easiest just to substitute in values of $j\omega$ and calculate the magnitude and phase of G. In other cases, we often measure $|G|$ and $\angle G$ experimentally, and the curves of Fig. 16.20 are obtained without ever having a transfer function which is a ratio of polynomials in s or $j\omega$.

(2) Next we construct the polar plot of $G(j\omega)$, as shown in Fig. 6.21. This is the *Nyquist plot* or *Nyquist diagram*. Values of ω are indicated along the curve, and the arrow shows the direction of increasing ω.

(3) Now we wish to interpret the effects on system stability when K is varied. By inspection of the equation for $G(s)$,

$$G(s) = \frac{K}{(s + 10)^3} \tag{16-22}$$

we see that G itself is stable. Therefore, the number of *RHP* poles of $T(s) = C/R$ equals the number of clockwise encirclements of the (-1) point by the Nyquist plot of Fig. 16.21. With $K = 1000$ (the plot shown), the system is stable. As we increase K, the curve is magnified until, with $K = 8000$, the plot passes through -1; the system is on the borderline of instability and poles of T are crossing into the *RHP* at $s = \pm j\,17$. Thus, the system is stable for

$$K < 8000 \qquad\qquad (16\text{-}23)$$

and, for K larger than 8000, unstable with two *RHP* poles.

The Nyquist plot shows the effects on stability when we vary the gain K. In this particular example, we could have accomplished the same task much more easily with the Routh test. The characteristic polynomial is

$$s^3 + 30s^2 + 300s + K + 1000$$

and the Routh table has the form

s^3	1	300
s^2	30	$K + 1000$
s^1	$\dfrac{8000 - K}{30}$	
s^0	$K + 1000$	

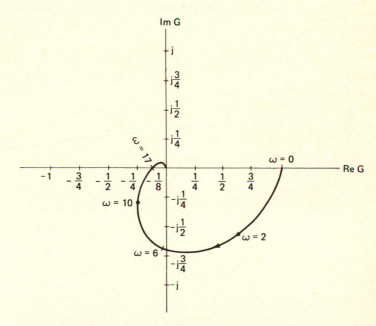

FIG. 16.21 Nyquist plot for

$$G(s) = \frac{1000}{(s + 10)^3}.$$

The system is stable for K positive and less than 8000. Certainly, this test is much simpler than the tedious procedure of constructing the Nyquist plot.

The Nyquist test is particularly appropriate when we have the measured open-loop frequency characteristics, when the transfer function is so complex that the Routh test is difficult, or when we want to design networks to add to G to shape the Nyquist plot so it avoids the -1 point.

16.7   GAIN AND PHASE MARGINS

The example of the last section emphasizes that the Nyquist plot not only determines stability, but also gives an indication of the margin of stability—in other words, how "far" from instability are we operating? In engineering work, it is rarely sufficient to be sure that a system is stable. We must also know that, if certain parameters change with time or environmental conditions, the system will remain stable. For example, the hydraulic steering system of an automobile is a familiar feedback control system which we certainly want to remain stable. During operation, the compressibility of the oil in the system changes radically with variations in temperature and changes in the amount of air in the oil. The system must be stable over the entire expected range of this compressibility parameter.

Figure 16.22 shows the Nyquist plot of the last section, for

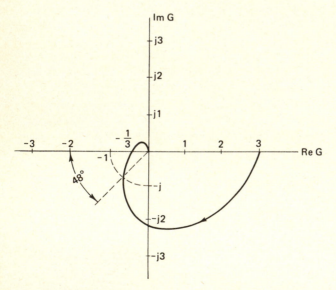

FIG. 16.22 Nyquist plot for
$$G(s) = \frac{8000/3}{(s + 10)^3}.$$

$$G(s) = \frac{K}{(s + 10)^3} \qquad\qquad\qquad (16\text{-}24)$$

redrawn for $K = 8000/3$. The system is stable, and indeed the Nyquist plot really does not come very near the -1 point.

In order to measure how "close" the plot is to encircling -1, engineers commonly use two quantities:

(a) The *gain margin*—how much in dB the gain must be increased to cause instability.

(b) The *phase margin*—how much in degrees the phase lag must be increased to cause instability.

For the system of Fig. 16.22, the gain can be multiplied by 3 (since the plot crosses the negative real axis at $-1/3$). This added gain of 3 corresponds to 9.5 dB. The gain margin is 9.5 dB.

The Nyquist plot can be rotated clockwise through 48° before the system becomes unstable. The phase margin is 48°.

Thus, the *degree of stability* is measured by a gain margin of 9.5 dB, a phase margin of 48°.

Gain and phase margins are the conservative safety factors for the feedback engineer. The construction engineer designing a bridge calculates the strengths required to support the maximum anticipated load, then uses support four or ten times as strong to be sure. Likewise, the systems engineer designs a stable feedback system, then makes sure he has a gain margin of at least 6 dB and a phase margin of at least 30°.

The gain margin measures directly how much K can increase before the system becomes unstable. The phase margin is equally important. If there are small time-constant terms that have been neglected in $G(s)$, the primary effects are increased phase lag.* More generally, the gain and phase margins together indicate how far the Nyquist plot stays away from the -1 point—in some sense, how much any of the parameters of the system can change without the system becoming unstable.

Conditionally stable systems

The concepts of gain and phase margins are not particularly useful when the system is conditionally stable. A possible Nyquist diagram for such a system is shown in Fig. 16.23. If we try to adopt our previous definition, we find that:

(1) The gain margin is 6 dB. The gain can be increased by 6 dB (a factor of 2) before instability appears.

---

*From the gain and phase plots, we recall that as we go up in frequency, the first effect of a real pole is a phase lag. The phase is affected a decade below the break, while the gain is not influenced until about 1/2 the break frequency.

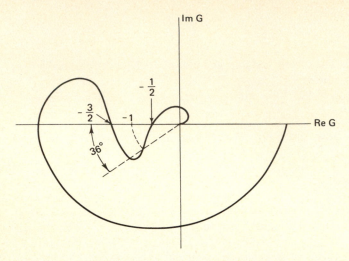

FIG. 16.23 Nyquist plot for conditionally stable system.

(2) The phase margin is 36°. The −1 point is 36° clockwise from the locus (the point on the locus which is a distance of one from the origin).

The gain and phase margins give relatively little information about the stability problem. They tell nothing essentially about the portion of the plot which lies in the second quadrant. For example, if the gain is reduced by 3.5 dB (a factor of 2/3), the system becomes unstable.

We might try to define gain margin to take account of either an increase or a reduction in gain to cause instability. Since gain and phase margins are really just conveniences to guide feedback system design, it is simpler when we have a conditionally stable system to avoid these "figures of merit" and work directly with the Nyquist plot.

16.8 CRITICAL POINT FOR NYQUIST PLOT

In all preceding sections of this chapter, we have been working with the system of Fig. 16.24 or its equivalent. In each case, the overall transfer function has the form

$$T(s) = \frac{G(s)}{1 + G(s)} \tag{16-25}$$

The poles of the system function $T(s)$ are determined by the zeros of the denominator, which has the form

[1 plus a transfer function]

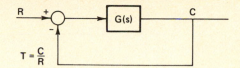

FIG. 16.24 Single-loop system with input comparator.

where the transfer function can be called $G(s)$. The Nyquist plot is for $G(s)$. Since the system stability is determined by the roots of

$$1 + G(s) = 0 \tag{16-26}$$

we are interested in encirclements of the point, $G = -1$.

In the literature on electronic circuits, we often find reference to the configuration of Fig. 16.25. Hence, the adder gives an output which is the sum of the two incoming signals. For this system

$$T = \frac{\text{Out}}{\text{In}} = \frac{A}{1 - \beta A} \tag{16-27}$$

Now the Nyquist plot is for the transfer function (βA) and is drawn in the (βA) plane. Since the poles of T are where $\beta A = +1$, we are then interested in the encirclements of the +1 point. Clockwise encirclements of +1 mean instability.*

In general, in order to decide what transfer function is to be plotted and what is the critical point, we simply look at the denominator of the system function. For Fig. 16.26,

$$T = \frac{C}{R} = \frac{G_1 G_2}{1 + G_1 G_2 H_2 - G_2 H_1} \tag{16-28}$$

Hence, the Nyquist diagram (or polar plot) should be constructed for the transfer function

$$[G_1 G_2 H_2 - G_2 H_1]$$

and we are interested in the encirclements of the -1 point in this plane.

*The directions (clockwise and counterclockwise) still have the same meaning.

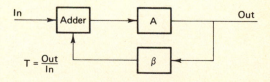

FIG. 16.25 Electronic feedback amplifier.

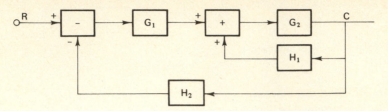

FIG. 16.26 Multiloop feedback system.

16.9 SIMPLE NONLINEAR SYSTEMS

We conclude this chapter with a brief indication of the extension of the Nyquist test to an important class of simple nonlinear systems. All real systems are nonlinear for at least some range of signals (as indicated in Chapter 10). For example, as the signals are made larger and larger, eventually amplifiers saturate or component characteristics change.

An important class of nonlinear systems can be represented as shown in Fig. 16.27. The open-loop system includes both a linear portion $G(s)$ and a nonlinear element n. If we could describe the nonlinear element by an equivalent gain K_n, the stability of the system would depend on the zeros of the denominator of C/R, the roots of

$$1 + K_n G(s) = 0 \tag{16-29}$$

Let us assume for the moment that we can indeed replace the nonlinearity by an equivalent gain K_n for purposes of stability analysis. Then we will be able to study stability not only in the simple configuration of Fig.16.27, but also in much more complex systems (Fig. 16.28). In each of these cases we can write the denominator of C/R as

$$1 + K_n G(s)$$

where $G(s)$ is an ordinary transfer function of a linear system. In case (c), the investigation of stability depends on two Nyquist plots: one for $G_2 H$ and the other for $K_n G(s)$, where

$$G(s) = \frac{G_1 G_2}{1 + G_2 H} \tag{16-30}$$

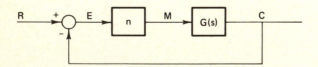

FIG. 16.27 Simple feedback system with one nonlinearity n.

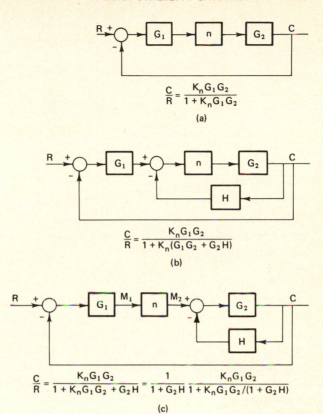

$$\frac{C}{R} = \frac{K_n G_1 G_2}{1 + K_n G_1 G_2}$$

(a)

$$\frac{C}{R} = \frac{K_n G_1 G_2}{1 + K_n (G_1 G_2 + G_2 H)}$$

(b)

$$\frac{C}{R} = \frac{K_n G_1 G_2}{1 + K_n G_1 G_2 + G_2 H} = \frac{1}{1 + G_2 H} \frac{K_n G_1 G_2}{1 + K_n G_1 G_2 / (1 + G_2 H)}$$

(c)

FIG. 16.28 More complex feedback systems with a single nonlinearity n represented by K_n.

Indeed, we can show that whenever the system possesses a *single* nonlinearity n which can be represented by a gain K_n, system stability depends on a Nyquist plot of

$$K_n G(s)$$

where $G(s)$ is the transfer function of a linear system.*

Equivalent gain K_n

Stability analysis of feedback systems with a single nonlinearity depends on replacing that nonlinearity by an equivalent gain, K_n. What do we use for K_n? Let us first consider the nonlinearity described in Fig. 16.29. This is simple *saturation*. When the input signal is less than 2 in magnitude, the gain is clearly 3:

*Actually, $G(s)$ is the transfer function from the output of n back to the input—from M_2 to M_1 in Fig. 16.28(c), for example.

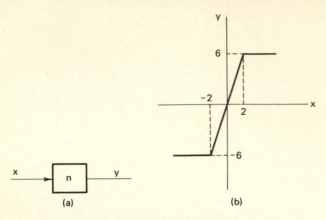

FIG. 16.29 A simple nonlinearity as a first example. (a) Nonlinear block; (b) nonlinear characteristic.

$$y = 3x \qquad |x| < 2 \tag{16-31}$$

When the input exceeds 2, the output remains constant at 6.

The essential characteristic of this nonlinearity is that the "gain" depends on the amplitude of the input signal. When the input signal is small (less than $|2|$), the gain is 3. Larger input signals cause outputs which are limited in size. If the input is *very* large, the "gain" is clearly near zero (a fixed output divided by a very large input). Hence, intuitively we might expect that we should represent the nonlinearity by a gain which varies with the input signal amplitude as sketched in Fig. 16.30.

We have developed Fig. 16.30 with qualitative reasoning. There is no basis for describing accurately how the gain should decrease for large input signals. We can refine this description if we focus attention on *sinusoidal* input signals for x.*

*We are interested ultimately in stability analysis using the Nyquist test. This criterion is based upon the sinusoidal ($j\omega$) characteristics of the various system components.

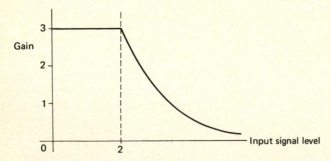

FIG. 16.30 General shape wanted for K_n.

When a sinusoidal signal is applied to the nonlinearity, the output y is three times the input as long as the input amplitude does not exceed two. With larger inputs, the output is clipped (Fig. 16.31). For large inputs, the output is a periodic signal, hence contains a fundamental component (at the frequency of the input) plus higher harmonics. One way to describe the "equivalent gain" of the nonlinearity is to use

$$K_n = \frac{\text{Amplitude of fundamental of } y}{\text{Amplitude of } x} \tag{16-32}$$

We assume that the higher harmonics can be neglected. For example, in the system of Fig. 16.32, if e is sinusoidal, m contains the harmonics. The system stability depends on the signal fed around the loop—hence the signal c which is fed back to e. If $G(s)$ tends to filter out the high frequencies (as often happens), an m which contains a fundamental plus higher harmonics will cause a c which is close to sinusoidal at the fundamental frequency.

Admittedly, this "justification" for Eq. (16-32) is weak. The only real argument for this definition of K_n is that it has worked in the stability analysis of a great number of real, nonlinear feedback systems. In the advanced control literature, many papers investigate the validity of this definition; in spite of all the research, however, the surprising aspect is that Eq. (16-32) almost always results in a stability analysis which describes the system with excellent accuracy.

Describing function

Let us summarize the above paragraphs. We saw that the saturation nonlinearity should be described by a "gain" K_n which depends on the amplitude of the input signal. (This is really the meaning of a nonlinearity.) We then restricted consideration to sinusoidal input signals:

$$x = X_1 \sin t \tag{16-33}$$

The resulting output is a fundamental plus higher harmonics. We define the equivalent gain K_n as

$$\frac{Y_1}{X_1}$$

the amplitude of the output fundamental divided by the input amplitude.

This gain is called the *describing function* of the nonlinearity and is usually indicated by the symbol N. In other words,

$$N \equiv \frac{\text{Amplitude of output fundamental}}{\text{Input amplitude}} \tag{16-34}$$

for a nonlinear device.

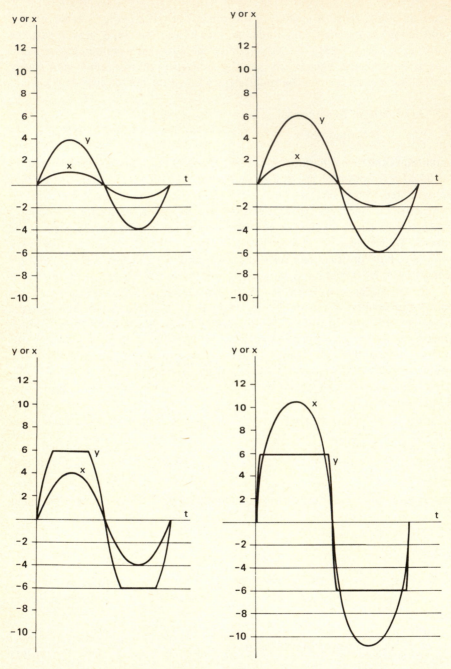

FIG. 16.31 Input and output of nonlinearity when x is a sinusoid of different amplitudes (4/3, 2, 4, and 10.6). Saturation occurs with $|x| > 2$.

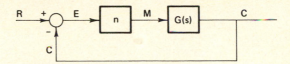

FIG. 16.32   Simple system with single nonlinearity.

Describing function for saturation

With this definition of the describing function, we can calculate this equivalent gain as a function of the amplitude of the input sinusoid. We need only find the output for any given input, then compute the amplitude of the fundamental component in the usual Fourier series analysis. Since this is nothing more than an exercise in Fourier series, we simply show the result in Fig. 16.33.

In this plot, the gain for small input signals is unity. If the actual gain is k, the vertical axis of (b) is simply multiplied by k. Likewise, the saturation starts at $|x| = 1$. If in the actual device saturation begins at $|x| = a$, we simply multiply by a all the numbers labeling the horizontal axis in (b). Thus, the one plot of Fig. 16.33 suffices to describe all saturation nonlinearities.

Describing functions for piecewise-linear nonlinearities

In Chapter 10, we saw how any odd, piecewise-linear nonlinearity can be represented as the sum of linear gains and saturating nonlinearities. Consequently, we can use Fig. 16.33 to calculate the describing function (equivalent gain) for any odd, piece-wise-linear characteristic. The procedure is illustrated in the problem at the end of the chapter.

16.10   STABILITY ANALYSIS WITH DESCRIBING FUNCTIONS

To complete this section, an example illustrates the use of the describing function in stability analysis. For the system of Fig. 16.34(a), if we can replace the nonlinearity n

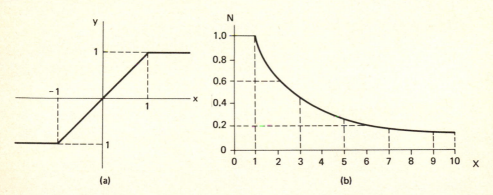

FIG. 16.33 N for saturation. The gain is unity for small inputs, and saturation occurs beyond $|x| = 1$. (a) Nonlinearity; (b) describing function.

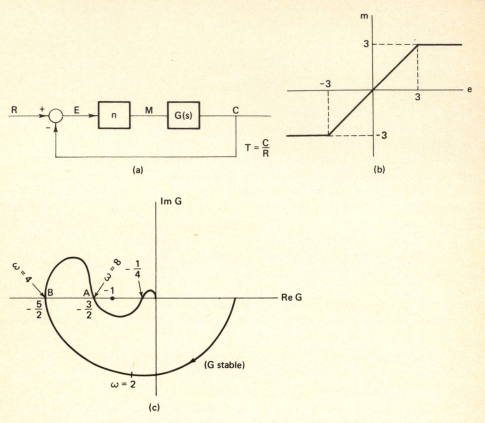

FIG. 16.34 Final example illustrating stability analysis of nonlinear system. (a) The system; (b) the nonlinear characteristic; (c) the Nyquist plot for $G(j\omega)$.

by its describing function N,

$$T = \frac{G}{1 + NG} \tag{16-35}$$

Stability depends on the zeros of $(1 + NG)$. Since N varies with the amplitude of the signal e, it is convenient to view the Nyquist criterion as depending on the encirclements of the point

$$G = -\frac{1}{N} \tag{16-36}$$

First we find the describing function N as it depends on E, the amplitude of the sinusoidal error signal. The characteristics of Fig. 16.34(b) show the small signal gain is unity; beyond $E = 3$, this gain falls off with E (Fig. 16.35 which is just 16.33 redrawn).

For small signals, N is unity and we are interested in encirlement of $-1/N$ or -1. The Nyquist plot of Fig. 16.34(c) shows that the system is *stable* for small signals (that is, when $E \le 3$).

As E increases beyond 3, the gain N starts to decrease and the critical point $(-1/N)$ moves leftward in Fig. 16.34(c). When the critical point reaches $-3/2$

$$-\frac{1}{N} = -\frac{3}{2} \quad \text{or} \quad N = \frac{2}{3} \tag{16-37}$$

The system is on the borderline of instability. Figure 16.35 shows that this occurs when E is 5.4.

If E continues to increase, N decreases, and the critical point crosses the Nyquist plot again when

$$-\frac{1}{N} = -\frac{5}{2} \quad N = \frac{2}{5} \quad \text{From Fig. 16-35, } E = 9.3 \tag{16-38}$$

For even larger values of E, the critical point is to the left of the Nyquist plot, and the system is stable.

What do we now know? For small E, the system is stable; as E exceeds 5.4, instability appears; beyond an E of 9.3, we are again stable (Fig. 16.36). There are two *equilibrium* points:

$$E = 5.4 \qquad N = \frac{2}{3} \qquad \omega_0 = 8$$

$$\tag{16-39}$$

$$E = 9.3 \qquad N = \frac{2}{5} \qquad \omega_0 = 4$$

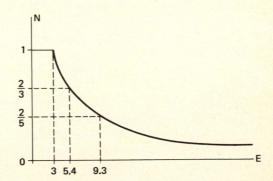

FIG. 16.35 Describing function for Fig. 16.34(b) .

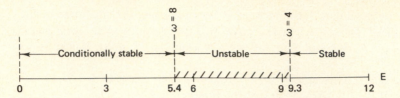

FIG. 16.36 Stability as a function of E, where E is the amplitude of the sinusoidal error signal.

These are equilibrium points because, if e is ever a sinusoid of amplitude 5.4 and frequency 8 rad/sec, the system will continue to oscillate indefinitely with these values.

Actually, this equilibrium point itself is unstable. If for any reason, E changes slightly, we do not return to the equilibrium point. For example, if E decreases slightly (perhaps because of noise), N increases, the critical point $-1/N$ moves to the right, the system is stable, and the oscillation tends to die out (E further decreases). If E increases slightly, N decreases, the critical point moves to the left, the system is unstable, and E increases further. Any slight motion away from the equilibrium point A in Fig. 16.34(c) causes even greater motion away.

In contrast, equilibrium point B is stable. A slight increase in E results in motion of the critical point to the left, the system is stable, and E decreases back toward its value of 9.3 at B. Likewise, a decrease in E makes the system unstable, and E grows again back toward 9.3.

Thus, stability analysis tells us we can operate at either A or B in Fig. 16.34(c). Only B is a stable equilibrium point. Consequently, if we actually build this system, we will find that it behaves as follows:

> As long as signal levels are small, the system is stable.
> If at any time the signal e becomes large (for example, greater than 5.4 for a cycle or two), the system thereafter operates with a constant oscillation for e. The amplitude is 9.3 and the frequency is 4 rad/sec.

This analysis is, of course, only approximate. We have replaced the nonlinearity by an equivalent gain. In actuality, we should then interpret the quantitative results liberally. We should not be surprised to find the oscillation amplitude 11 instead of 9.3, or the frequency 4.5 instead of 4.

The describing-function concept does permit us to estimate the stability of feedback systems with a single nonlinearity. Essentially, the analysis represents an extension of the Nyquist critierion to an important class of nonlinear systems. Within this class fall many practical oscillators—systems specifically designed to give constant-amplitude, sinusoidal outputs—as well as feedback systems with saturation, dead zone, backlash, and other familiar nonlinearities.

16.11 WHAT CAN THE NYQUIST TEST DO?

The Nyquist stability criterion

(1) Indicates stability of the closed-loop system from the open-loop frequency characteristics;

(2) Is useful when the open-loop characteristics are measured experimentally or known analytically;

(3) Suggests how the open-loop characteristics should be modified to obtain stability;

(4) Indicates relative stability, for example through the gain and phase margins; and

(5) Permits stability analysis of systems with a single nonlinearity, when that nonlinearity is represented by a describing function.

Thus, the Nyquist test complements the Routh test and thereby broadens the arsenal of approaches available to the engineer concerned with system stability.

PROBLEMS

16.1   For the feedback system shown, two different Nyquist plots are given. In each case, what values of K cause instability?

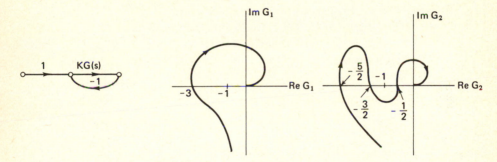

16.2   For the system shown, determine $T(s) = C(s)/R(s)$ in terms of G_1, G_2, and H. We wish to plot a Nyquist diagram to determine the overall system stability before we close the loop at A. What transfer function should be measured experimentally for the test (where should we apply the input and take the output)? What is the Nyquist criterion in terms of this plot?

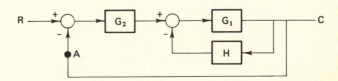

16.3   For the system shown, we plot the Nyquist diagram when ω varies from zero to infinity and obtain the curve as given (the arrow shows the direction of increasing frequency). What can be said about the nature of $G(s)$—that is, how many poles and zeros, behavior as s tends to zero, behavior as s tends to infinity, and so forth? What can we say about the stability of the closed-loop system as K is increased from zero?

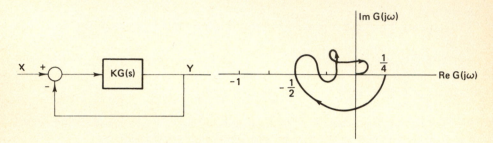

16.4   (a) For the transfer function

$$G(s) = \frac{K}{s(s + 2)^2}$$

determine the frequency at which the phase shift is $-180°$.

(b) When this $G(s)$ is used in the system shown, sketch the Nyquist diagram to determine the stability of the total system.

(c) How large can K be for the system just to be stable?

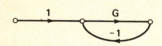

16.5   For the same system shown in Prob. 16.4,

$$G(s) = \frac{K(s + 1)}{s^2(s + 3.5)(s^2 + 0.5s + 10.25)}$$

(a) Construct the Nyquist diagram.

(b) Determine how large K should be to make the system marginally stable.

(c) Determine the frequency of oscillation when K has the value of (b).

(d) Compare this analysis with the Routh test. What are the advantages and disadvantages of the Nyquist and Routh criteria?

(e) The poles and zeros here were chosen to be convenient numbers. How would the analysis change if more practical numbers were used—for example, each pole and zero multiplied by 10^5 above?

16.6 Three different Nyquist plots are given, each referring to the system of Prob. 16.4. In each case, $G(j\omega)$ has been measured experimentally by applying a sinusoidal signal of fixed amplitude and adjustable frequency, with ω varied from very small to very large values. The arrow shows the direction of increasing ω. For each of these Nyquist plots, answer the following questions.

(a) What is the behavior of $G(s)$ as s tends to zero? In other words, does G have a pole, zero, or constant value at $s = 0$? If a zero or pole, what is the order?

(b) What is the behavior of G as s tends to infinity? What can we say about the degrees of the numerator and denominator polynomials?

(c) Indicate the form of the transfer function G. In other words, is it something like

$$G(s) = \frac{s + a}{(s + b)(s + c)}$$

(d) In tandem with $G(s)$ we now insert a variable gain K. What conditions must K satisfy for the closed-loop system to be stable?

(e) With $K = 1$, what is the gain margin?

(f) With $K = 1$, what is the phase margin?

(g) Resketch the Nyquist diagram and then indicate the new form if we multiply $G(s)$ by $10(s + a)/(s + 10a)$, where $3.2a$ is chosen equal to the lowest frequency at which the phase of G is $-180°$.

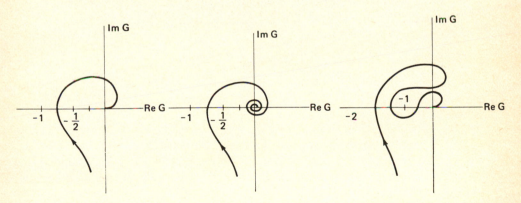

16.7 In the single-loop system shown,

$$G(s) = \frac{K}{s(s + 1)^2}$$

Sketch the Nyquist plot. Notice that the contour γ_s in the s plane cannot pass through $s = 0$, since there is a pole of G at that point. As we come up the $j\omega$ axis in the s plane, we detour around $s = 0$ by a semicircle to the right of the pole. What path in Γ_G, the Nyquist plot, corresponds to this very small semicircle of γ_s?

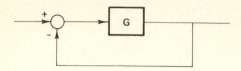

16.8   For the system of Prob. 16.7, we wish to add a transportation lag of 2 sec to the open-loop transfer function. Sketch the new Nyquist diagram. How does a transportation lag affect the Nyquist plot? (Notice that the lag does not really complicate the stability analysis; this is an advantage of the Nyquist test over the Routh criterion, since the Routh test is basically inapplicable unless the transfer functions are ratios of polynomials in s.)

16.9   Sketch the describing functions for the two nonlinearities shown in the figure.

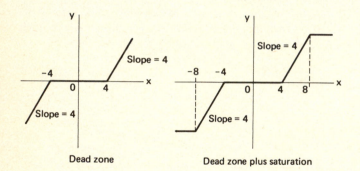

Dead zone   Dead zone plus saturation

16.10   For the system shown, the G is that given in Prob. 16.7, and we wish to consider separately each of the nonlinearities of Prob. 16.9. In each case, determine whether the system can be unstable; if so, find the approximate frequency and amplitude of the oscillation. Is the oscillation itself stable or unstable?

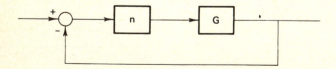

INDEX

Accelerometer, 108, 443
    (P 12.11)
Active network:
  building blocks for, 216–217
    (P 6.14)
  sensitivity in, 411–415
Active RC oscillator, 172–176
Adaptivity of man, 88
Adder, 19, 168
Admittance, short-circuit
    transfer, 295
Adrenal gland, 435
Aircraft force simulator,
    250–251 (P 7.12)
Algebra of signal flow graph,
    42
Aliasing error, 454
All-pass network, 382–383
All-pass system function, 381
Amplifier:
  difference, 299–306
  differential, 290–292
  instrumentation, 291
  logarithmic, 340–341
    (P 10.7)
  non-ideal, 189–190
  operational, 165–217
  (*See also* Op-amp)
Amplitude modulation, 454n
Amplitude scaling, 197–202
Analog computer:
  amplitude scaling in,
    197–202
  for nonlinear system,
    320–329
  parameter choice with,
    278–284
  potentiometer in, 188
  programming of, 165–217
  time scaling in, 191–197
Analog computer program, 177
Analog, electrical, 282
Analog simulation, 165–217
  with difference amplifier,
    299–306

Analog simulation (*Cont.*)
  initial conditions for
    285–289
  nonlinear, 311–338
Analytic linearization, 318
Antenna position control,
    429–433
Apartment system, 15–17
Asymptote:
  high frequency, 349
  low-frequency, 349
  for root loci, 142
Asymptotic gain calculation,
    354
Asymptotic gain plot, 345–356
Asymptotic phase characteris-
    tic, 373–374
Asymptotic stability, 514, 543

Backlash, 82
Ball-and-disk integrator,
    129–130
Band:
  pass, 204
  stop, 204
Band-elimination filter, 75
Band-rejection filter, 75
Bandwidth:
  definition of, 72
  evaluation of 72–40
Bandwidth sensitivity, 414
Barkhausen stability criterion,
    555
Bel, definition of, 385
Blending system, 271–275
Block diagram, 17–25
  construction of, 23
  for electronic amplifier, 22
  for nonlinear system, 313–
    316
  of radar tracking system, 51
  and signal flow diagram, 50
Bode, Hendrik, 345

Bode plot, 345–356
  for complex poles, 360–363
  correction for, 356–363
  measurement from, 363–368
Break frequency, 349
  for complex factor, 360
Buffalo population model,
    479–486
Butterworth filter, 204–211

Car resonance, 95
Cardiac pacemaker, 7
Cause-effect relation, 17
Characteristic equation, 139
  definition of, 58
Characteristic, phase, 369–377
Characteristic polynomial,
    definition of, 58
Centroid of pole-zero pattern,
    143
Circuit parameter, choice of,
    281–283
Clipping system, 339–340
    (P 10.5)
Closed loop, 396
Code:
  Hamming, 511–512 (P 14.8)
  zip, 5
Coding, 505
Coefficient, choice of, 279–
    281
Comb filter, 75–77
Comparator, 19, 290–292
Compensation, feedback, 425–
    429
Complex pole:
  Bode plot for, 360–363
  correction for, 362
  phase for, 377–379
Complex system, 40–42
Computer, analog, program-
    ming of, 165–217
Condition, initial, 285–289

Conditional stability, 151, 530, 563–566
Constant current source, 309–310 (P 9.10)
Constant-R network, 249–250 (P 7.9)
Control:
  of cortisol secretion, 436
  disturbance, 416–417
  dynamics, 422–429
  expressway, 312
  noninteracting, 153–257
  of water level, 399
Control system:
  blending, 271–275
  with human operator, 86
  man-machine, 87
  pupillary, 394–395 (P 11.17)
  state model for, 233–236
Controllability, 238–243
  of multidimensional system, 268–270
  test for, 242
Controller, 22
Converter, D/A, 309 (P 9.9)
Correction:
  for Bode plots, 356–363
  for linear factor, 358
  for quadratic factor, 362
Cortisol, 435
Counter-intuitive behavior, 12n
Criterion:
  Barkhausen, 555
  Lyapunov, 542
  Nyquist, 554–585
  Routh, 516–526
Critical frequencies, 58
Critical point for Nyquist criterion, 574–575
Crystal, piezoelectric, 391 (P 11.4)
Cut off frequency, 206
  scaling of, 209
Cycle, limit, 538

D/A converter, 309 (P 9.9)
d'Alembert's principle, 281
Damper, vibration, 281–283
Damping ratio, relative, 116, 137

dB gain, 385–390
  for linear factor, 348–351
Dead time, 77
Dead zone, 338–339 (P 10.3), 588
Decay rate from z transform, 472–473
DeciBel, definition of, 385
Delay line, tapped, 77
Delay, time, 116, 371
Describing function, 339 (P 10.4), 577–581
  for dead zone, 588
  for piecewise linear characteristics, 581–582
  with stability analysis, 581–584
Determinant of feedback configuration, 34
Diagram, Nyquist, 570
Difference amplifier, 299–306
Difference equation, 505
Difference, return, 417–422
Differential amplifier, 290–292
Differential equation:
  model of, 219
  and transfer function, 55
Differentiator, 100–108, 297–299
  filter, 103
  gain for, 102
  network, 129
  numerical, 107
Digital filtering, 489–490 (P 13.12)
Digital simulation, 504–508
Diode, ideal, 321–322
Direct transmittance, 34
Directed branches, 25
Discrete data processors, 464–470
Discrete systems, 464–470
Distortionless transmission, 369–372
Disturbance control, 416–417
Driven system response, 229–233
Duality of signals and systems, 69
Dummy output variable, 37
Duration of transient, 133

Dye dilution test, 124
Dynamics, 11
  control of, 422–429

Elimination:
  of a node, 42
  of a self loop, 44
Energy function, 543
Epidemic model, 510–511 (P 14.7), 553 (P 15.14)
Equation:
  characteristic, definition of, 58, 139
  Lotka-Volterra, 553 (P 15.15)
Equilibrium point, 316, 537, 583
  stability of, 541
Equivalent gain, 577–581
Error, aliasing, 454
Error correction, automatic, 511–512 (P 14.8)
Expansion, partial fraction, 63–66
Expressway control, 312

Feedback:
  in accelerometer, 443–444 (P 12.11)
  in body temperature control, 434
  compensation, 425–429
  to control sensitivity, 403–405
  to control unstable system, 95
  definition of, 396
  for disturbance control, 416–417
  for dynamics control, 422–429
  after hemorrhage, 435–440
  homeostatic, 433–440
  measure of, 417–422
  negative, 421
  positive, 421
  purpose of, 400–402
  in radar tracking, 429–433
  of state variable, 237
  for time-constant control, 423–425

Feedback loop, 28
Feedback system:
  human, 399
  z transfer function for, 474
Feedback, transportation lag
    in, 81
Filter:
  band elimination, 75
  band-rejection, 75
  Butterworth, 204–211
  comb, 75–77
  definition of, 75
  differentiating, 103
  digital, 489–490 (P 13.12)
  low-pass, 135
  notch, 74–75, 443 (P 12.10)
  for sampled system, 495–
    499
Force simulator, 250–251
    (P 7.12)
Forced response, 57, 230
Forrester, Jay, W., 15
Fourier series for impulse
    train, 452
Frequency:
  break, 349
    for complex factor, 360
  critical, definition of, 58
  cutoff, 206
    change of, 209
  in a measured signal, 487
    (P 13.5)
  natural, definition of, 59
  resonant, 70, 72
  sampling, definition of, 450
  undamped natural, 116
Frequency measurement of
    transfer function, 363–
    369
Frequency multiplexing, 488
    (P 13.10)
Frequency scaling, 193
(*See also* Time scaling)
Friction:
  rotor, 330
  static, 337–338
Function:
  describing, 339 (P 10.4),
    577–581
  energy, 543
  Lyapunov, 543–544
  system, 54 (*see* Transfer
    function)

Function: (*Cont.*):
  transfer, 54–99
  weighting, 508–509 (P 14.1)

Gadd severity index, 132
Gain:
  asymptotic, 354
  in dB, 385–390
  for differentiator, 102
  distortion from, 369–372
  equivalent, 577–581
  for linear factor, 348–351
  linearized, 318
  for resonant system, 71
  along root loci, 152
  of sampler, 492–493
Gain margin, 572–574
Gain-phase relations, 380–383
Gain plot, asymptotic, 345–
    356
(*See also* Bode plot)
Gain realization, 169
Graphical analysis of non-
    linear circuit, 315–316
Graphical linearization, 316–
    318
Growth rate from *z* transform,
    472–473

Hamming code, 511–512
    (P 14.8)
Harmonic generation, 541
Harvesting policy, 481
Hearing, human, 343–344
Heart block, 8
Heart pacemaker, 7
Hemorrhage, 435–440
Hilbert transform, 380–383
Hold circuit, 495–499
Homeostasis, 433–440
Human feedback system, 399
Human operator:
  in control system, 86
  simulation of, 214–215
    (P 6.11)
Human pilot controlling two
    loops, 90 (Fig. A3.4)
Human reaction time, 77
Human sampling, 458
Human transfer function, 85–
    91

Hurwitz polynomial, 549
    (P 15.2)

Ideal differentiator, 102
Ideal voltage amplifier, 166
Identification:
  in frequency domain, 363–
    368
  from impulse response, 124–
    128
  system, from step response,
    117–119
Impedance:
  series, 119
  shunt, 119
Impedance level, change of,
    208
Implantable pacemaker, 8
Impulse:
  for initial condition, 285
  in time scaling, 196
Impulse response, identifica-
    tion from, 124–128
Impulse sampling, 452
  Laplace transform of, 459–
    464
Impulse train, Fourier series
    for, 452
Incremental linearization, 318
Industrial dynamics, 15*n*
Inert systems, 83
  definition of, 55
Inertial navigation system,
    108, 444 (P 12.11)
Initial condition, 285–289
Input-output model, 218
Instrumentation:
  feedback in, 422–429
  sampling in, 458
Instrumentation amplifier,
    291
Integration:
  numerical, 112
  system for, 111
  z operator for, 509–510
    (P 14.4)
Integrator, 108–112, 168,
    170–172, 294
  ball-and-disk, 129–130
  mechanical, 129–130
Integrator network, 128–129
    (P 4.1)

Inversion of state model, 288
Isolation of one element, 44

Jarmain, W. E., 15n

Ladder network, 119–123, 302
Lag, transportation, 72–82
Leakage transmission, 444 (P 12.12)
Letter frequency, 4
Limit cycle, 538
  stable, 539
Linear factor, correction for, 358
Linear system, stability in, 514–516
Linearity, 82
Linearization, 316–320
  algebraic, 272–273
  analytic, 318
  graphical, 316–318
  with partial derivative, 274
  by Taylor series, 300
Linearized gain, 318
Loading reduced by sampling, 458
Logarithmic amplifier, 340–341 (P 10.7)
Logarithmic gain, 385–390
Loop, closed, 396
Loop gains, 34
Lotka-Volterra equations, 553 (P 15.15)
Low-pass filter, 135
Lumped system, 55, 83
Lyapunov criterion, 542
Lyapunov function, 543–544
Lyapunov stability, 541–548
  and Routh test, 546–548

Man-machine system, 87
  stress in, 91
Margin:
  gain, 572–574
  phase, 572–574
Mason's theorem, 33
  for block diagram, 38–39
  for complex system, 40-42

Matrix:
  exponential of, 228
  rank of, 265
  residue, 265
  unity, 226
Measurement, temperature, 308–309 (P 9.8)
Mechanical integrator, 129–130
Minimum-phase system, 380–383
Minor-loop controller, 407
Model:
  buffalo population, 479–486
  differential equation, 219
  epidemic, 510–511 (P 14.7), 553 (P 15.14)
  input-output, 218
  predator-prey, 552–553 (P 15.13), 553 (P 15.15)
  rumor, 511
  state, 177–191, 218–151
    definition of, 221
Modeling, 11
Modulation, 337
  amplitude, 454n
Motor, 71, 329–330
Multidimensional systems, 252–277
  controllability of, 268–270
  noninteraction in, 253–257
  observability of, 270
  order of, 264–267
Multi-input system, 257
Multiloop system for sensitivity control, 410–411
Multi-output system, 257–259
Multiple poles:
  evaluation of, 66
  synthesis for, 187n
Multiplexing:
  frequency, 488 (P 13.10)
  time, 455–458
Multiplier, 340–341 (P 10.7)
Multivalued response in nonlinear system, 540

Natural frequency:
  definition of, 59
  undamped, 116

Navigation system, inertial, 108
Negative feedback, 421
Network:
  all-pass, 382–383
  constant-R, 249–250 (P 7.9)
  differentiator, 129
  integrator, 111, 128–129 (P 4.1)
  ladder, 119–123
  prediction, 131–132
  predistortion, 383
  two-port, 293
Network synthesis, 165–217
New York City housing, 18 (Fig. 2.1)
Node, 25
  elimination of, 42
  properties of, 26
  sink, 27
  source, 27
Node splitting, 41
Noncompetitive pacing, 9
Noninteracting control, 253–257
Nonlinear circuit, 314–316
Nonlinear simulation, 311–338
Nonlinear system:
  driven, 540–541
  linearization of, 316–320
  Nyquist criterion for, 576–581
  reduction of, 313–316
  stability of, 536–541, 576–581
Nonlinearity:
  even zero-memory, 328–329
  product, 337
  simulation of, 320–329
  in three variables, 318
  with transfer function, 329–335
  zero-memory, 322
Nonminimum-phase system, 380–383
Non-touching loops, 34
Normal form, 184–187
Notch filter, 74–75, 443 (P 12.10)
Northeast Corridor, 2
Numerical integration, 112

Nyquist criterion:
  critical point for, 574–575
  gain margin in, 572–574
  for nonlinear system, 576–581
  phase margin in, 572–574
  proof of, 566–567
  and Routh test, 572
  statement of, 557
  with transportation lag, 588 (P 16.8)
Nyquist diagram, 570

Observability, 243–246
  of multidimensional systems, 270
Offset, 318
Op-amp:
  analysis of, 172–176
  definition of, 166
  with difference amplifier, 299–306
  differentiating, 297–299
  for even nonlinearity, 328–329
  frequency dependence of, 191
  general, 292–297
  initial condition for, 285–289
  nonlinear, 323–328
  potentiometer in, 171, 188
  practical problems in, 188–191
  related to state models, 222
  time scaling in, 191–197
  with two inputs, 191, 213 (P 6.8), 289–292
  use of, 168
Op-amp building blocks, 216–217 (P 6.14)
Op-amp clipper, 339–340 (P 10.5)
Open-loop system, 401
Operational amplifier, 165–217
(*See also* Op-amp)
Optimization, 12
Order, minimum, 164–267
Organization, signal flow diagram for, 52

Oscillator, 584
  design of, 214 (P 6.10)
  RC, 172–176
Overshoot, 332
  versus ζ, 114

Pacemaker:
  cardiac, 7
  implantable, 8
  transvenous, 8
Parameter:
  isolation, 44
  and stability, 527
Parameter choice from simulation, 278–284
Parameter dependence of system function, 158
Partial fraction expansion, 63–66
  of residue matrix, 265
  of z transform, 470–471
Partial fraction realization of $T(s)$, 184–187
Partitioning of signal flow diagram, 52
Pass band, 204
Periodic sampling, definition of, 450
Phase:
  asymptotic, 373–374
  for linear factor, 372–377
  for quadratic factor, 377–379
  related to gain, 380–383
  for transportation lag, 383
Phase, distortion from, 369–372
Phase characteristic, 369–377
  for nonminimum-phase system, 381–382
Phase margin, 572–574
Phase portrait, 539
Piecewise-linear characteristic, 322, 581
Piezoelectric crystal, 391 (P 11.4)
Pituitary gland, 435
Plot:
  Bode, 345–356
  Nyquist, 570

Point, equilibrium, 316, 537, 583
Poles:
  of Butterworth characteristic, 207
  gain for, 348–351
  Importance of, 95–99
  phase for, 372–379
  related to stability, 60, 516
  of transfer function, definition of, 58
Pole-zero pattern, centroid of, 143
Policy, harvesting, 481
Polynomial:
  characteristic, definition of, 58
  evaluation of, 95
  Hurwitz, 549 (P 15.2)
Population model, buffalo, 479–486
Positive feedback, 421
Potentiometer:
  in analog computer, 188
  in op-amp, 171, 188
Power from human being, 86
Predator-prey model, 552–553 (P 15.13), 553 (P 15.15)
Prediction, 101, 131–132
Predistortion network, 383
Probability and signal flow graph, 53
Process, 22
  definition of, 398
Processor, discrete data, 464–470
Product nonlinearity, 337
Programming:
  analog, 283 (*see* Simulation)
  of analog computer, 165–217
  simulator, 283 (*see* Simulation)
Proper loop, 28
Pupillary control system, 394–395 (P 11.17)

Q, 73:
  maximum, 137
  sensitivity of, 412

Quadratic factor:
  correction for, 362
  phase for, 377–379
Quality factor, 73
Quantization, 505
Quarter-squaring operation, 341

Radar tracking system, 429–433
  block diagram of, 51
Rank of matrix, 265
RC oscillator, 172–176
RC synthesis, 296
Reaction time, human, 77
Recursion equation, 505
Reduction, 33
  of nonlinear system, 313–316
Redundancy for error correction 511–512 (P 14.8)
Relative damping ratio, 116, 137
  and overshoot, 114
Relative stability, 116, 572–574
Relay system, 551–552 (P15.12)
Remnant, 88
Reservoir feedback system, 399
Residue, evaluation of, 64–66, 95 (P 3.11, 3.12)
Residue matrix, 265
Resonance, 70–73
  bandwidth of, 72
  in car, 95
  motor, 71
  Q of, 73
Resonant circuit, sensitivity of, 441 (P 12.6)
Resonant frequency, 70, 72
  sensitivity of, 412
Response:
  driven, 229–233
  forced, 230
  steady state, 67–70
  step-function, 112–119
  total, 62
  undriven, 225–229
Return difference, 417–422

Rise time, 116
Root loci:
  asymptote of, 142
  definition of, 137
  for negative K, 159
  Routh test for, 530
  rules for, 140–143, 152, 154
  use of, 160–161
Routh test, 516–526
  and Nyquist criterion, 572
  related to Lyapunov, 546–548
  in root-locus plotting, 530
Rumor model, 511

Sample value, sequence of, 449
Sampled signal, equivalence of, 491–494
Sampler, gain of, 492–493
Sampling:
  in analysis, 499–504
  artificial, 494
  to control loading, 458
  definition of, 448
  frequency, definition of, 450
  with hold circuit, 495–499
  human, 88, 458
  impulse, 452
  in instrumentation, 458
  Laplace transform in, 459–464
  periodic, 455n
    definition of, 450
  summary of, 506–507
Sampling theorem, 451–455
Saturation, 313, 319, 338–339 (P 10.3), 577
  torque, 329–332
Scaling:
  amplitude, 197–202
  of cutoff frequency, 209
  frequency, 193
  of impedance, 208–209
  time, 191–197
    signals in, 195–196
Second-order system, step response of, 113–117
Seismograph, 108, 132, 393 (P 11.13)

Self-loop, 28
  elimination of, 44
Sensitivity:
  in active network, 411–415
  bandwidth, 414
  definition of, 403
  to forward element, 403
  frequency dependence of, 405
  of Q, 412
  of resonant circuit, 441 (P 12.6)
  of resonant frequency, 412
  and return difference, 419–421
  and stability, 405
  zero, 410–411
Sensitivity control with minor loops, 407
Sensor, 331
Series impedance, 119
Servomechanism, 130, 233–236, 331, 429
  definition of, 234
  design of, 236
Settling time, 116
Severity index, Gadd, 132
Shannon sampling theorem, 451–455
Short-circuit transfer admittance, 295
Shunt impedance, 119
Signal:
  sampled, equivalence of, 491–494
  in time scaling, 195–196
Signal flow diagram, 25–33
  algebra of, 42
  compared to block diagram, 50
  for complex system, 40–42
  determinant of, 34
  for ladder network, 121
  for nonlinear system, 313–316
  for op-amp circuit, 173
  for organizational arrangement, 52
  partitioning of, 52
  and probability, 53
  state transition of, 53

Simulation:
  analog, 165–217
  of clipping, 339–340
    (P 10.5)
  with difference amplifier,
    299–306
  digital, 504–508
  of even zero-memory non-
    linearity, 328–329
  of human controller, 214–
    215 (P 6.11)
  initial condition for, 285–
    289
  of multidimensional system,
    264–267
  nonlinear, 311–338
  of nonlinearity plus transfer
    function, 329–335
  parameter choice from, 278–
    284
Sink node, 27
Sinusoidal response, 67–70
    (see Steady state re-
    sponse)
Slope change realized by
    diode, 323–324
Small-signal model, 318
Source node, 27
Speech, infinitely clipped,
    339–340 (P 10.5)
Splitting, node, 41
Square root circuit, 342
    (P 10.11)
Squarer, 340–341 (P 10.7)
Stability, 60
  asymptotic, 514, 543
  conditional, 151, 530, 563–
    566
  definition of, 514–516, 541
  with differentiator, 298
  Lyapunov, 541–548
  of nonlinear system, 313–
    314, 536–541, 576–
    581
  and parameter, 527
  relative, 572–574
  from Routh test, 516–526
  and sensitivity, 405
  and step response, 113
  from transfer function, 515
  from z transfer function,
    471–472, 490
    (P 13.14)

Stability analysis with describ-
    ing function, 581–584
Stability criterion:
  Barkhausen, 555
  Nyquist, 554–585
State:
  definition of, 223–224
  initial, 285
State model, 177–191, 218–
    251
  for control system, 233–236
  definition of, 221
  inversion of, 288
  related to transfer function,
    270
State-transition matrix, 227
State variable feedback, 237
Static accuracy, 130
Static friction, 337–338
Steady state component, 68
Steady state response, 67–70
Step response:
  of second-order system,
    113–117
  system identification from,
    117–119
Step for initial condition, 285
Stop band, 204
Stress in human task, 91
Subharmonic generation, 541
Subtractor, 19
Synthesis, network, 165–217
System:
  blending, 271–275
  clipping, 339–340 (P 10.5)
  conditionally stable, 151
  conditionally static, 563–566
  controllable, 238–243
  differentiating, 100–108
  discrete, 464–470
  homeostatic, 433–440
  inert, 83
  inertial navigation, 44
    (P 12.11), 103
  for integration, 111
  linear, 82
  lumped, 55, 83
  multidimensional, 252–277
  multi-input, 257–259
  multi-output, 257–259
  noninteracting, 153–257
  nonminimum-phase, 380–
    383

System: (Cont.)
  observable, 243–246
  open-loop, 401
  predicting, 101
  radar tracking, 429–473
  relay, 551–552 (P 15.12)
  telephone, 6
  time invariant, 83
  two-port, 55
  of urban apartments, 15–17
System determinant, 34
System engineering, definition
    of, 2
System function:
  all pass, 381
  gain of, 345–356
  (See also Bode plot)
  phase of, 369–377
System, nonlinear:
  linearization of, 316–320
  reduction of, 313–316
  stability of, 536–541
System, order, minimum, 264–
    267
System response:
  driven, 229–233
  undriven, 225–229
System, stability, 514–516

Tackometer, 107
Tapped delay line, 77
Taylor-series linearization, 320
Telephone system, 6
Temperature control, body,
    434
Temperature instrument, 422
Temperature measurement,
    308–309 (P 9.8)
Test, Routh, 516–526
Theorem, sampling, 451–455
Time:
  dead, 77
  rise, 116
  settling, 116
Time-constant control with
    feedback, 423–425
Time delay, 116, 371
Time invariance, 83
Time multiplexing, 455–458
Time scaling, 191–197
  signals in, 195–196

Torque:
  developed, 330
  saturation, 329–332
Tracking system, radar, 429–433
Transfer function, 54–99, 279–281
  critical frequencies from, 58
  definition of, 55
  forced response from, 57
  frequency measurement of, 363–368
  gain of, 345–356 (*see* Bode plots)
  human, 85–91
  importance of poles in, 95–99
  limitation on use of, 82–85
  with nonlinearity, 329–335
  phase of, 369–377
  poles of, definition of, 58
  related to state model, 270
  resonance in, 70–73
  stability from, 60, 515
  steady state response from, 67–70
  total response from, 62
  for transportation lag, 77–82
  uses of, 218
  zeros of, 58
Transform, z, 461
Transient duration, 133

Transmission:
  distortionless, 369–372
  leakage, 444 (P 12.12)
Transportation lag, 72–82, 588 (P 16.8)
  in feedback loop, 81
  phase for, 383
Transvenous pacemaker, 8
Two-port network, 293
Two-port system, 55
Typewriter keyboard, 4

Uncontrollability, 238–243
Undamped natural frequency, 116
Undriven system response, 225–229
Unicycle suspension system, 250 (P 7.10)
Unstable system, contol of, 95
Urban dynamics, 15

Vibration damper, 281–283
Visual acuity measurement, 394–395 (P 11.17)
Voltage amplifier, ideal, 166

Water temperature during shower, 19

Water-level control, 399
Weighting function, 508 (P 14.1), 508–509
Wiener, Norbert, 101

z, definition of, 461
z transfer function:
  for feedback system, 474
  realization of, 476
  stability from, 471–472
z transform, 461
  decay rate from, 472–473
  growth rate from, 472–473
  partial fraction expansion of, 470–471
  of recursion equation, 466
  table of, 464
Zero:
  gain for, 348–351
  in ladder transfer function, 122
  phase for, 372–379
  of transfer function, definition of, 58
Zero-memory nonlinearity, 322
Zero-order hold circuit, 495–498
Zip code, 5
Zone, dead, 338–339 (P 10.3)